D0065292

THE HUMAN BLUEPRINT

BOOKS BY ROBERT SHAPIRO

Origins: A Skeptic's Guide to the Creation of Life on Earth

Life Beyond Earth: The Intelligent Earthling's Guide to Life in the Universe
(with Gerald Feinberg)

The Human Blueprint: The Race to Unlock the Secrets of Our Genetic Script

THE HUMAN BLUEPRINT

THE RACE TO UNLOCK THE SECRETS OF OUR GENETIC SCRIPT

ROBERT SHAPIRO

St. Martin's Press New York

 For information, address St. Martin's Press, 175 Fifth Avenue, New York, N.Y. 10010.

Production Editor: Mark H. Berkowitz
Design by Judith A. Stagnitto

Library of Congress Cataloging-in-Publication Data

Shapiro, Robert, 1935–
The human blueprint : the race to unlock the secrets of our genetic script.
p. cm.
Includes index.
ISBN 0-312-05873-X
1. Human Genome Project—Popular works.
2. Human gene mapping—Popular works. I. Title.
QH445.2.S53 1991
573.2'12—dc20 90-29153

First Edition: September 1991
10 9 8 7 6 5 4 3 2 1

To my parents, and theirs, and theirs,
back to the start of humankind.

Know then thyself, presume not God to scan;

The proper study of Mankind is Man.

Sole judge of truth, in endless Error hurled,

The glory, jest and riddle of the world.

—*Alexander Pope,* AN ESSAY ON MAN, 1733

CONTENTS

PREFACE

Imagine that a huge space-traveling rocket was under construction in Florida. Books and articles described the selection of the crew, the construction problems, and the likely cost. However, little attention was paid to its destination or mission after it arrived. This was not what happened, of course, when the Apollo project landed men on the moon for the first time. However, I think that the analogy works very well in describing another large government-funded effort that is taking place in a different area today: the attempt to read the entire genetic DNA text of billions of characters that describes the heredity of a typical human being. This undertaking has been named the *Human Genome Project.*

The media has been aware of the immense burst of activity that has appeared in the field of human genetics. Some good books appeared at the end of the 1980s that dealt with recent advances in medical genetics. In some cases, they dealt with the problems of gene mapping, which is an early chore of the project, but not its final goal. A number of articles have been written that cover particular aspects of the project, particularly those related to human conditions and diseases such as cystic fibrosis, Alzheimer's disease, and alcoholism. In addition, the funding of this project in a period of government budgetary constraint has caused a good amount of debate among scientists, which has been reported in the media. No work has appeared, however, up to the start of 1991 that deals with the Human Genome Project in terms of its full scope, historical background, and ultimate meaning for humanity.

I have written this book in the hope that it would fill the gap and bring the public more closely in touch with the enormous events that are taking place in that area of science where chemistry borders human genetics. I have traced the roots of current events into the past, back to the first efforts of humans to understand why inheritance works as it does. The separate investigations of many scientists in different countries gradually converged to produce the insight that our genetic plan was written in script and that it was possible to read and interpret that information. From this new awareness, the Human Genome Project was born.

I continue by describing the present state of the project and then try to anticipate the benefits and choices that will follow its completion in sections that deal with the immediate and more distant future. The very act of learning how our bodies function at their most basic level will in itself bring satisfaction to the scientists involved. In addition, the progress in reading our own genetic text will provide first hope and ultimately relief for humans who suffer from thousands of hereditary diseases. Advances in this area will be among the earliest and most visible results of the project.

In much of the book, however, I move into new territory and try to anticipate the impact of this effort on "normal," reasonably healthy human beings. I indicate the type of information that each of us may gain about our biological strengths and weaknesses, our relatedness to others, and our genetic history. One of the great strengths of the Human Genome Project, in my view, is that its results will be very relevant to the everyday lives of many people on this planet. It has potential then for rekindling an interest in science for many individuals who are not engaged by abstract questions or interested in discoveries about objects that are incredibly tiny or are very remote from us. One body exists, however, that has an almost immediate claim on our attention: our own. I hope that many readers will move from this account of scientific progress in understanding the basic plan of the human body to a deeper appreciation for the way that science works more generally.

I have used a number of devices in constructing this book in order to make it as accessible as possible to readers who are not scientists. Our genetic plan is written in text, and whenever possible I have tried to illustrate its working by making an analogy with a different form of script with which every reader will be familiar: the English language. I have also avoided discussion of some worthy, but complex subjects that diverge from the central theme of the book. I say little about the functions of RNA or how proteins are assembled in the cell, for example, to keep more focus on the major idea that DNA is text.

In the same spirit, I have avoided technical terms as much as I can. Some genetic and biochemical words that have gained common recognition, or cannot be described simply in common language, such as *DNA*, *gene*, and *chromosome*, are retained in this book. Others, however, are unfamiliar and cumbersome and can be represented by simple English equivalents. In such cases, I present the technical name once and thereafter use the simpler one. For example, the term *restriction enzyme*, which describes a substance that severs DNA at the site where a particular sequence of letters appears, is replaced by the simpler "text cutter."

In following these practices, I may introduce some minor inaccuracies in places, but I feel that this is a lesser evil than reader incomprehension. Those who are inspired to probe more deeply into the subject will find many sources that can provide them with a full discussion of the rich complexities of our biochemistry.

A great number of scientists, both past and present, have contributed to the advances in knowledge concerning human biochemistry and genetics that I describe in this book. I have selected only a few of them for discussion to allow the narrative to flow smoothly and keep the size of the book within reasonable limits. Many others who made important discoveries or introduced significant concepts have been mentioned briefly or omitted altogether, for example, Archibald Garrod, George Beadle, E.L. Tatum, Joshua Lederberg, J.B.S. Haldane, Francois Jacob, Jacques Monod, Barbara McClintock, Torbjörn Caspersson, Stanley Cohen, Herbert Boyer, Leroy Hood, David Botstein, Thomas Caskey, Louis Kunkel, Maynard Olsen, Raymond White, and many, many more. Although they have received little or no space here, their scientific contributions are acknowledged with deep respect.

The subject of this book has enormous magnitude, and I drew upon the help of many individuals in writing it. I am very grateful to the following who provided in-depth interviews and discussions: James Watson, Frederick Sanger, Walter Gilbert, Nancy Wexler, Charles Cantor, Cassandra Smith, Stuart Fischer, Allan Wilson, Svante Pääbo, Mary-Claire King, Gabriel Dover, Alan Coulson, Erika Hagelberg, Bryan Sykes, Robert Hedges, Robert Moyzes, Henry Erlich, Randy Saiki, Kary Mullis, and David Schwartz.

I also wish to thank Lise Hazen and Gerald Feinberg for making my visit to the site of Thomas Hunt Morgan's lab possible; Jan Sapp for providing material on Gregor Mendel; Benjamin Barnhart, Thomas Lee, Leonard Lerman, Richard Kouri, and Marcia Lewis of Bios Corporation; the staff of the DNA Museum in Cold Spring Harbor; and the research groups of Allan Wilson, Gabriel Dover, and Charles Grunberger for helpful discussions and demonstrations.

Finally, I am indebted to my agents John Brockman and Katinka Matson for their role in making this book possible. I wish to thank my editor, Robert Weil, and the staff of St. Martin's Press for their encouragement, advice, and assistance.

PROLOGUE

With some trepidation, the knight threw his weight against the old rusty gate. He had crossed many levels of the dungeon to reach this barrier in a quest that had lasted as long as he could remember. Beyond this gate, said the voice that accompanied him (he did not know whether it came from outside of or within him), lay many of the answers that he had been seeking: who he was, what history he had had, what powers and weaknesses lay within the body in which he found himself, and what his future might be. He had also learned from the voice that some dreadful obstacle must be overcome beyond the gate before these answers could be learned. When he faced this obstacle, one of his powers would be tested to the utmost.

He was surprised when the gate swung open easily, revealing a huge and dimly lit chamber. It was a library, containing scrolls stored on shelves in an array that extended back as far as his eye could see. A table stood nearby, with a scrap of paper and a supply of candles, one already lit, upon it. A smaller room adjoined the library. Within it were a bed, freshly made, and a number of cases of preserved food, sufficient to sustain a man for an indefinite time.

A feeling of dread and unease made the knight return to the library. He cautiously raised his sword and shield and searched for a time among the many rows of shelves. No dragon or other monster appeared. Finally, he seized one of the scrolls and carried it back to the table. No marking had existed on the outside of the scroll or on the shelves to guide his choice. Opening it, he found that it contained unfamiliar characters. The paper, however, contained a scheme that allowed him to translate it to his own language. Unfortunately, the message still made no sense. It was full of terms he did not understand with references to other scrolls whose locations were not given.

The knight yelled out in anger: "Where are the answers I was promised?"

"In the scrolls, of course," the voice responded, almost immediately.

"But which scroll must I read then?" the knight asked, in desperation.

"All of them, all of them," came the answer.

Then the knight realized, all at once, that the quality that would be tested was not his quickness or his bravery, but his patience.

He looked down the length of the vast room again yearning for a dragon. Only a timeless silence greeted him. With a deep sigh he removed his armor, stored it with his weapons in a corner, and, seating himself as comfortably as he could, began to decipher the first scroll.

The knight represents us: the humans who are alive and those who have lived until now. His ignorance is ours. We have all risen to awareness in bodies that we did not choose within a world that we did not make. With the help of the others that came before us, we have learned to use our bodies and explore the world. As we have searched in it, we have developed the need to improve it: to convert an often frightening and painful realm into one better for humans to live in. I have termed this search the Quest. Its direction is clear; the final goal is not.

Our ancestors found their way by trial and error—the world itself was their teacher. As they gained skill and experience, they understood more clearly the method they were using—it evolved into modern science. Science differs from worldviews based on faith. Ideas are put forward to be tested and cast aside if they do not measure up to the verdict of experience. One feature above all recommends science to each new generation—it works.

Most of the triumphs of science to the present age brought improvements in our control of the world outside of us. Inventions and discoveries from the wheel and hammer to the factory and telescope have helped us to learn about our outer environment and cope with it. We have found that our universe extends much further in time and space than we imagine. Our vision, in what I will call the Outer Quest, has extended for billions of light years and back to a time within a second of the birth of the universe.

The other direction of science, the Inner Quest, has lagged behind until now. We have known more about the movements of distant planets and galaxies than the cycles within our own cells. Suddenly, this has changed. We ourselves are the new frontier for exploration. We are about to come into possession of our body plan, the human blueprint. In the scheme of our experience, only the human mind, our consciousness, is more fundamental. That will be a story for a later day. Our generation will be known, for all of history, as the first to read our own text. A vast and ancient document, compiled over billions of years, lies open for the first time before us.

Our understanding in this area has grown slowly. For millennia, a combination of magical thinking and clever, but wrong, guesses held the field. A potent and sacred fluid, our own blood, appeared to control the way that we function and reproduce. No philosopher dreamed that our plan was simply written down. This notion emerged over a century from the uncoordinated efforts of a number of gifted men. A train of thought that began with an obscure Austrian monk who raised peas in a garden culminated suddenly with flash of insight by two impetuous young men who played with cardboard and metal constructions in an English physics laboratory, James Watson and Francis Crick. Finally, in the mid-1970s, Frederick Sanger in England and Walter Gilbert in the United States developed methods that allowed us to read the text if we chose.

No huge generators or atom smashers were required. One person with simple equipment could decipher a page of the script in a day or two. However, there were millions of pages. Yet the messages written within were so compelling that hundreds of separate laboratories began to search and rummage around the huge text. Vital passages were found that told of the working of unsuspected mechanisms deep within us, of the cause of long-known and cruel diseases, and of the history of humans and other life forms on our planet. First in a trickle, then in a flood, passages of the human plan emerged and were stored in computers, until finally the idea took hold: Why not gather all of it? Let us now, in this generation, capture the entire plan. We could only begin the task of interpreting its message and applying it to enrich human existence, but the plan itself would be our legacy, our inspirational gift, for the future.

Thus, the human genome (genetic text) project was born—first as an idea, then as an effort carried out by working scientists in their laboratories. New funds were appropriated by the United States Congress and governments abroad, and committees were formed to manage the growing effort. One of the two men who, in 1953, had discovered the language of the human text was put in charge, and a date, 2005, was set by which time the job of reading it should be done.

So large an undertaking could not come to pass without some discussion and controversy, but the sources of complaint were surprising: They did not arise from those parts of the public who usually show antipathy to science, but rather from other scientific laboratories. Many biological scientists, accustomed to working on their own, were not comfortable with a plan that would require coordinated effort. They would have preferred to see the funds diverted to themselves or their colleagues for their own projects. Quarrels of this type have most likely come along with all large human endeavors, from the raising of cathedrals to the establishment of an independent United States of America. The most startling objection of all, however, concerned a particular feature of the human plan.

If the workings of the human body were to be captured and described in text, we would have hoped for a document that would flatter us. Perhaps it would be rather short, but rich in phrase and deep in meaning, like the Gettysburg Address or the Declaration of Independence. Or possibly it would be longer, poetic and rich with the experience of the ages, like the Bible. At the least we would anticipate something like an auto repair manual or anatomy text: a no-nonsense, nuts-and-bolts description of parts and functions. What we found instead, in our early samplings, resembled more the collected writings of an elementary school class, including the contents of the wastebasket. Doodles, discarded drafts, ramblings, and interpersonal notes were present in abundance with their bulk dwarfing that of the completed essays.

Some scientists only wanted to read the essays in the human text, ignoring the remainder as "junk." As such, they resembled a schoolmaster whose interest in his pupils ended when he learned how well they had learned their math or could put "What I did last summer" into correct English prose. Another type of teacher, interested in the pupils as individual human beings, might also be fascinated by the material not handed in. In the same sense, the junk represents a treasure chest of information on human identity, relationships, and history.

The most astonishing objection, in my view, concerned the way in which the project had been presented to the public. To some, the presentation was an exercise in "hype" and "public relations." The objectors must presume, then, that science already relates well enough to the public, that support and enthusiasm for science are sufficient, and that matters outside of the science laboratory are running so well that no further input of scientific attitude and methods are needed.

For those of us who disagree with this analysis, the new Inner Quest offers a lustrous opportunity to reunite the excitement of science with the concerns of the human race. For much of history, the objects studied by scientists were part of our visible daily experience: birds, falling objects, river valleys. In the past decades, the focus has shifted to the remote and invisible: collisions of obscure molecules, magnetic fields of distant planets, differences in the tiniest objects that are so subtle that only instruments of colossal size can probe them. Finally, one area of science has returned to ask vital questions that affect us all: How do our bodies work? What makes each of us different from our fellows? What makes us ill? Why do we grow old and die? The answers lie written in the human text, and whether we do it more rapidly or slowly, we shall surely read that text. All of it.

We shall first want to read the plan for a prototype, a typical individual. The chapters and phrases that govern every human function will be located, recorded, and indexed. No single person shall provide it all; it shall be taken, in bits and pieces, from many of us. When this common goal has been attained, the army of scientists and scholars will divide and proceed in two different directions.

The biological scientists will seek to interpret the text. Perhaps one hundred thousand separate passages will have been uncovered, each of which describes a substance involved in human function. The purpose of each must be determined, and then the complex interactions between them untangled and understood. At the end of this effort, which may take centuries, we shall understand our bodies as we do an automobile and be skilled in maintenance, repair, and even redesign.

Long before the text is fully understood, the individual spelling differences in it will fascinate many of us. Anthropologists will find them a vital tool for tracing the development of races and ethnic groups. Historians will probe the relationships of famous individuals. Population biologists will trace the history of genetic diseases. Everybody will have a chance to learn about his or her own ancestry and personal variations. *The Book of Humanity*, as some have called it, is not reserved for specialists.

Many somber and depressing pictures have been portrayed in literature, such as Aldous Huxley's *Brave New World*, of the possible effects of this new understanding of human life. Governments have performed atrocities on entire populations during this century, citing imaginary principles of science in partial justification. These fictional and real abuses no more represent the essence of science than the Inquisitions, persecutions, and wars of the past represent the ethical teachings of the great religions.

The Nazi tyranny and the society pictured in *Brave New World* operated to eliminate human variety and choice. Their actions reflected repugnant decisions by governments. Science works to increase our control over nature and so enhance possible human choices. We may choose to drive a car, or we can walk; we can suffer a disease or be cured of it. It is good to have the decision in our own hands.

Our understanding of the human plan will bring many new choices to individuals and to our spacies as a whole. Some will involve the long-term future of the human race. Many ethical questions will arise. Our experience thus far has indicated that, even with the best of intentions, we will disagree on the answers. In my view, we will be wise to accept these disagreements with tolerance for the other point of view and allow each group to follow its own destiny. In doing so, we will create a future that will be the antithesis of *Brave New World*. Choice and variety will proliferate. The human race will emulate Lord Roland who, in a tale by the Canadian humorist Stephen Leacock, "flung himself upon his horse and rode madly off in all directions."

I myself have ridden briefly over the ground I wish to cover. The material begs for closer inspection, and we shall do so in this book. I have divided our tour into sections that deal with Yesterday, Today, Tomorrow, and After Tomorrow.

PART I
YESTERDAY

1

OF TEXT AND BLOOD

The answer seems so tricky 'til you catch it in a flash. When I work certain problems in biochemistry class, I cannot understand why the students are so perplexed. One, for example, involves a chain of atoms that is cut through. Yet only a single piece, not two, are produced. It is obvious to me—the atoms are connected in a circle. Yet most of the students continue to stare until I provide the answer. I feel the same way when I watch a mystery film for the second time. The identity of the murderer is clear. How did I miss it when I first saw it?

The secret of human identity now seems equally clear. The general human plan is stored in a language of four letters. Some modest spelling variations provide for our individual differences. We each carry two copies of the plan: One is taken from our mother and one from our father. When we reproduce, we shuffle and splice our two copies as if we were shuffling a deck of cards, then we donate half of the mixture to our child.

All of this was missed by our ancestors for millennia. The document was well guarded; few hints escaped that a plan existed. The greats who speculated on heredity, from Pythagoras to Darwin, never dreamed that it was preserved in a row of individual letters of the alphabet, much like the Bible, or the words in which they wrote their theories. And what a text it is! Like a massive encyclopedia, it stretches for forty-six volumes (scientists call them *chromosomes*), or rather a double set of twenty-three volumes, each set differing one to three percent in spelling from the other. However, each set is immensely larger than the ones we see in the library. Our heredity is not written in words separated by spaces, so a comparison is best made in terms of letters. The *Encyclopedia Britannica* has perhaps 280 million letters; each of our double sets has more than ten times that amount—about 3 billion letters for each.

For a more exact comparison, think of the largest volume that you have seen in the library. Most likely it is that mammoth unabridged dictionary that stands open on a special table or stand, so that you do not have to haul it about. The one in my local library has over three thousand pages of tiny print. I estimate that it holds about 76 million letters that would match the content of one of our smaller chromosomes (volumes). The largest one would be four times the size. If it were printed as a book, any user who tried to move it would risk a rupture!

The plan for each of us is kept in forty-six volumes of that type. Even though it is immense, many copies of it have been "printed." With few exceptions, every cell in our bodies has its own edition. In other words, I have a few trillion copies of my own text; you have as many of your own slightly different version, and so on. No human best-seller ever sold a fraction of this number of copies. Every human on the planet would have to buy a dozen copies for his or her own use to justify a printing of such enormous size.

The idea that our plan is stored in letters has been around for a little more than a generation. Remnants of a much older idea are still preserved in our language as a type of verbal fossil: that heredity is preserved and transmitted by our blood. The thought has become so familiar that we do not even pause when we see such phrases as "royal blood," "bad blood," "blood relative," "blue blood," and "mixed blood." The blood theory was first devised by the ancient Greek philosopher Aristotle (384–322 B.C.) and others of that era. They felt that the male semen was made of purified blood. The thickest part of the blood in our veins was absorbed by our bodies, while the remainder was transformed into the warm, thick, liquid substance familiar to us.

According to this theory, the male parent had the most important influence in heredity (guess which sex Aristotle and his colleagues belonged to!), which was expressed by the motion and "formative power" of his semen. The female contribution to heredity was present in her menstrual blood; its major effect was to hinder the energy of the semen. Anomalies or deformities could be one result of this interaction. To quote Aristotle: "He who does not resemble his parents is already, in a certain sense, a monstrosity; for . . . nature has simply departed from the type. Indeed, the first departure occurs when the offspring is female rather than male. This, however, is a natural necessity."[1]

Apart from its unflattering and mistaken view of woman's role, this theory had a deeper defect: It was not built on careful observations and experiments. It came from an extension of the idea held by humans through the ages: that blood was the very stuff, the soul of life. Francois Rabelais, a noted French satirist, humorist, and physician of the sixteenth century, said it all: "Life consisteth in blood; blood is the seat of the soul." In medieval philosophy, blood was one of four humors that comprised the human body. The phrases *hot-blooded* and *sanguine* survive to describe a warm and passionate temperament, one dominated by blood. Blood could be spoiled as well; bloodletting was used until the nineteenth century as a means of ridding the body of fouled fluids. How natural (and how wrong) then to conclude that blood was the hereditary stuff.

For those who believed this idea, inheritance involved a blending of parental qualities, as in the mixing of two different liquids. As an example, let's think about the Bloody Mary, a nifty mixture of vodka and tomato juice (and also various spicy ingredients, which I will ignore in the following discussion). Once stirred together, the ingredients mix evenly. The vodka does not curl up in one corner, leaving the remainder of the space to the juice. Suppose I decided that the drink was too alcoholic. I could take half of it and mix it with an equal volume of tomato juice. I could expect that the "offspring" had half of the vodka. If I again mixed the product of this "cross" with tomato juice, the third generation would still have twenty-five percent of the original alcohol. Even after a dozen generations, the product would not be entirely free of alcohol.

Our experience has also taught us that tomato juice and alcohol, once blended, stay blended. Suppose I combined two "first-generation" Bloody Marys and divided the mix into four portions. I would expect to get four drinks, identical in taste and composition, if not in size. I would never get a mixture that held two Bloody Marys, one glass of tomato juice, and one of vodka. However, this result is exactly what you get in certain cases of real inheritance.

Linear text is different. It can be spliced, but not blended. As an example, let us imagine a book: *Great Works by British Poets; The Nineteenth Century*. It would contain poems by Keats, Tennyson, Shelley, and additional writers, but each item would be separate from the others. The poems would not be blended. Similarly, a collection from the twentieth century might include Auden, Graves, and Thomas, among others, with each contribution distinct from the others. If a book of the same size were wanted that covered both centuries, it would be easy to put one together. We would just select some of the contents from each volume. If the final volume were arranged alphabetically by author's name, nineteenth- and twentieth-century works would be intermingled, but each poem would still keep its separate identity. Some authors from the original two volumes might have vanished in the final product, however.

Biological heredity works by the poetry book model, not the Bloody Mary one. The child receives a selection of components from both parents. Some remain intact, while other traits from a parent may be lost entirely. If we take the analogy further, we can see other interesting effects emerge.

Suppose that another editor wants to use the combined nineteenth- and twentieth-century book as source material for a work of modern poems in the English language. He selects some of the poems and combines them with works by American writers. If this "second-generation" anthology featured recent works, all of the nineteenth-century English works would be left out. The contribution from one "grandparent" source would be lost entirely.

In the same way, the contribution of a human grandparent could disappear entirely after two generations. He would be biologically unrelated to his grandchild. In practice, this is very improbable, as the human blueprint has far more entries than any poetry compendium. As we shall see later, though, the possibility of total loss of an ancestor's contribution gets much larger as we go back through the generations.

The blood-is-heredity idea hardly matched the observable facts, yet it persisted for thousands of years. It came accompanied with other stray baggage, as no accepted way existed to get rid of bad ideas. Another long-term loser was the pangenesis theory launched by Hippocrates (450–377 B.C.), a Greek physician viewed as the "Father of Medicine." More than two thousand years later, the idea was still alive and had found a new champion in Charles Darwin.

Hippocrates felt that the whole body controlled inheritance, not just the blood. The shape of my fingers would somehow affect the shape of my son's, my spine would control his spine, and so on. Let us quote the gentleman himself:

> I say that the semen comes from the whole body, from the solid parts as well as the soft parts and from all the humors which are in the body. . . . It is the most active part which

> separates off. Here is the proof: after coition, the evacuation of such a small quantity of semen renders us feeble.[2]

The testes were then just an assembly point at which the troops gathered before they shipped out.

This theme was taken further by the nineteenth-century French biologist Georges Buffon:

> The male seminal fluid . . . and the corresponding female fluid . . . are both equally active substances and equally loaded with organic particles, endowed with engendering properties. . . . When the two seminal fluids mix, . . . each organic particle becomes immobile at an appropriate location, this being none other than that which it occupied previously in the body. Thus, all the particles which came originally from the spinal cord will be relocated in a comparable way.[3]

Charles Darwin endorsed and extended these ideas. In his "Provisional Hypothesis of Pangenesis," he called the hereditary particles *gemmules*. In Darwin's words:

> Gemmules are supposed to be thrown off by every cell or unit, not only during the adult state, but during all the stages of development. . . . It is not the reproductive elements, nor the buds, which generate new organisms, but the cells themselves throughout the body.[4]

Darwin agreed with other workers in that his gemmules migrated to the genitals at an appropriate time. Gemmules passed from cell to cell during cell division. Normally, they went from parents to offspring, but they could remain in a dormant state for many generations and then develop. This provided a tidy explanation for the observation that hereditary features could skip a generation and then reappear: The gemmules were asleep.

Such ideas could explain some things that we saw, but did not tell us what was going on at a deeper level. A finger gemmule is obviously smaller than a finger, yet it has the power to make a finger. Questions of this kind would bother us today, but they did not trouble our ancestors very much for a good reason. Even though they could see that some parental qualities were handed down, they felt that at the bottom line, reproduction involved supernatural forces.

In his history of genetics, Nobel laureate Francois Jacob wrote, "Living beings did not reproduce; they were engendered. Generation was always the result of a creation which, at some stage or other, required direct intervention by divine forces."[5] For like to produce like was the most common result, but God and His agents could change the rules whenever they pleased. Thus, the most remarkable events were possible.

For this reason, most people from ancient times until the twentieth century believed in spontaneous generation: Living things could arise directly from nonliving material. Thus, worms could be produced from timber, beetles from dung, and mice from river mud. Lepidus said to Antony in Shakespeare's *Antony and Cleopatra*: "Your serpent of Egypt is born of the mud, by the action of the Sun, and so is your crocodile." If mud could substitute for both your mother and father, then there was not much point in constructing a family tree or wondering where you got that particular shape for your nose. Anything could happen.

Such tales did not just supply gossip in the kitchen or marketplace. They penetrated into academic channels as well. For example, Jan Baptiste Von Helmont, a Flemish biologist, published the following procedure in 1667:

> If you press a piece of underwear soiled with sweat together with some wheat in an open mouth jar, after about twenty-one days the odor changes and the ferment, coming out of the underwear and penetrating through the husks of the wheat, changes the wheat into mice. But what is more remarkable is that mice of both sexes emerge (from the wheat) and these mice successfully reproduce with mice born naturally from parents. . . . But what is more remarkable is that the mice which came out of the wheat and underwear were not small mice, not even miniature adults or aborted mice, but adult mice emerge![6]

In addition to spontaneous generation, the idea that outside events can change heredity was also widely accepted. The story of one such attempt can be found in the Bible (Genesis 30): Jacob makes an arrangement with his father-in-law, Laban. He agrees to look after Laban's herds, and in return, he is allowed to keep the less attractive animals. For example, he can claim all of the spotted and speckled goats, but Laban keeps those of a solid color. Jacob then decides to tilt the odds more to his favor. He cuts strips of bark from rods made of green poplar, hazel, and chestnut branches, so that white markings appear on them. These rods are placed in the gutters of the watering troughs, which stand near a favorite mating place of the animals. Jacob places his rods only when the stronger specimens come to mate, however, and not the feebler ones. The striped rods exert their effect, and the stronger goats produce speckled and spotted offspring for Jacob's flock, whereas the weaker ones produce feeble solid colored goats for Laban. Jacob was lucky. He was able to fool around with the colors of the goats by using rods, but their strength continued to breed true.

History is studded with many such anecdotes: The Greek philosopher Empedocles (fifth century B.C.) told of a mother who fell in love with the physical beauty of some statues. Her children acquired some of that beauty. A French gentleman of the same profession, Nicolas de Malebranche, reported some two thousand years later about a woman "who gazed too long at the picture of St. Pius . . . and she gave birth to a child who perfectly resembled the picture of the Saint." This idea had lasting power.

The notion that acquired characteristics can be inherited is closely related to the above. Again, we can call on Aristotle for his power of observation:

> It has happened that the children of parents who bore scars are also scarred in just the same way and in just the same place. In Chalcedon, for example, a man who had been branded on the arm had a child who showed the same branded letter, though it was not so distinctly marked and had become blurred.[7]

The most colorful products of imagination and inaccurate observation were the mythical beings that were produced by unlikely crosses. For example, Ambrose Paré reported in 1573: "Nature always tries to create its own likeness: A lamb with a pig's head was once seen because the ewe had been covered by a boar."[8] Such occurrences were reported commonly; a man was seen with the hands and feet of an ox, and a dog with the head of a fowl. Creatures of these types were born when the action of some person or animal violated the intended order of the world. Such beasts no longer come up in everyday conversations, but they have found a new home: mermaids, centaurs, and other fanciful creatures now inhabit advertisements and commercial films.

When the microscope was invented and perfected, in the fifteenth through seventeenth centuries, human fantasies could be extended into a whole new realm. One pioneer in that area was Antonie van Leeuwenhoek (1632–1723). He first described human sperm and decided that each one had a complete miniature being within it. Physician-biologist Marcello Malphigi, a physician and biologist of the same era, observed egg yolks under the microscope and came to a similar conclusion. Thus arose the theory of *preformation*. Believers in this theory felt that all future generations to come had been made at the time of Adam and Eve and nested in one another like a set of Russian dolls. Each doll represented a complete creature-in-miniature. Fertilization served only to turn on some growth mechanism, so that it could enlarge itself to a size appropriate for the outside world.

The preformation theory disposed of all questions concerning the inheritance of features. All of that had been settled by divine action at the very beginning of Creation. Only one issue remained to divide the advocates of this idea: Was it the sperm or the egg that held the little person and all of his or her descendents? Thus, began the hundred years' war between the spermists and the ovists.

Van Leeuwenhoek, quite understandably, favored the "animalcules" that *he* had discovered. One colleague who agreed claimed that each animalcule "contains a little male or female animal of the same species hidden under a tender and delicate skin." Leeuwenhoek added that the animalcules "differ in sex . . . and male and female can be distinguished." Why invent creatures needed for generation in the egg? They could be seen swimming around in male semen. This primary role for the male was felt to be in keeping with his dignity. The job of the female, as usual, was that of providing a nest and nourishment.

Malphigi, of course, preferred to award first prize to *his* discovery. His feelings, and those of the other ovists, were summarized by Francois Jacob:

> As to man, everyone knew that the liquid he releases with so much pleasure does not penetrate the uterus, but runs out of it as soon as it is discharged. Plenty of cases were known of girls becoming pregnant without even letting the man's fluid penetrate them.[9]

Preformation in both its guises was eventually dispatched through more careful observation of developing animal embryos in the eighteenth and nineteenth centuries. Thus, we returned to the starting point or at least to pangenesis.

Given this confusion of theories, we might include that no progress would have been made by humans in coping with questions of heredity over the millenia. Fortunately, another more practical set of humans was paying little heed to the teachings of philosophers and priests and was attempting by trial and error to make things work. Any observant child could see that like begets like, that humans give birth to baby humans and horses to baby horses, and that human–horse combinations were never to be seen in the real world. Farmers and herdsmen learned that within species like could sometimes give birth to something slightly different. Individual cases were not predictable, and exact rules hard to come by, but in the longer haul events could be controlled, and improved species could be produced.

Thus, from 8000 B.C. on, wolves were slowly domesticated into dogs. Pigs were gradually bred from wild boars over a stretch of time beginning about 6000 B.C., and a number of other animals were altered by selective breeding. Cultivated maize was developed from wild maize in ancient Mexico between 5000 and 3000 B.C. Many other plants and animals were developed for human use in this way in the distant past. Skulls and other remains provide a record of these ancient projects, as do the Babylonian accounts and Assyrian reliefs that describe and display the intentional cross-pollination of date palms thousands of years ago.

Our ancestors understood that humans were also governed by the rules of heredity. Ancient Hindu religious writings advised that "in choosing a wife a man must make sure that she has no illness which could be inherited and that her family is free of such illnesses" and also that "a man of base descent inherits the bad characteristics of his father or mother or both; he can never escape his origin." Similarly in Homer's *Odyssey* it was assumed that physical features, strength, courage, and ability were handed down within noble families.

Plato's *Republic* echoes these same themes.

> But the god who fashioned you added gold in the composition of those of you who are to become rulers (which is why their prestige is greatest); he put silver in the auxiliaries, and iron and bronze in the farmers and other artisans. As you are all of the same stock, children will mostly

> resemble their parents. Occasionally, however, a silver child may be born of golden parents, or a bronze child of silver parents, and so forth.[10]

Alas, Plato also thought that the mental, physical, and moral condition of their parents during the intercourse in which conception took place was also inherited by their children. With hindsight, we can see that a mixture of correct and incorrect inferences lay side by side in Plato's conception of his ideas and that without the systematic practice of modern science there was no way to sort out one from the other.

With that practice, we have seen much, much more revealed in a century and a third than our forbears dreamed of in ten thousand years. We stand now poised at the threshold of the largest leap of all.

Many scientists contributed to this spectacular advance, and their stories have been told in histories such as *The Eighth Day of Creation*, by Horace Freeland Judson. I do not want to add to these scholarly accounts here. Rather, I would like to capture some of the spirit of these events by selecting a few of the more inspirational insights and setting them against the background of the year or years in which they occurred. Some of the key individuals were rewarded with almost immediate fame; others earned obscurity. Together they illustrate the remarkably different ways in which great science can be done. So let us enter the time machine of our minds for a trip back to the years 1865, 1910, 1953, and 1975–77.

2

1865: THE PEA PATCH

BRÜNN, AUSTRIA–HUNGARY, FEBRUARY 1865

On a clear, cold evening in February 1865, in a small city in central Europe, a middle-aged monk left his monastery to address the monthly meeting of the local scientific society. Somewhat heavy, with broad shoulders, modest stature, a receding hairline with dark hair brushed to one side, and gold-rimmed glasses, Gregor Mendel presented a very ordinary appearance. At the meeting, he told what he had learned by growing peas for eight years in a corner of the monastery garden. We can imagine that the audience listened politely, but the minutes record no discussion or questions.

A month later Mendel returned to finish his presentation and gave a mathematical analysis of his results. Again, no audience response was recorded. In keeping with the practice of the society, the talks were published the next year as a paper in its own journal. This series, *The Proceedings of the Brünn Society for the Study of Natural Science*, although a minor one, was colleced by over a hundred libraries. Mendel's work, however, was mentioned only four times in print by other scientists in the next thirty-four years. Yet the presentation was an immortal moment in the history of biology.

We may judge that Mendel's colleagues were short-sighted, but I will suggest that he might do no better today. To illustrate my point,

I have arranged a fantasy: Imagine that a study section of the U.S. National Institutes of Health has fallen into a time warp and been swept back to the Austro-Hungarian Empire in 1865. Their current scientific knowledge has been erased and replaced by its 1865 equivalent, but their attitudes have been left intact: They are attempting to conduct their usual business in this new setting. They have convened to review a group of grant applications. The chairperson of the panel speaks first, and the members respond.

"What about the application from this guy Mendel? Something to do with peas."

"It must have been sent here by mistake. Lets ship it over to Agriculture."

"Wait. He wrote in the Significance section that he is going to work out the general laws of heredity that apply to all living things."

"A bit grandiose, isn't he. Look, I have inside information that Darwin is working on that problem. When what's his name's track record matches Darwin's, we can give him more attention. Ready to vote?"

"Hold on. We have to write a pink sheet, so we need some comments."

"Okay. To start, how can he work with peas? *Nobody* is working with peas these days. Not even the plant botanists. Now if he wants an interesting plant system, he should work with hawkweed. A lot of good work is being done with hawkweed. Von Nageli's stuff, for example."

"Good. Now how about his Preliminary Studies section. He sent us a copy of a manuscript of his. Has anybody read it?"

A moment of awkward silence passes. Then another board member, hitherto silent, speaks up: "Oh, I know all about that paper; it is of no importance. It is nothing but numbers and ratios, ratios and numbers. It is pure Pythagorean stuff; don't waste any time on it, forget it."

"Pythagorean? What do you mean?"

"You know. The idea that you can explain everything in terms of the properties of mystical numbers."

"Let's get back to the job. We have to say something about the Biographical Sketch, and the Resources and Environment."

A pause follows while the committee thumbs through the grant application to refresh their memories. Then there is a sudden explosion of voices:

"He must be kidding. He wants to carry out his research at a monastery?!"

"Well, perhaps he has some special deal. The use of novices for inexpensive labor. Greenhouses, lab space, lots of land in which to grow his specimens. I know that monastery. They have very large gardens."

"No, they've only given him a strip that measures 120 by 20 feet and no assistants to help him. He has to work alone."

"What about his credentials. He may be an eccentric genius of some kind. Has he affiliated with any well-known places? Where did he train?"

"I don't think he has any scientific degrees at all, though he fooled around at the University of Vienna for a couple of years. He teaches science part-time at the local high school, but without proper credentials. He flunked the qualifying exam for teachers twice."

"I think we've heard enough!" said the chairperson. "I can fill in the rest. But we can't even give this one a priority score. We have to reject it outright." No voice of dissent was heard from the section. One member did speak up, however.

"I agree. But we have to say *something* encouraging. After all, the guy *has* put in a lot of work. He counted almost thirteen thousand pea plants over eight years. We have to encourage public interest in science, or our funding sources may dry up."

"*If* he put in that work. There's something that I don't trust about his ratios. They're too good to be true. They shouldn't work out so exactly at three to one and so on."

"We can't get involved in that. We're not funding him anyway, so nobody can point a finger at us. We'll encourage him to keep on working on his own. What if we say this: 'Your experiments with Pisum (peas) are only beginning.' " The committee agreed and moved on to the next grant.

The above proceedings, of course, have been invented by me. However, I have tried to give in modern terms the response of nineteenth-century biologists to Mendel's work. He sent a copy of his paper to Dr. Karl von Nageli, a noted Munich botanist, from whose response I have lifted the comments about the experiments only beginning.[1] Von Nageli also made the suggestion that Mendel switch to his own species of interest, hawkweed.

The comment that Mendel's work was made of numbers and ratios without significance also dates from the nineteenth century.[2] In the present century, statistician Sir Ronald Fisher has made the opposite claim—that his numbers have too much significance. Fisher has felt that authentic experimental data would not come so close to the exact results expected. Something funny was going on according to him. Whatever the explanation for Mendel's exceptional accuracy, it cannot dent the essential fact that he was *right*.

What had he accomplished? He simply captured a fragment of twentieth century science and deposited it into the literature 34 years ahead of its time. There it lay unnoticed but not entirely forgotten, until biology was ready to catch up with it. Suddenly, in 1900, three separate European botanists rediscovered his work and found it to be valid and vital. Mendel had shown that heredity worked through message units, not fluids, and the units were passed on intact from generation to generation.

A message might say, for example, "make smooth peas" or "make wrinkled peas." Each individual would get *two* copies of each message, however, one from "mom" and one from "dad." (I put in the quotes because I am still talking about peas, but it holds for us as well, despite what the grant board said.) Mendel developed plant strains that produced smooth peas, generation after generation, when bred with one another. Each plant had two copies of the "smooth" message. Strains of wrinkled peas that he interbred did equally well, as each had a double copy of the "wrinkled" instructions. What happened, then, when smooth was crossed with wrinkled? In that case, smooth won. All of the peas of the next generation were fully smooth; none were observed that were even mostly smooth with a few wrinkles. The messages had not blended. A geneticist would say that smooth was dominant and wrinkled was recessive. The wrinkled message had not been lost in this case, however; it had only gone into hiding. Each descendent had one copy of the smooth message and one of the wrinkled one. The situation could be revealed by crossing two members of this hybrid generation.

In the second generation, both plants bearing smooth peas and plants bearing wrinkled ones were seen. The latter were now pure in this respect, having thrown off all vestiges of the smooth message that had been present in two grandparents. The smooth group could be subdivided, however. One contained only the smooth instructions, while the other still carried a mixed message (this could be detected by breeding further generations).

Why hadn't anyone else noticed this before the mid-nineteenth century? Some scattered observations had been made, but not on a large enough scale so that the general rules could be extracted. Further, Mendel was either incredibly clever or incredibly lucky. The messages (or *genes*, as they were later named) do not usually control obvious features, such as pea shape, any more than the congressional representative from your district decides the final size of the federal deficit.

Genes determine submicroscopic agents called *proteins*, and the interaction of many proteins in a complex manner determines most of the traits in our everyday world. For example, 63 genes work together to control the color of a mouse's coat. Most features of interest such as the size of cattle, their milk production, and the speed of race horses were also under complicated control and could not be easily analyzed. Another species with complicated genetic properties was hawkweed, which the famed botanist Karl von Nageli urged him to study. Mendel got so bogged down in that system later in his career that he quit plant research.

At the start of his research, however, Mendel had not yet received bad advice on what to study, and he selected peas. In particular, he chose to follow seven obvious features, height, flower color, and unripe pod color, for example, as well as pea shape, all of which were controlled by single genes. In every case, they followed the scenario I described earlier, where one gene dominated and the other hid. Many properties are now known where things do not work this way. Lucky Gregor. Further, he noticed that each property behaved as if it were completely independent of the other, in the same way that the redness or greenness of a traffic signal acts independently of the weather. We now know that this is usually true, but there are some important exceptions. Yet this rule held for every case studied by Mendel. Lucky, lucky Gregor.

Wrinkled peas for example have that shape because they contain more sugar than normal ones at a certain point in their development. For that reason they absorb additional water and swell up more than usual. When they mature, this water is lost, and they wrinkle.

Peas of that type have excess sugar to start because they have lost an ability to build this sugar into starch that normal pea plants possess.[3] The genetic text that provides the instructions for the job (construction of a protein called *starch-branching enzyme*) has been ruined in both gene copies. Once, many generations earlier, a migrant piece of text, eight hundred letters long, spliced itself into the normal pea text in a copy of that gene, ruining it. That large typographical error was then handed down faithfully from generation to generation, spoiling the text in each case. Two ruined copies must come together in a single plant, however, for it to have wrinkled peas. It is rare that visible traits follow the fate of single genes, but this was one such case. Mendel did not know any of this, of course, but he knew how to follow the generations and count wrinkled peas.

Even with this luck, or skill, it is still surprising that the job got done where and when it did. Of course, a nineteenth-century monastery, and in particular the ancient Augustinian monastery of St. Thomas in Brünn (now Brno, Czechoslovakia), was not as unlikely a place for science as a twentieth-century study section might imagine. Scientists were then studying items that they could see and handle and had not yet escaped into the realms of extremely large and incredibly small things. An amateur, with his own hands and eyes and using inexpensive equipment, could make useful observations on plants, minerals, and the weather. Monks, with their food and lodging guaranteed and duties often not very demanding, had enough time for such activities. In particular, Mendel's monastery "pulsed with artistic and scientific energy,"[4] with almost every resident pursuing his own project. Mendel's principal biographer, Hugo Iltis, in *Life of Mendel*, has painted an engaging picture:

> On the farther side of the main building, where the library window and the little clock-tower look down on the garden, there is a special strip of garden cut off from the rest by a hedge and a path. This strip, adjoining the monastery wall is only one hundred and twenty feet long and a little over twenty feet wide. . . . Here there were to be seen, clinging to staves, the branches of trees and stretched strings, hundreds of pea plants of the most various kinds, with white and with violet blossoms, both tall and dwarf, some destined to bear smooth and others wrinkled peas. . . . During the fifties and sixties of the nineteenth century, any one passing this way might have seen, on fine spring days, a vigorous, short rather sturdily built and somewhat corpulent man engaged in a laborious occupation which would have been puzzling to any uninstructed observer. . . . He carefully took pollen from one plant and dusted the pollen onto another, then wrapped the treated flower in a paper bag to prevent any bee from pollinating the treated plant. . . . The investigator's patience was indefatigable. . . .
>
> When visitors came to the monastery, he would walk them around the gardens and then say, in a serious tone "now I am going to show you my children." He would smile at their astonishment, and then lead them to the peas.[5]

The benign and industrious, if slightly impish, person pictured in this account coincides with the one remembered by his students, who described him as kind and conscientious. No surprise, then, that he was elected prelate of the monastery later in the 1860s, and was described by the local newspaper upon his death in 1884 as "a man of the noblest character, one who was a warm friend, a promoter of the natural sciences, and a natural priest." Many men who might fit this obituary have walked the earth, however, but few have left behind a scientific legacy of this magnitude. Unless we choose to believe that Mendel was a disguised extraterrestrial, sent on a mission to move earthly biology onto schedule, then we must probe more deeply into his circumstances. Other accounts have left a less flattering, but more human picture.

He was born into a mixed German–Czech farming family in a century in which farmers had to struggle for survival. Two of his four sisters died at an early age, and his father had to sell the farm after an accident in which a tree fell on him. Young Johann (he changed it to Gregor when he began his religious studies) was an apt pupil, but suffered from bad health that was related to his extreme nervousness. His decision to opt for the priesthood arose more from his poverty than any obvious deep calling to religion. Mendel wrote of himself: "Now that he had been relieved of that anxiety about the physical basis of existence which is so detrimental to study, the respectful undersigned acquired fresh courage and energy."[6]

The tone of that fragment gives us some hint of his personality. According to his biographer,[7] he not only shunned all relationships with women, "holding strictly to his monastic vows," but in addition "he even found it difficult to enjoy any sort of intimacy with his clerical or monastic brethren," a sacrifice not required by his vows. Nor was he comfortable with the usual duties of a priest: "He was seized with unsurmountable fear at the sickbed and the sight of invalids, and he became dangerously ill."[8]

For this reason, Mendel was relieved of his pastoral duties. He turned his courage and energy rather to the plant and rock collections of his monastery and to teaching. The monastery had contracted to teach mathematics, philosophy, and theology in the local schools, and Gregor could at least help in these ways. Life as a teacher was less stressful, except on the occasions when he tried to pass the qualifying exams. On his second try, he became ill and quit after the first question.

In this life, Gregor found just the conditions of quiet and isolation that he needed to carry out his intensive eight-year-long experiments with peas. He was intelligent with a background in farming and some training in science. His fellow monks and his colleagues in the school took a lively interest in botany, although they were more interested in classification and crop improvements than the laws of heredity. Some of Mendel's hybrid peas demonstrated their worth on the monastery dining table long before they attracted scientific attention. Mendel's ample curiosity was not limited to peas. His students later claimed that he crossed grey and white mice in his rooms, although Mendel never mentioned it in public. Biographer Hugo Iltis speculated that "in the eyes of many clericalist zealots, it was sufficiently improper to take any interest in the natural sciences at all, and some persons must have regarded breeding experiments with animals as positively immoral." So, publicly, he may have avoided mice "for the less suspicious field of botany and the crossing of flowers."

External forces eventually ended Mendel's plant-breeding experiments. His specimens were devoured by weevils, he received bad advice from established botanists, and, finally, in 1868, he suffered the ultimate setback: election as prelate of the monastery. With this honor, he had at his disposal the wide spaces of the monastery gardens rather than the meager patch of land allotted to him by his predecessor, "but he was now hampered by lack of time instead of lack of space." Suddenly, he acquired many luxuries, but also many duties. He now lived in the prelate's apartments, a suite of rooms with costly furniture and paintings by old masters. He could enjoy the cuisine that made the monastery famous. However, he was expected to serve on the governing boards of banks and other local institutions, and he got involved in a long and exhausting struggle with the government over the taxation of monasteries.

He no longer had the immense amounts of time and energy needed for the plant experiments, although he could still observe sunspots through a telescope and make detailed observations on a tornado that uprooted trees and shook the monastery. He had become a pioneer in another role: that of the scientist whose career is effectively ended by promotion to an administrative position.

In the closing years of his life, Mendel "was overshadowed with gloom." He became fat and smoked up to twenty cigars a day. Kidney disease set in in 1883, and he died the next year. He continued to make scientific observations until two days before he died. His biographer wrote: "At the time of his death it had occurred to no one—least of all in the monastery—that he was a man of mark; and naturally enough, the few documents he had left behind were heedlessly torn up or committed to the flames" by his successor. Only well-bound books were retained. "Twenty years after his death his personality had become no more than a vague memory almost effaced by time." He was reported, however, to have said to a colleague on at least one occasion: "My time will come."[9]

What had he accomplished? We will have to jump forward to the twentieth century to find out. Before we do that, we will take note of certain other events of the year 1865. Mendel's theory and the following happenings represent threads that were separate at that time, but they were knit together closely by subsequent history.

PARIS, FEBRUARY 1865

The French Academy of Sciences endorsed a report of its own committee that ended the spontaneous generation controversy in France. A series of brilliant experiments by Louis Pasteur had made it clear that living things, even microorganisms like yeast, do not arise from sterile broths, milk, or urine, let alone wheat and sweaty underwear. New germs arise only from parent germs. In Pasteur's own words: "There is no circumstance known today in which one can state that microscopic beings are produced without germs." These results, said he, "carry a final blow to the doctrine of spontaneous generation."[10] Now, more than a century later, we can understand that bacteria are so complex that 4 million characters of genetic text are needed to describe them. They cannot easily arise by chance.

Heredity worked through a mechanism, not by magic, but that mechanism was still a mystery. To penetrate it, followers were needed to carry on Mendel's work.

FORT DELAWARE, PENNSYLVANIA, FEBRUARY 1865

The year in which Mendel presented his papers to the Brunn society was also the one in which the American Civil War was drawing to a close. On the days when peas were discussed in Austria–Hungary, General Sherman's armies were marching through the Carolinas. On April 9, General Robert E. Lee surrendered at Appomattox Court House, Virginia. Related to these events was the release in February of the Confederate officer Charlton Hunt Morgan from a prisoner-of-war camp at Fort Delaware. A resident of Kentucky and a U.S. diplomat before the war, Morgan had cast his lot with the Confederacy. He served as a raider in the cavalry unit headed by his more famous older brother John Hunt Morgan, "the Thunderbolt of the Confederacy," and had assisted his brother in a daring escape from prison camp.

It would have taken extraordinary powers of precognition for any individual in 1865 to see a connection between Gregor Mendel's presentation and the freeing of the confederate raider, other than that they happened at the same time. Yet the release of Charlton Hunt Morgan was an obvious prerequisite to his marriage on December 7 to Ellen Key Howard in Baltimore, Maryland. That event led in turn to the birth of their first child in Lexington, Kentucky, on September 25, 1866. The year in which Mendel's ideas were first presented to an uninterested world was also the year of the conception of the man who, more than any other, would bring them to prominence. We shall see in Chapter 3 how Mendel's factors found their place in Morgan's maps.

Many years later, in 1934, friends celebrated a visit to New York of Nobel laureate Thomas Hunt Morgan by bringing out a bottle of 1865 brandy. According to his biographers, Morgan "appreciatively cradled it in his arms like a baby and remarked on the appropriateness of the date." He was asked whether he had been born in 1865 and answered "No, that he had been born in 1866—but 1865 was the year he was laid down." In two ways, then, modern genetics was laid down in 1865.

EAST HAMPTON, LONG ISLAND, NEW YORK, 1865

George Huntington intended to follow in his father's footsteps as a physician. Six more years of study were still needed before he was to graduate from Columbia's College of Physicians, but he already carried within him the memory of an event that was to shape his career and perpetuate his name. As a child he liked to accompany his father on his horse and buggy rounds. One day they came upon a striking sight. He could remember it years later:

> I recall it as vividly as though it had occurred but yesterday. It made a most enduring impression on my boyish mind which was my very first impulse to choosing chorea

> as my virgin contribution to medical lore. Driving with my father through a wooded road leading from East Hampton to Amagansett we suddenly came upon two women, mother and daughter, both tall, thin almost cadaverous, both bowing, twisting, grimacing. I stared in wonderment, almost in fear. What could it mean? My father paused to speak with them and then we passed on. . . . [With this] my medical education had its inception.[11]

Huntington was to ensure his fame by publishing a paper in 1872, "On Chorea,"[12] which summarized observations he made during his medical education. In his own words: "The name chorea is given to the disease on account of the *dancing* propensities of those who are affected by it and it is a very appropriate designation." He first described another disease that had these symptoms then continued to discuss the one that had impressed him years earlier: "*hereditary* chorea." The disease was later to be known as "Huntington's chorea," or "Huntington's disease."

The symptoms of hereditary chorea first appeared in adult life. The disease then asserted itself, according to Huntington, "coming on gradually but surely, increasing by degrees, and often occupying years in its development, until the hapless sufferer is but a quivering wreck of his former self." The symptoms included an "irregular and spasmodic action of certain muscles," which increased as the disease progressed, as well as a tendency to insanity and suicide. Not all bodily functions ground to a halt at once, however. Huntington disapproved of the behavior of two afflicted men ("they can hardly walk, and would be thought by a stranger to be intoxicated") who yet "never let an opportunity to flirt with a girl go past unimpeded" and were "constantly making love to some young lady."

As the adjective *hereditary* implied, this disease ran in families. If a parent suffered from it, the child was likely to get it as well, but if the offspring went through life without it, the thread was broken and future generations were free of the blight. Although Huntington could not have known about it, this genetic pattern had been described and analyzed mathematically by Gregor Mendel. Hereditary chorea acted as a dominant trait in the same manner as the monk's smooth peas. The normal condition was recessive and behaved like the wrinkled pea trait. The rules for peas could also apply to disease.

BASEL, SWITZERLAND, 1865

The studies of another physician-in-training, Friedrich Miescher, were interrupted by an illness that lasted for a year. The son of a respected Basel family, serious and talented, but shy by nature, Miescher had wanted to be a priest, but this choice was vetoed by his father. Upon finally completing his medical training three years later, Miescher opted for a life in research. His attention was captured by a prominent biologist who had located his laboratory in Tubingen, Germany: Felix Hoppe-Seyler.

Many workers of that age believed that it was pointless to isolate and study the individual components of living cells. Their life-giving properties would change when they were separated from the cell, and the results would be meaningless. Hoppe-Seyler disagreed. He had investigated blood and crystallized the molecule hemoglobin that gave color to it. Further, he had shown that it could serve as a carrier of oxygen gas. Miescher, however, on arriving in Hoppe-Seyler's lab in 1868, selected another substance for investigation. Certain cells in pus resembled white blood cells: Pus would be the subject of his attention.

The bandages discarded by a nearby clinic would afford him all of the raw material he needed, but first he had to find some way to strip the pus off of them. He started to explore various treatments with alkali and acid and accomplished more than he expected. He first found that he could separate the nuclei of the cells in pus from the remainder of the material and then that he could isolate a whitish material from the nucleus. He named this new substance *nuclein.*[13]

Protein was the best appreciated of life's components at that time, and Miescher expected that his nuclein was another protein. It was so in part, but held another component as well, one that was rich in the element phosphorus. So surprising was this latter discovery that Hoppe-Seyler held up publication for several years until he could check it with his own hands. Miescher, in the interim, returned to Basel and continued his own study of nuclein. He found that salmon sperm provided a better source of the material than pus. He pursued his studies despite the suggestion of Karl von Nageli that the phosphorus was simply a contaminant of the protein. Miescher, however, showed that it was bound in a large molecule, which another worker named *nucleic acid.* Today we recognize two subdivisions of nucleic acids, called *DNA* (deoxyribonucleic acid) and *RNA* (ribonucleic acid).

Miescher continued with a life devoted to relentless hard scientific work. It took its toll, and he died of tuberculosis at age fifty-one in

1895. Three years before his death, however, he made a remarkable speculation in a letter to his uncle. He suggested that large biological molecules could carry the message of heredity "just as the words and concepts of all languages can find expression in twenty-four to thirty letters of the alphabet." He offered only proteins as examples of the carrier, however. He did not recognize that the hereditary substance was the one he had discovered with his own hands: DNA.

LONDON, 1865

Queen Victoria sat on her throne, twenty-eight years into a reign that was to last until 1901 and make her name an eponym for England in the late nineteenth century. For the past year she had been slowly emerging from the seclusion that she had adopted following the death of her husband, Prince Albert. She began to appear in public again and take a more active role in matters of state.

Among the matters that concerned her was the health of her twelve-year-old son Leopold. He was the eighth of nine children born to her, but the only one to suffer from the disease hemophilia. Leopold was thin at birth, frequently fell sick, bruised easily, and suffered from internal bleeding. He was also troubled by a chronically injured knee. Apart from the disease, he was an intelligent and affectionate, but willful child. His father had felt that he would outgrow his malady, whereas his mother protected him as much as she could. Despite many crises, he was to survive to adulthood, marry, and have a daughter. In 1884, he reinjured his weakened knee and died within a day. Queen Victoria believed for a time that his malady was "not of our family." It had not been so in the past, but it was from that time on.

A happier note was struck on June 3, 1865, with the birth of a new grandson of Victoria. The father was her oldest son and heir to the throne, Albert Edward ("Bertie"), the Prince of Wales. As the second son of his father, the new arrival, George, did not stand directly in line for the throne. He was, however, described as "a jolly little pickle." He did not suffer from hemophilia.

In that same year, in England, *Macmillan's Magazine* published a two-part article by Francis Galton, a cousin of Charles Darwin. Galton's ideas were later expanded in a book published in 1869, *Hereditary Genius*.[14] Galton was born in the same year as Gregor Mendel, but approached heredity in a very different way. He had observed the results of selective breeding in plants and animals and, inspired by his cousin's work, advocated that humans should direct their own future evolution by judicious marriages. He later gave the name *eugenics* to the field that he founded.

In his own work, Galton collected pedigrees and biographical data of accomplished individuals, noting that characteristics such as musical ability and political success often ran in families. Although such observations may have been valid in a number of instances, the conclusions that he drew from them went far beyond his data. He presumed that the effects he saw were entirely due to heredity, neglecting environmental and cultural influences. Such complex attributes as musical ability and other forms of achievement were thus equated with the height of pea plants and the color of their flowers.

As we have seen, genetics from Aristotle on has been littered with a host of incorrect theories that were stated with profundity and at great length. These fancies were brought to an end in the decade of the 1860s, when the experiments of Pasteur, Mendel, and Miescher laid the foundation for the new science of genetics. Ironically, in that same decade, Francis Galton planted the seeds of new fantasies concerning inheritance that were to have far more devastating effects on humans than the previous set. Both his approach and the scientific one would develop slowly during the remainder of the nineteenth century and come to full flower during the twentieth century.

3

1910: THE FLY ROOM

BRÜNN, AUSTRIA–HUNGARY, 1910

On October 2, 1910, a memorial to Gregor Mendel was unveiled in the center of Brünn. The suggestion that such a monument be erected initially caused some astonishment in the town. Biographer Hugo Iltis noted that "twenty years after his death his personality had become no more than a vague memory almost effaced by time."[1] The scientific community had come awake as if to a thunderclap, however, when the biological vanguard suddenly arrived at the height that Mendel had staked out. In the spring of 1900, six papers from three different research groups resurrected his results and affirmed their importance. These developments did not penetrate readily into the consciousness of the Brünn burghers. Upon viewing a picture of Mendel displayed in a bookshop in town and learning that he was somehow connected with inheritance, they concluded, according to Iltis, that he was being honored because he had left the town a large bequest!

Other townspeople, feeling themselves modern thinkers, objected for a time to the display of a monument to a priest. Their objections and those of others failed, and the memorial was commissioned. The protesters may have been mollified when they saw the actual work. The upper part did display a statue of a man in priestly robes, with a beatific expression on his face. Below the statue, however, were the words:

TO THE INVESTIGATOR
P. GREGOR MENDEL
1822—1884
ERECTED 1910 BY THE FRIENDS OF SCIENCE

Most noteworthy of all was the illustration on the center of the pedestal, which showed a nude man and woman, kneeling, with hands clasped. The message was clear. In his work, Mendel had commented on more than peas. He had shed a deep light upon the vital human processes in which he, as a priest, could not participate directly.

LONDON, MAY 20, 1910

The magnificence of the event was captured years later by Barbara Tuchman in her book *The Guns of August*:

> So gorgeous was the spectacle . . . that the crowd, waiting in hushed and black-clad awe, could not keep back gasps of admiration. In scarlet and blue and green and purple, three by three the sovereigns rode through the palace gates, with plumed helmets, gold braid, crimson sashes and jeweled orders flashing in the sun.[2]

She had described the funeral of King Edward VII of England. As Victoria's oldest son, Prince Albert Edward, he had been known for his sporting and gaming and for his extramarital affairs. In the nine years since accession to the throne, however, he had proved so skilled in statesmanship that Kaiser Wilhelm of Germany (a grandson of Victoria on his own) commented "He is Satan. You cannot imagine what a Satan he is." Through personal diplomacy, Edward had created understandings with France and Russia and improved relations between England and other European nations. The Kaiser felt encircled.[3]

Edward's successor was George, the "jolly pickle," who became upon his accession King George V. He had moved into the line of succession when his older brother died in 1892. His cousin, the Kaiser, was more approving of George than of his father. He had earlier termed George "a very nice boy." Now Wilhelm added: "He is a very thorough Englishman and hates all foreigners, but I do not mind that as long as he does not hate Germans more than other foreigners."

ST. PETERSBURG, RUSSIA, 1910

Another granddaughter of Victoria was contending with a problem that her grandmother had faced, but which had bypassed George, Wilhelm, and their families. The concerned woman was Alexandra, empress of Russia, wife of Czar Nicholas II. Her son, Czarevitch Alexis, suffered from hemophilia.

Alexis was the fifth child, but only the first son, of the royal couple. He stood first in line for the throne, and his birth in 1904 had been particularly welcome. Shortly thereafter, however, he began to suffer from the bruises, internal hemorrhages, and unchecked bleeding that typify the disease. As nosebleeds and minor injuries could place his life in danger, he was kept under constant watch: Two sailors were assigned to keep him constantly in their sight as he played.

Alexis was destined to die at a young age, but not of hemophilia. He and his entire family were murdered in a basement in Ekaterinburg, Siberia, on July 31, 1918, during the Russian Revolution. According to historical accounts, Alexis, his parents, and all but one of his sisters were killed in the first volley of bullets. The youngest sister, Anastasia, had only fainted. When she awakened and screamed, however, she was dispatched with bayonets.

Robert K. Massie, in his book *Nicholas and Alexandra*,[4] has suggested that Alexis's hemophilia played an important role in bringing about these tragic consequences. The child's suffering led his mother to call on the notorious monk Gregory Rasputin for assistance. Rasputin could ease the child's pain through hypnosis or psychological assistance, but his advice on matters of state is said to have contributed to the downfall of the monarchy.

WASHINGTON, D.C., MAY 5, 1910

The twenty-fifth meeting of the Association of American Physicians was held in the New Willard Hotel with a distinguished roster of physicians in attendance. Dr. James Herrick of Chicago chose this occasion to present a case study on a patient that he had started to treat six years earlier and resembled no other condition that he had ever encountered. Herrick was no routine practitioner, but a professor at his alma mater, Rush Medical College. A historian has described him as "a scholarly and perceptive clinical observer."He had sharpened his skills by undertaking a project in the laboratory of Nobel Prize winner Emil Fischer, a brilliant master of chemical analysis. Herrick, forty-nine in 1910, was to go on to win a number of medals and awards in his medical career.

His case concerned a twenty-year-old black West Indian who had come to the United States in 1904 for professional study. He had been troubled since his arrival with a cold, coughing, and fever, which eventually led him to seek medical help. Herrick did not treat him routinely, but took his history and performed a thorough examination. The student had a low red blood cell count, enlarged heart and lymph nodes, and a yellowish tinge in his eyes and suffered from dizziness and a shortness of breath. He had been disinclined to exercise for the past three years. Earlier in life he had suffered from ulcers and lesions on his legs, which had left many scars. His most noteworthy and unique symptom, however, was the presence of red blood cells of unusual shape: "What especially attracted attention was the large number of thin, elongated, sickle-shaped and crescent-shaped forms." This appeared to be a key observation: "Some unrecognized change in the composition of the corpuscle itself may be the determining factor."

Little more than rest and nourishing food could be prescribed for the condition. The patient improved, but required additional treatment over two years later. At that time he remained in the hospital for two months with a malaise that included pain in his back and limbs. After some additional time, Herrick decided to publish his observations: "Peculiar Elongated and Sickle-Shaped Red Blood Corpuscles in a Case of Severe Anemia," in the *Archives of Internal Medicine*. He noted that "not even a definite diagnosis can be made."[5] Before publishing the case, however, he presented it orally at the Washington meeting.

According to one historian, "there was no discussion of Herrick's paper. None of his colleagues recognized the unique disorder that he described." Herrick himself apparently did not consider the matter one of great importance. In his own autobiography, written in 1949, "there is barely a passing reference to the disease."

The condition had great importance, despite its lack of medical impact at that time. For centuries it has been, and it continues to be, the most significant genetic disease afflicting black populations in Africa and the New World: a killer of multitudes, particularly in infancy. Herrick was the first to describe its central feature, the one that gave the disease its name: sickle-cell anemia.

HEIDELBERG, GERMANY, 1910

A torchlight parade was held to commemorate the award of the Nobel Prize to Albrecht Kossel, professor of physiology in Heidelberg University. Kossel, another pupil of Felix Hoppe-Seyler, had continued the investigations on the chemistry of the cell nucleus that Friedrich Miescher had pioneered. Like Miescher, he felt that the proteins were the gene stuff with nucleic acids having some lesser role. They were there, however, and just as Mount Everest, they had to be conquered. Because of their complexity, the task would require over eighty years of excruciating hard work.

Although many chemists in different laboratories were to participate in this task, one scientist in each period was foremost in the effort. In the closing years of the nineteenth century, it was Kossel.

Among his discoveries were several whose significance he could not imagine in his wildest dreams. The chemist's job was to determine, atom by atom, how the nucleic acids were put together. The best way available then was to knock them apart into their component pieces, identify each piece, and then figure out how they were put together. Kossel was fated to collect the four substances, called *bases*, that have immortal importance in biology. Their names can be found in the schoolbooks and newspaper columns of today: adenine, cytosine, guanine, and thymine, more often just abbreviated as A, C, G, and T. Together, they make the alphabet of the genetic text and carry the hereditary message within DNA.

One of this famous foursome had been identified earlier in a less glamorous context. Guanine is a principal component of guano, the bird droppings that are so prominent on Pacific coastal rocks and elsewhere. It also serves to give fish and reptile scales their shiny appearance. The other bases were discovered by Kossel himself. Kossel could not have forseen that the first initials of these routinely assigned names would be listed within journals in arrays of thousands and ultimately be stored in binary code within computers by the millions and billions.

NEW YORK CITY, 1910

Kossel's successor in the line of great nucleic acid chemists was at work in his laboratory in the Rockefeller Institute for Medical Research. Phoebus Aaron Levene had been born in Russia and received a medical degree there, but fled to the United States in 1891 to escape persecution as a Jew. He joined the newly founded Rockefeller Institute in 1905 and worked there until his death in 1940.

Miescher had noted the presence of phosphorus, or phosphate, in nucleic acids, and Kossel had identified the bases. Levene discovered the remaining components, members of the class we call *sugars*. Two types of nucleic acid exist in nature, and each has its own sugar to identify it. Their chemical names, deoxyribose and ribose, are of importance to us here because the first initial of each has become immortalized in the abbreviations DNA (for *d*eoxyribo*n*ucleic *a*cid) and RNA (for *r*ibo*n*ucleic *a*cid).

With the pieces identified, the job that remained for chemists in the decades ahead was to deduce their connection. Much of it was done by Levene. No glamour was attached to the work in those days: The media paid little attention to DNA and RNA. The discovery of their roles in biology lay far in the future. Ironically, an incorrect theory of Levene helped delay that discovery. Eager to complete the effort, he suggested that DNA was a relatively small molecule with only four subunits: Each molecule contained exactly one A, C, G, and T. He was off by a factor of millions.

CHICAGO, ILLINOIS, SEPTEMBER 19, 1910

Clarence B. Hiller had been disturbed by the entrance of an intruder into his home. He strugged with the stranger at the top of the staircase. Both fell down, and two shots rang out. Hiller was killed and his murderer escaped.

Shortly thereafter, two off-duty policemen apprehended a suspect in that neighborhood. The man, Thomas Jennings, had a wound on his arm and bloodstains on his clothing. Unfired cartridges in his home matched those found in the Hiller house. Jennings, however, vigorously denied having been near that house.

In the trial of Jennings the next year, a new type of evidence was introduced that had not yet been considered in courts in the United States, although the English had used it since 1901: fingerprints. Prints that had been left in fresh paint on the Hiller porch matched those of Jennings. As reported in the book *Fingerprinting*, by Eugene B. Block, "the defense made vigorous objections, arguing that fingerprinting, if in reality it was a science, was in no way recognized in the laws of

Illinois."[6] Experts in the procedure argued in favor of its use. Jennings was found guilty and sentenced to death.

When the Supreme Court of Illinois reviewed the case, it upheld the procedure and the verdict, noting that "when photography was first introduced it was seriously questioned whether pictures thus created could properly be used in evidence." These events were to be echoed more than seven decades later, when another revolutionary scientific technique for determining human identity was introduced in the courts.

The new method depended on DNA.

NEW YORK CITY, MAY 1910

Thomas Hunt Morgan was fed up with *Drosophila.* This tiny insect, more commonly known to us as the fruit fly, had become fashionable in biology laboratories within the last decade. Fruit flies bred rapidly, requiring ten days for a generation, lived happily on mashed, rotting bananas, and could be housed by the thousands in small milk bottles. No good had come to Morgan out of their presence in his Columbia University laboratory, however. "There's two years work wasted," he had commented to a colleague. "I've been breeding these flies for all that time and have got nothing out of it."[7]

Morgan had put considerable effort into one line of work. He wished to test again the old, somewhat discredited idea that acquired traits could be inherited. To do this, he bred generation after generation of *Drosophila* in the dark to see if descendents would be born with atrophied eyes. As reported by his biographers, Ian Shine and Sylvia Wrobel,

> The sixty-ninth generation emerged momentarily dazed . . . but the flies soon recovered and flew to the window as if nothing had ever happened. . . . Morgan often jokingly said that he did three kinds of experiments—those that were foolish, those that were damn foolish, and those that were worse than that.[8]

His genetic experiments with these insects also seemed to be headed for one of the above categories. He had exposed them to a variety of conditions—acids, alkalis, radium, and X rays—in an effort to produce some new mutations, new traits not present in the parents that would be transmitted to the offspring. After two years, no new mutants had been seen. One problem, of course, was that he did not know what he was looking for. The field was new—even the term *genetics* had only been coined in 1906, and *gene*, in 1909. The word *mutant* may have brought to the mind of Morgan the type of gross deformities that the same word suggests to film viewers today.

One day in May 1910, his luck changed. He peered into one of his milk jars and a white-eyed male fly looked weakly back at him. As *Drosophila* normally have red eyes, this was the mutant he had been waiting for. As reported by Shine and Wrobel:

> The fly was feeble. Morgan is said to have carried it home at night to sleep in a jar by his bed, returning it to the laboratory during the day. There it mustered enough strength to mate with a normal red-eyed female before dying, leaving the mutated gene in what was to become a prodigious family line.[9]

With the descendents of "white-eye" safely enlisted in his fly army, Morgan could now march on his main objective: a critical examination of the ideas of Gregor Mendel and his followers. As a man dedicated to experimental science, he distrusted the hypothetical factors of Mendel, suspecting that they had little basis in reality. His own experience in trying to verify Mendelian principles using the coat color of mice (hardly the best choice; recall that sixty-three genes are involved) had not magnified his enthusiasm for this topic. The previous year, in 1909, he had attacked Mendelian ideas in an "almost bitter fashion" at a meeting of the American Breeders Association. Now he could put them to another test.

Fortunately, he had written three years earlier in his own textbook, *Experimental Zoology*: "The investigator must . . . cultivate also a skeptical state of mind toward all hypotheses—especially his own—and be ready to abandon them the moment the evidence points the other way."[10] Further, Morgan practiced what he advocated. Through his fruit fly studies he became Mendel's inheritor, rather than his detractor: the man who, more than any other, confirmed the basic ideas and extended them far beyond their previous limits. In the words of biographer Garland Allen:

> It was to be Thomas Hunt Morgan and his closely knit group of young colleagues who, in the period after 1910,

> would give a physical basis to Mendelian theory by demonstrating the structural relationship between genes and chromosomes. . . . It would be his role to exploit the new vistas of heredity that Mendel had opened up.[11]

The first Morgan *Drosophila* paper was sent to the journal *Science* that July and was followed by a procession that almost resembled a cloud of fruit flies. Over the next five years, Morgan, with three close collaborators, Alfred H. Sturtevant, Calvin B. Bridges, and Hermann J. Muller, bred generation after generation of flies with white eyes, vermillion eyes, pink eyes, bent wings, rudimentary wings, yellow bodies (brown was normal), and other mutational variations. They found that the traits almost always followed Mendel's rules. Remember, Mendel had discovered that his factors acted independently. A wrinkled pea could be yellow or green and so could a smooth one. This was not always the case with the flies. Pink-eyed males and females appeared in equal numbers, but white-eyed boy flies vastly outnumbered the white-eyed girls. Some traits were unrelated, à la Mendel, but others definitely had their destinies linked.

Mendel's factors appeared as so many items in a grocery shopping cart. Chance determined which got packed together and which ended up in separate grocery bags. In Morgan's lifetime, however, the chromosomes had been discovered. (Their name is derived from *chroma*, color, and *soma*, body.) These tiny structures exist in double sets in the nuclei of cells, one set derived from each parent. They were now plausible sites for the location of the factors, or genes. *Drosophila*, for example, had four chromosome pairs. If each trait belonged to a particular chromosome, then they should travel in four groups, like four separate classes of school children visiting a museum at the same time. White eyes, miniature wings, and a yellow body traveled together in the same group. A fly with those three features was likely to pass them on together to his or her descendents.

Yet this exception to the Mendel laws had an important subexception. Occasionally traits switched groups and separated from one another. Let's get back to our vision of classes of school children in a museum. Imagine that two teachers are responsible for each class, and that each teacher moves with half of the class following behind him or her in a tight single file. Every so often, the two files line up alongside one another, select some arbitrary place in the line, and switch all members as a unit behind that place. If you were a child with a close friend in the same line, you would be happy if she were next to you in the line, for you would get separated only if the splitting point was exactly between you. The worst possibility would be for one of you to be at the very head of the line and the other at the end, for you would be separated into different groups the very first time that switching, or "crossing-over," took place.

The Morgan group, with their hordes of flies, could follow many linked traits (each equivalent to the child and his friend) for many generations. The rule was simple: The more often a linked pair got separated, the further apart they were on their chromosome. Thus a map could be made for each chromosome, listing the traits in a linear order, with rough distances between them (the unit of distance was later named the *morgan* by biologist J.B.S. Haldane). This "map" was, of course, more like the plan for an unbranched railroad or subway line than a highway map with its three-dimensional network of roads.

During the next few years, dozens of traits were mapped to *Drosophila* chromosomes by the Morgan group (the number of genes located by all scientists had grown to more than three thousand by 1988). In 1915, the four co-workers published their classic book on genetics, *The Mechanism of Mendelian Heredity*,[12] and the flow of recognition and honors began. They included, for Morgan, the Darwin Medal, the Copley Medal, presidency of the National Academy of Sciences, numerous honorary degrees, and, ultimately, the Nobel Prize in physiology or medicine in 1933. In the words of his biographers, Shine and Wrobel, "Morgan was established as the twentieth-century Mendel and attracted scientists and visitors from all over the world prepared to kneel at the shrine of some scientific deity" (certainly in that respect he differed from the poor monk).

Those who arrived on such a pilgrimage were likely to be appalled by the conditions at the shrine. The "fly room," as Room 613, Schermerhorn Hall, came to be called, was a mess. Eight desks and a small preparation table (for mashing the bananas) were crammed into a small area of 23 by 27 feet, which also held hundreds of half-pint cotton-stoppered milk bottles "borrowed" from the Columbia University cafeteria. The room also housed many thousands of fruit flies, both enslaved and free. The captives were bred, counted, and ultimately killed and thrown into a jar of oil called "the morgue." The lucky escapees could leave, but many chose to join a group of native free flies that lived in the room. They gathered around a stalk of bananas that hung conspicuously near the entrance to the room, perhaps to plot revolution. The natives had bred, without scientific help, in the laboratory garbage can that was never fully cleaned. For an escapee to tarry involved some risk, however. On one occasion, Morgan's wife Lilian, herself a trained biologist, found that an important mutant had slipped away. Unruffled, she simply inspected the hordes loose in the laboratory and located her desired quarry by the window.

Adjacent to the fly room was Morgan's own small office, which also housed an assistant. His own roll-topped desk was also a mess. Much of his data were kept on the backs of envelopes and on old scraps of paper. He shared the repetitive, endless work of counting flies, but with one difference: when he was done with the flies, "he squashed them on the porcelain counting plate that was already covered with the mold of last week's previous counts." Occasionally the assistant, often a graduate student's wife, would clean the plate, but he would take little notice and squash on with a vengeance. This attitude did not stop at the limits of the laboratory, but could carry over onto his own personal appearance.

Photographs and accounts portray him as a healthy, striking man, six feet tall, although slightly stooped, with a full beard and moustache, dark hair, and "startling" blue eyes. They show him in the conventional professorial dress of jacket, tie, and white shirt. No hint of eccentricity appears. His originality in dress displayed itself in the way he handled minor crises. When he could not find his belt, for example, he would tie his pants with a string. A hole in his shirt was pasted over with white paper. When his collar was torn off before a lecture he said, "Oh never mind, Bridges will fix it with adhesive tape."[13] On more than one occasion, he was mistaken for the janitor.

This style of dress is a familiar one to me. In the neighborhood where I grew up, there could be found academic *schlemiels*—misfits who had never learned to care for themselves. Their combination of emotional denseness, physical unpresentability, and a sharp intelligence that was limited to abstract matters ultimately steered them into the only position suitable for such a combination—a professorship. Morgan was definitely not of that type, as he was fully aware of the image he projected. In Shine and Wrobel's words: "There is no doubt that Morgan enjoyed exactly this kind of laboratory. He was by nature a careless and sloppy man, but he also delighted in shocking others, in playing the imp."[14]

The chaotic and even filthy laboratory that Morgan preferred would seem a tremendous handicap in a science that depended on keeping track of thousands of flies over many generations with a careful mathematical analysis of the results to follow. Despite this, it worked for him. The lab bred cockroaches, but also a sense of informality, cameraderie, and ease of communication. Morgan's integrity, fatherly qualities, lack of pomp, sense of humor, and joy in his work made him an ideal leader for the closely knit group that he gathered for his most memorable achievements.

He first encountered Sturtevant and Bridges in a Columbia biology class in 1909. They were both teenagers and undergraduates at the time. They were to remain with him for the remainder of his working life, accompanying him during summers at Woods Hole, on sabbatical at Stanford, and, subsequently, in 1928, on his move to the California Institute of Technology.

Bridges attracted Morgan's attention on one of his early visits to the lab by spotting the first vermillion-eyed mutant, without the aid of a magnifying glass. Sturtevant, who was color blind, could not match that feat. He earned his honors in other ways, such as by developing the first genetic maps. The two were hired in 1910 as bottle washer and research assistant, respectively, but stayed on to complete their undergraduate and doctoral degrees and take on faculty positions. The fourth associate, Muller, also started as a Columbia undergraduate. He was less closely linked and departed and returned several times. He left permanently, in 1920, for the University of Texas. Muller followed a stormy and politics-ridden path that subsequently led him to Germany and the Soviet Union. His career as an academic vagabond ended at the University of Indiana, where he received the Nobel Prize in 1946 for his work on the mutational effects of X rays.

Thus, we have the environment in which modern genetics was laid down: one middle-aged professor and three eager young helpers with two future Nobel Prize winners among the four. According to Sturtevant:

> There was an atmosphere of excitement in the laboratory and a great deal of discussion and argument about each new result as the work developed rapidly. . . . This group worked as a unit. Each carried out his own experiments but knew exactly what the others were doing, and each new result was freely discussed. There was little attention paid to priority or to the source of new ideas or new interpretations. . . . I think we came out somewhere near even in this give-and-take and it certainly accelerated the work.[15]

Thus, Morgan could provide his depth of experience, but be aided by the others in areas such as the mathematical analysis of data, which was not his strength.

In this environment, research was the center. Morgan disliked administrative duties and had little interest in teaching: "Excuse my big yawn, but I just came from one of my own lectures," he once commented. Morgan "would bear no discussion of a man's politics, religion or private life." He defended the rights of others to lead their lives as they wished. Morgan, for example, was uninvolved in politics and "almost puritanical in matters of sex,"[16] but shared the laboratory comfortably with Bridges, who was a Bolshevik sympathizer and an advocate of free love. The work intensity spread to other students who joined the laboratory. They would often take their bottles home and score the mutants on the kitchen table. When asked what his father did for a living, one child responded, "My daddy counts flies for Columbia University."

In its peak period, the work of this group proceeded without an aid that is considered essential by its modern counterparts: external funding. Large-scale federal grants did not exist. Salaries for teaching assistants and some basic equipment were provided by the university. Most of the remaining costs came from Morgan's own pocket, including meals for the fruit flies, which initially cost a dime a day. The one essential ingredient was Morgan himself with his scientific tenacity and his unique personality. He was able to combine dignity and reserve with casualness and informality. He brought these qualities together in a personality that provided leadership and inspired loyalty. What were the factors that shaped this man?

We have seen his close connection at birth to heroes of the Confederacy, a connection that followed him through his life. At a post-Nobel celebration in his hometown, for example, he was billed as the nephew of the "Thunderbolt of the Confederacy." Prior to the Civil War, his family had been as American as "The Star-Spangled Banner." In fact, his great-grandfather *wrote* "The Star-Spangled Banner." Another great-grandfather was a Revolutionary War hero and governor of Maryland.

The solid sense of identity that can come from well-defined roots undoubtedly helped him build a life in which he could follow his deepest interests and be comfortable in his own personality, even in its contradictions. He combined, for example, a deep generosity in important matters with a stinginess on minor ones. His Nobel Prize money was divided among Bridges, Sturtevant, and his children.[17] When his assistants ran into financial difficulties, he would help them and even pay their salaries out of his own pocket. On the other hand, he was reluctant to spend institutional funds for equipment, preferring lenses to microscopes ("Morgan thought fancy equipment was a kind of decadence"[18]); he would destroy files rather than purchase another filing cabinet; and he bought his Christmas trees on Christmas Eve, because prices were lower at that time.

His passion, from his earliest days, was nature. As a boy, he organized collecting expeditions, hunting for fossils, butterflies, minerals, and other curiosities. He studied natural history at the State College of Kentucky and emerged as valedictorian (of a class of three). His career followed the rising curve that might be expected of a bright, dedicated, and well-connected young biologist: a Ph.D. at Johns Hopkins; summers at the Woods Hole Marine Biological Laboratory (establishing a tradition that was to become a family ritual for much of his life); an associate professorship at Bryn Mawr; a year at the Zoological Station in Naples, Italy; and finally, in 1904, appointment as professor of experimental zoology at Columbia.

In that same year he married the Bryn Mawr biologist Lilian Sampson, who devoted the next decades primarily to managing their household and raising their four daughters. Although he was thirty-eight when he married, biographers Shine and Wrobel noted that "this seems to have been his first romance."[19] Until then, there had been at least one feature of his life that he shared with Gregor Mendel.

The marriage was, to all observers, a happy lifelong one, and he entered a comfortable routine, centered about his work. For most of the year the Morgans lived in a house near Columbia. He put in a full day in his laboratory seven days a week, taking his meals at home and playing with his children after dinner. Sunday differed only in that he omitted a usual predinner handball game with other faculty. He was indifferent to religion. In the summer, the family moved to another large house they owned at Woods Hole, and the work continued at the laboratory there.

Morgan seemed to be quite comfortable financially at the time, with income from stocks and bonds. (One biographer noted that he was kin to financier J.P. Morgan.) If any frustration had entered his life during the first decade of this century, it undoubtedly came from his failure to learn much about his main interest, the development of embryos. This proved to be to his advantage, however, as it redirected his interests to fruit flies and genetics, where he made his most important discoveries. Only now, decades later, are biologists coming to grips with the problems of development, and one of the best organisms for this study is the fruit fly.

Morgan spent the last part of his career as head of the new Division of Biology at the California Institute of Technology, surrounded by equipment that he was reluctant to use. He had hope to the end that he could make some fundamental discovery in development. He died in 1945, his lifetime spanning the period from Mendel's paper to the discovery that genes are made of DNA.

POSTSCRIPT: NEW YORK CITY, 1990

I set out on a pilgrimage across the Columbia campus with my friend Gerald Feinberg, a physics professor at that university. We wanted to locate the remnants of Morgan's laboratory, if any existed. Our prospects were not encouraging. Schermerhorn Hall still stood, of course, with its traditional Columbia red bricks and green roof. It formed one side of a small ornate plaza in the northeast area of the campus that must have looked much as it did in Morgan's day. The Biology Department had left for more modern quarters, however, over a decade ago. My friend had learned that Art History now occupied the top floor. The former fly room had been partitioned and renumbered, and was now Room 920, Schermerhorn Hall. A woman named Lise Hazen had the key, however, and could meet us and show us what there was.

Our arrival on that floor was inauspicious. It had been elegantly decorated, with ornate pillars and lighting and reprints of art works hung on the corridor walls. There was one hopeful sign, however; an old, thick, wooden door standing ajar, Room 920, beckoned at the end of the corridor. Within was an old, although not old enough, biology lab. Its dimensions, 18 by 14 feet, were smaller than the original fly room, but represented a surviving fragment. The workbenches lining the sides of the lab were of recent origin, but the two old wooden windows opposite the door were likely to be those that Morgan and his group had used. As no plaque existed, I felt that they and the door would have to stand silently as monuments to his work.

Once again I was wrong. A better marker was in place. Lise Hazen arrived; no indifferent keeper of the keys was she, but a highly dedicated, white-coated lab technician. Lise led us into an adjoining small preparations room and threw open the door of the incubator (it looked like a refrigerator to me) that stood there. Inside were half-pint milk bottles filled with swarming *Drosophila* living on mashed banana!

A fragment of biology still survived atop Schermerhorn. The remnant of the fly room and another much larger room on that floor were used for laboratories in introductory biology. The undergraduate inheritors of Bridges, Sturtevant, Muller, and Morgan now repeated their fruit fly breeding and gene mapping experiments as part of their normal training. I was no Calvin Bridges. I could recognize *Drosophila* as the familiar specks that I would shoo away from the pineapple in the supermarket, but I could not even make out their eyes, let alone the color. Lise put one normal colony temporarily to sleep with ether and placed them under the microscope. Suddenly, I could see one magnified to full picture-book glory with its eyes an unmistakeable bright ruby red. A bottle of white eyes was also present. They were less striking with their eyes actually a muted pale ivory color.

This was 1990, however, and not 1910. No stalk of bananas hung by the door. The rounder milk bottles were authentic and marked "half pint," but alongside were ones that were squarish in shape and marked "*Drosophila*." The *Carolina Drosophila Manual* now offered a choice of fly foods with or without bananas. Vials full of vigorous white- or vermillion-eyed *Drosophila* could be purchased, or if those colors did not meet your decorator requirements, apricot-, yellow-, brown-, cinnabar-, scarlet-, and claret-eyed mutants were also in stock.

The 1990 *Drosophila* models came with a variety of body and wing shapes and colors as well, affording much more consumer choice than 1910. You were advised to anesthetize your flies with FlyNap (registered trademark of Carolina Biological Supply Co.) rather than the more hazardous ether. If you use too much ether, your flies become sterile; a bit more than that and they are dead. A drawing showed a content, smiling fruit fly in a bed, snoring away. Another section of the manual provided tips on how to keep your flies healthy, free of mites and bacteria, and the proper way to cook their meals if you did not want to bring food in.

Gary, Lise, and I toured the large student lab, as well as the small one. Old doors, wall charts, and specimen bottles with pickled fish and octopuses remained as survivors of the old era. Finally it was time to go, but I wanted to linger. I felt that if I only waited a bit, a distinguished old gentleman with dark, greying hair and a grey beard and moustache would show up. He would roll up his sleeves and comment that things have been neglected here for too long, and, grabbing the nearest fly bottle, he would start to work.

COLD SPRING HARBOR, LONG ISLAND, NEW YORK, OCTOBER 1, 1910

A day before the Mendel monument was dedicated, the Eugenics Records Office started its operations at this peaceful site on the north shore of Long Island. It stood next to an existing Center for Experimental Evolution, run by the Carnegie Institution of Washington, D.C. Also nearby was a Biological Laboratory, affiliated with the Brooklyn Institute of Arts and Sciences. The person in charge of these enterprises and responsible for the establishment of the Eugenics office was a former Harvard biology professor and eugenics advocate named Charles B. Davenport.

In February of that year, Davenport had convinced Mrs. E.H. Harriman, the wife of the railroad magnate, to provide support for an office of this type. He wrote in his diary that it was "A Red Letter Day for humanity." Red is, however, also a color used to enter losses into accounting books. Davenport's gain proved to be a loss for humanity. Historian Daniel Kevles in his book *In the Name of Eugenics* summarized the outcome:

> If so much negative eugenics was written into law in America, it was probably not because American eugenics activists could draw upon the publications or allies of the Eugenics Records Office—the principal scientific and authoritative institution in its field.[20]

Sir Francis Galton, the founder of eugenics, was still alive, although he had been born in the same year as Gregor Mendel. His reputation had flourished, and he had been knighted the previous year. He would pass away January 1911, but his ideas, like Mendel's, would continue. In fact, the rules discovered by the monk had been seized upon enthusiastically by advocates of eugenics such as Davenport and used to furnish an updated, expanded mythology. The list of human characteristics controlled by heredity now contained not only such valid items as hemophilia and Huntington's disease, but also very dubious ones such as pauperism, feeblemindedness, nomadism, shiftlessness, and eroticism. Also included were some medical conditions such as pellagra, now recognized to be due to a vitamin deficiency. Davenport, according to Kevles, "thought in terms of single Mendelian characters, grossly oversimplified matters, and ignored the force of environment." As the mathematical eugenicist Karl Pearson remarked to Galton, "our friend Davenport is not a clear strong thinker."

Thinking of this sort promoted dreadful science. Far worse, however, were the effects of additional claims that linked the above hereditary traits to particular ethnic groups. In Davenport's 1911 book, *Heredity in Relation to Eugenics*, for example, he commented that Italians tended to "crimes of personal violence," whereas Jews showed "the greatest proportion of offences against chastity and in connection with prostitution, the lowest of crimes."[21] In the same vein, another believer, Madison Grant, wrote in 1916:

> Whether we like to admit it or not, the result of a mix of two races in the long run gives us a race reverting to the more ancient, generalized and lower type. . . . The cross between a white man and a Negro is a Negro, . . . and a cross between any of the three European races and a Jew is a Jew.

These claims appeared in a book *The Passing of the Great Race*, which *Science* called "a work of solid merit," according to historian Mark H. Haller.[22]

As we have seen, there was legitimate scientific work on genetic linkages going on at the time. By the application of ingenuity and hard work, Thomas Hunt Morgan and his colleagues had established that white eyes and miniature wings were linked in the fruit fly. They would do more with that fly, but the establishment of the first human genetic maps lay decades ahead in the future. The eugenecists of that time were not really interested in science, however; they had another agenda: bigotry. Their science was based on the ideas of the ancient Greeks that heredity was rooted in blood.

For example, Albert Wiggam, a eugenic popularizer, wrote in 1924:

> Heredity has cost America a large share of its labor troubles, its political chaos, many of its frightful riots and bombings, the doings and undoings of its undesirable citizens. Investigation proves that an enormous proportion of its undesirable citizens are descended from undesirable blood overseas. America's immigration problem is mainly a problem of blood.[23]

If good blood is mixed with bad, it becomes adulterated.

The harvest of this nonscience and bigotry for America was a series of ill-advised laws governing immigration, sterilization of the feebleminded, and miscegenation. Abroad, the results were far worse. In 1910, a young German biologist, Dr. Eugen Fischer, was involved in studies of people of mixed descent in the colony of Southwest Africa. As cited in Benno Muller-Hill's *Murderous Science*, his subsequent recommendation was "we should provide them with the minimum amount of protection which they should require, for survival as a race inferior to ourselves, and we should do this only as long as they are useful to us."[24] As the century moved along, so did Fischer's ideas. In

1939, as rector of the University of Berlin, he would write, "When a people wants, somehow or other, to preserve its own nature, it must reject alien racial elements, and when these have already insinuated themselves, it must suppress them and eliminate them." His daughter later characterized him in an interview with Muller-Hill: "My father was a kindly man. So sensitive. He was brave too. And unassuming. A piece of sausage, a roll, and a glass of wine, he wanted nothing more."[25]

Millions would die before this rubbish could be swept away. World War II was needed to dismantle the vicious regimes that had incorporated doctrines derived from eugenics into their philosophy and used them to justify their murderous behavior. Benno Muller-Hill's *Murderous Science* describes the doctrines and those who carried them out. Many who endorsed the policies held titles and received salaries as scientists, and some data were published in papers that bore scientific titles. Yet the doctrines were not science.

The process that defines science requires skepticism, debate, and doubt before new theories are accepted. During that stage, the theories are subjected to tests in which they can lose and be discarded. This was not done for the doctrines of eugenics. Further, science can describe conditions, but does not make value judgments. Modern laboratory techniques can show that the genetic plan of one individual differs from that of another, but declarations that one plan is superior and another inferior do not belong within science.

In 1910 and subsequent years, much data concerning human families and behavior were collected in Cold Spring Harbor. Some of the statistics may have been accurate, but they led nowhere in terms of understanding genetic mechanisms. They were of no more use in describing how genes govern human characeristics than data on planetary movements has been in predicting everyday events on earth. Thomas Hunt Morgan and his colleagues by studying fruit flies were arriving at understandings that ultimately could be used to diagnose and cure human disease. By the last decades of the twentieth century, Davenport and his work had become sad episodes of history, and his old grounds at Cold Spring Harbor had become a vital center of activity for the new genetics.

Much patient, hard work was needed to move events forward in this direction. In our brief tour, we cannot give these efforts the attention they deserve. Having said this, we will now leap forward past the world wars and their destruction to 1953, the year in which another extremely fruitful insight was grasped by two young scientists.

4

1953: THE EAGLE PUB

LONDON, JUNE 2, 1953

On a chilly, wet, and windy day, Elizabeth II was crowned Queen of England in Westminster Abbey by the Archbishop of Canterbury. She had actually succeeded to that position upon the death of her father, King George VI (son of George V), sixteen months earlier. During the ceremony, according to biographer Elizabeth Longford,[1] the queen smiled at a "small apparition with satin and silky hair, sitting alert and attentive between his grandmother and aunt." It was her four-year-old son, Prince Charles, who "was brought in by a back entrance in time to see the crowning." By the law of succession, Charles now stood next in line for the throne. It was likely to be a long wait, however, as his mother, only twenty-six at the time of her coronation, seemed likely to preside over the empire for a time that rivaled in length the reign of her great-great-grandmother, Queen Victoria.

MOSCOW, MARCH 5, 1953

Joseph Stalin, the inheritor of the Revolution that had deposed Nicholas and Alexandra, died suddenly in the Kremlin. The departure of this brutal dictator was most convenient, as his "Doctor's Plot" earlier that year appeared to signal a new round of terror.

Stalin's treatment of biologists, rather than his ordinary political opponents, concerns us most here. Power over Soviet genetics was put in the hands of an ignorant plant breeder named Trofim Lysenko. Lysenko felt that heredity was controlled by the entire cell and by the environment as well. He denied the importance of chromosomes and genes. This belief agreed with the Socialist view that an improved environment would lead to a better form of human. If a Socialist state, through its very existence, could defeat unemployment and illiteracy, why shouldn't it do so for hereditary diseases and the more corrupt aspects of human nature as well?

Lysenko did not just argue his view in scientific journals. In 1948, he denounced Mendel–Morganist genetics at a meeting, declaring it to be abstract, Fascist, racist, and incompatible with the views of Mendel's contemporary, Karl Marx. His followers echoed him in articles, producing such choice quotes as

> Mendelism–Morganism has fully exposed its gaping emptiness; it is rotting from within and nothing can save it now.
>
> The followers of Virchow, Weissman, Mendel and Morgan, talking of the immutability of the gene and denying the effect of the environment, are preachers of pseudoscientific tidings of bourgeois eugenecists and of various distortions in genetics, which provided the base for the racist theory of fascism in capitalist countries.[2]

Those who disagreed were not simply denied grants. They were fired, jailed, and sent to labor camps. Books were banned, and laboratories were shut down. Some celebrated foreign scientists who were Socialist sympathizers endured this behavior as long as they could and then bolted. One of them was the distinguished British population biologist, J.B.S. Haldane, who broke with the Communist party in 1949 announcing "I am a Mendelist–Morganist." Another was a person we met earlier in the fly room, Hermann J. Muller.

Muller had actually moved his laboratory to the Soviet Union in the early 1930s, only to watch the ascendancy of Lysenko with dismay. He considered Lysenko to be a fraud and a gangster and made these views public. Fortunately, he was able to get away from the Soviet Union in the late 1930s, before the situation deteriorated more.

Lysenko's power survived the fall of Stalin for a number of years, but ultimately he fell out of favor and his ideas were discredited. Gene and chromosome theories could return to the Soviet Union by the

1970s. However, much damage had been done. The Lysenko era was recently summarized in an article in *Nature* by Valery N. Soyfer:

> Russian genetics, which had produced splendid research recognized by the entire world, ceased to exist. Darkness fell on biological sciences in the Soviet Union. Nobody at that time could know how long that night would last—or whether he would survive until dawn. . . . [Lysenko's activities] contributed to the transformation of the Soviet Union from an exporter to an importer of grain.[3]

Chronic grain shortages have been an important cause of the current downfall of the Communist economic system in the Soviet Union. We saw earlier how genetic influences exerted through the hemophilia of Czarevitch Alexis and Rasputin's resulting influence on the czar may have contributed unwittingly to the rise of communism in Russia. Years later, another set of genetic circumstances helped pull it down.

CAMBRIDGE, ENGLAND, SATURDAY, FEBRUARY 28, 1953

Francis Crick (played by Timothy Pigott-Smith) raced out of the gateway of the old Cavendish Physics Laboratory, dodged the parked bicycles in narrow Free School Lane, and bursting into the nearby Eagle Pub shouted to the lunchtime crowd, "We've done it!" One step behind him and equally breathless, Jim Watson (played by Jeff Goldblum) echoed his "We've done it!" A celebration followed with laughter, kisses, and the raising of mugs of beer in toasts to mark the event. This scene represented a culminating moment of the film *The Race for the Double Helix*, although the race in question was not Watson versus Crick, as suggested above, but rather Watson and Crick versus the opposition. The latter group included master scientist Linus Pauling, X-ray crystallographer Rosalind Franklin, and others who may not even have been aware they were in a contest.

What had Watson and Crick accomplished? They had constructed in their office in the Cavendish a cardboard-and-metal-rod model of the atoms in a small portion of DNA, the stuff from which genes are made. Pauling had also prepared a model of DNA and announced it in the journal *Nature* a week previously. His was wrong, however. The

Watson–Crick version was the right one and revealed deep and abiding truths of biology.

Watson, in his best-selling autobiography of the period, *The Double Helix*, wrote:

> There was also the obvious fact that the implications of its existence were far too important to risk crying wolf. Thus I felt slightly queasy when Francis winged into the Eagle to tell everyone within hearing that we had found the secret of life.[4]

Crick has written a more restrained ("I am reluctant to write about close personal relationships with people still alive") and non–best-selling memoir, *What Mad Pursuit*, in which he denies the event:

> According to Jim, I went into the Eagle, the pub across the street where we lunched every day, and told everyone that we'd discovered the secret of life. Of that I have no recollection.[5]

Later that day he did tell wife Odile of their big discovery. His wife's remembrance is that "you were always coming home and saying things like that so naturally I took no notice of it."

The crowd in the Eagle also may not have taken the announcement seriously, if, in fact, Crick ever made it. According to Watson:

> Often he came up with something novel, would become enormously excited, and immediately tell it to anyone who would listen. Almost everyone enjoyed these manic moments, especially when we had time to listen attentively and to tell him bluntly when we lost the train of his argument. . . . He could not refrain from subsequently telling all who would listen how his clever new idea might set science ahead. . . . A day or so later he would often realize that his theory did not work.[6]

In fact, Watson and Crick had proposed a different model for DNA a bit more than a year earlier. They had based it on remembered, but unwritten, details captured by Watson during a brief lecture given by

Rosalind Franklin. His recollection was incorrect, and their scheme fell into ruins. Erwin Chargaff, a Columbia University biochemist whose own experimental data played an important role in the proceedings, had this comment: "It is a pity that the double helix was not discovered ten years earlier; some of the episodes could have been brought to the screen splendidly by the Marx Brothers."[7]

The Marx Brothers missed their chance, and there were also too many of them. So, we are left for the time being with the existing film version. My own feeling was that if it did not happen that way, it should have done so. The moment of discovery deserved it, as did the aftermath in which a succession of awed visitors, including the losers in the race, marveled at the model (much as the onlooking crowd gaped at the extraterrestrials in *Close Encounters of the Third Kind*), and trumpets sounded in the background. Whatever the style and circumstances of the discovery may have been, it was a magnificent moment for science. It was the time when two great, but separate exploratory efforts, each launched more than eighty years earlier, suddenly converged and joined as they attained their common goal. Physical science and genetics, together at last, had discovered the deepest secret of heredity.

The stage, of course, was perfectly set. The existing state of development of each field could not have been more ideal and the two individuals involved, bright, hungry, and inquisitive, one a geneticist by training and the other a physicist, could not have been better chosen.

THE GENETICISTS

By the early 1950s, bacteria and the even simpler viruses that preyed upon them had become the preferred targets for research upon the nature of genes. A new generation of fruit flies could be raised in ten days, but one of bacteria could be had in as little as twenty minutes. Bacteria, once thought to be simple blobs of life stuff that needed no heredity at all, were now recognized to be complex, sexual at times, and worthy subjects for genetic research programs. Bacterial genes could be mapped in great detail, but this effort would still give no clue as to what genes were made of. Once again, a key insight came primarily through the curiosity and persistent effort of one individual.

The hero in this case was Oswald Avery, a small, thin, bespectacled bachelor, cautious, gentle, and reserved, who led a life that centered about science. Avery had labored on and off for sixteen years in his laboratory at the Rockefeller Institute in New York to learn the nature of *transforming principle.*

This mysterious substance could change the heredity of bacteria. Its target was a natural "smooth" strain that could infect and kill mice and other laboratory animals. A mutant "rough" version of the germ existed that had lost that deadly power, but it could be reconverted back to the deadly smooth form by adding transforming principle. The offspring of the transformed germs remained smooth and deadly.

Transforming principle was prepared by heating and killing the "smooth" bacteria and extracting something from them. The extract was harmless on its own, but it had the ability to convert the rough mutants back to the natural smooth form. Something in the extract had the power to change heredity, and Avery set out to find out what it was.

Many scientists felt that the effect was caused by some error in procedure or represented a special situation that applied to certain bacteria, but not elsewhere in the realm of life. One who disagreed with this diagnosis was Hermann Muller, one of Thomas Morgan's key assistants. Muller, unlike Avery, was willing to put forward an explanation before all the data were in: "There were, in effect, still viable bacterial chromosomes, or parts of chromosomes, floating free in the medium used. These might, in my opinion, have penetrated" the living bacteria and have "taken root" within them.

He was exactly right—the free-floating substances in transforming principle were genes.

Avery published his crucial results on the chemical identity of the genes, the transforming principle, on February 1, 1944.[8] Most biologists at that time, if they had to guess at the nature of the genes, would have said that they were protein. Protein molecules were diverse and carried out a dazzling variety of tasks in living organisms. They were oxygen carriers, hormones, building materials, weapons, and, above all, enzymes—the mysterious substances that could selectively permit certain chemical changes to take place. Further, protein molecules were large and made out of many subunits. A vast number of proteins could exist, just as a staggering number of paragraphs can be written in English text using twenty-six letters and a few punctuation marks. A mindset, or paradigm, had developed that genes were made of protein.

Avery was more inclined to follow the dictates of his evidence, however, rather than public opinion. The best tests that he could apply to transforming principle, after he and his co-workers had purified it rigorously in the laboratory, indicated that it was not protein but DNA, a substance that had hitherto been assigned a minor supporting role in hereditary events. "Who could have guessed it?" he wrote in a private letter to his brother, "sounds like a virus—may be a gene." Ever cautious, he also added, "It's lots of fun to blow bubbles,—but it's wiser to prick them yourself before someone else tries to."

Although no flaw could be seen immediately in it, Avery's work met with disbelief. The chemists who had had the greatest role in determining what DNA was, Miescher, Kossel, and Levene, all felt that protein was the hereditary substance. Levene worked in the same institute as Avery and, until his death in 1940, could provide in-house discouragement as the DNA story unfolded. As was mentioned earlier, Levene had culminated his own studies with the idea held that DNA was a relatively small molecule (compared to proteins) and that each unit of DNA had exactly four letters: one each of A, C, G, and T. In English, you can get a certain amount of feeling across in a single four-letter word, but you can hardly pack in much information. As one noted biologist, Max Delbruck, put it, "At that time it was believed that DNA was a *stupid* substance, a tetranucleotide which couldn't do anything specific." If this were so, then Avery's results were due to contamination of his DNA by protein or to some special effect.

Not everyone shared this belief. Erwin Chargaff was working on another type of biochemical at that time in his laboratory at the medical school of Columbia University. In his words:

> Avery's discovery certainly made an impression on a few, not on many, but probably on nobody a more profound one than on me. . . . I saw in dark contours the beginning of a new chapter of biology. . . . Avery gave us the first text of a new language, or rather he showed us where to look for it. I resolved to search for this text.[9]

He immediately changed his research to work on DNA. Among the others who were impressed by Avery's results were James Watson and Francis Crick.

THE CHEMISTS AND THE PHYSICISTS

In 1944, the year of Avery's announcement, the celebrated physicist Ernest Schrödinger published a short philosophical book *What Is*

Life?.[10] The book offered no lasting solution, but was noteworthy in that it assumed that life processes, in general, and genes, in particular, could be described in physical terms, atoms and molecules. Among the scientists and would-be scientists who were profoundly influenced by the book were, again, Watson and Crick.

The chemists who worked with DNA from 1869 to 1940 had underappreciated its importance, but had nonetheless managed to put most of its chemistry in order. The basic subunits had all been collected and identified. They included the four bases, the substance phosphate, normally a component of rocks and bones, and a sensitive and elusive sugar component, deoxyribose, which was finally tracked down by Levene in 1929–30. Levene had also showed the manner in which base, phosphate, and sugar were put together to form one unit of DNA. What remained was to establish how the units (whether four or many, many more) were connected to build the entire molecule. This task was taken on by Glasgow-born Alexander Todd. By 1952, the job was essentially done: The culminating paper was published by Todd with another Scot, Dan Brown. The sugar and phosphate subunits served to form a backbone from which the bases dangled like so many charms from a bracelet.

Unlike many of his predecessors who either labored at a time when DNA was obscure or expired like Avery before they could get their Nobel Prize, Todd was not underrewarded. He was knighted in 1954 and received the Nobel Prize in chemistry in 1957. Two years later, I arrived to spend a year in his laboratory on Lensfield Road in Cambridge. Todd by that time was a tall, clean-shaven, and stocky man with a full head of white hair. His manner was formal, cordial, but without warmth. We agreed that I should work directly under one of his lieutenants, Colin Reese, an arrangement that was very agreeable to me. Colin and I had worked together in a lab at Harvard, and he was my friend. I was to meet Todd one-on-one on only two further occasions. A year after the end of my stay, however, a paper was published that included Colin, Todd, myself, and two others as authors. By that time, he had been promoted again, from knight to baron, and was now Lord Todd of Trumpington and a life peer. The listing of authors in our joint paper was alphabetical and ended ". . . , Robert Shapiro and Lord Todd." Although I was still quite young, I felt that it was as close as I was likely to get to a lord.

When a chemist investigates a substance extracted from nature, a vitamin, hormone, dye, or whatever, his or her job is usually done when the structure, the exact way that the various atoms in it are linked to one another, has been revealed. By the late 1940s, recognition was dawning that another feature was also crucial for large molecules in biology, such as proteins. The atomic connections did not automatically reveal the three-dimensional shape that a protein would take in space, and shape was vital to the way it functioned. Imagine an instruction sheet for a Tinkertoy, Lego set, or other child's construction kit that clearly explained what should be connected to what, but gave no illustration or explanation of what you would have when you finished it. You had no idea whether the plan was for a truck or a roller coaster.

When Alexander Todd completed the DNA project that many chemists had worked on for over eighty years, this exact situation was recreated. The chemical structure of DNA offered no clue as to how it held itself in space: A vast number of solutions were possible. No principles even existed to guide scientists from atomic structure to three-dimensional structure.

In 1951, however, Linus Pauling had published details of the alpha-helix, a corkscrewlike spiral shape that a protein or part of it could take on. This shape was one of several (the others were not yet known) that could be present or not present in a particular protein. To understand completely how a protein twisted itself in space, more information was needed, and the most likely way to get it was through the use of a tool called *X-ray diffraction analysis*. This was the domain of the physicists.

With some skill and much labor, an investigator could get a pattern from a substance, a complex series of spots and lines produced at that time on photographic paper. Analysis of the pattern, usually by elaborate mathematical calculations, would afford some clues about the shape of the molecule. If the shape of the molecule being investigated was symmetrical and simple, the investigator could attempt to build a model to see if it would fit the data.

By the end of the 1940s, the medical Research Council of Great Britain had established a unit dedicated to the use of X-ray techniques to solve the three-dimensional structure of important biological molecules. It was placed within the Cavendish Physics Laboratory at Cambridge, which was located on Free School Lane, perhaps a block from Todd's laboratory.

Historian Horace Freeland Judson in *The Eighth Day of Creation* has written that "biology is not the business of the Cavendish."[11] The century-old, nondescript grey and yellow structure, connected to others to enclose a courtyard, had been the site of some notable physics. The existence of the neutron had been predicted, the properties of the electron measured, and the first atom-smashing experiments performed all at that site.

The head of the Cavendish at that time was Sir Lawrence Bragg, who had won the Nobel Prize, together with his father, for formulating the basic laws of X-ray diffraction. Working under him was Austrian-born Max Perutz, who as a graduate student in 1936 had undertaken the X-ray analysis of hemoglobin, the substance that gives blood its red color.

As was mentioned earlier, the oxygen-carrying function of hemoglobin had been identified in Mendel's time by Miescher's supervisor, Felix Hoppe-Seyler. One excuse after another has continued the association of blood and genes, first linked in the speculations of the ancient Greeks, into modern times.

Perutz was not to complete his project until 1959; this was obviously not one of the easy tasks. Also in the unit was John Kendrew, who was working on a related molecule. As Kendrew's choice was only one-quarter the size of Perutz's, he was able to start later and finish earlier than Perutz. Both arrived at the Nobel ceremony at the same time, however.

THE WATSON–CRICK CONNECTION

Into this pregnant situation we now insert the two main characters, whose names are fated to be linked together permanently. They were also linked by certain characteristics that they shared. Both came from middle-class environments, although Crick's in the English Midlands was prosperous, while Watson's, on the South Side of Chicago was more impoverished. Both had become disenchanted with religion at about age twelve and taken an interest in nature; Watson, in birds, Crick, wildflowers. Both were bright and loved to speculate on important scientific questions. Each, impressed by Avery, Schrödinger, and Pauling, had concluded that DNA was the material of the genes, and the time was ripe for an attack on a three-dimensional structure of DNA in the manner of Pauling.

Their convictions were strengthened during the time of their collaboration by a noteworthy experiment performed by Alfred Hershey and Martha Chase at the Cold Spring Harbor Laboratory. Hershey and Chase showed that in certain viral infections of bacteria only the DNA and not the protein made its way into the infected cell. The DNA thus carried the heredity of the virus. Avery's conclusion that DNA was the genetic material no longer stood as one isolated case.

Many items of personality and history distinguished Watson and Crick. Watson was an early achiever, one of the original radio Quiz Kids. He entered the University of Chicago at age fifteen and earned his Ph.D. from Indiana University at twenty-two. At Indiana, he had passed by the opportunity to work with Hermann Muller in order to study bacterial viruses under Salvator Luria. A future Nobel Prize winner was preferred to one who had just received that award. "It seemed natural that I should work with Muller but I soon saw that *Drosophila*'s better days were over and that many of the best young geneticists . . . worked with microorganisms," remembers Watson.[12]

After he had earned his Ph.D., Watson erred by accepting a National Research Council Fellowship to study in Copenhagen. The work that he was given there was excessively chemical, a subject that bored him at the time. By various subterfuges, he arranged to have his fellowship transferred to the Cavendish and arrived in Cambridge in 1951 as a socially awkward, untidily dressed, nervous, ambitious, and intense young American Ph.D.

Francis Crick captured one aspect of Watson's physical awkwardness some years later. In his autobiography, *What Mad Pursuit*, Crick described how he toyed with the idea of presenting his own version of the events leading to their discovery, but written in the style of Watson.

Watson had opened *The Double Helix* with the words: "I have never seen Francis Crick in a playful mood." Crick proposed the following counter: "I did get as far as composing a title (*The Loose Screw*) and what I hoped was a catchy opening ('Jim was always clumsy with his hands. One had only to see him peel an orange')." The writing project was not carried further.[13]

Francis Crick had taken a more relaxed and meandering approach to his career. He appreciated science, but above all he enjoyed conversation, "especially with pretty women." This was another avocation that he shared with Watson. When Watson arrived at the Cavendish, Crick, at thirty-five, was still seeking his Ph.D. He was not fully responsible for this plight, however. When World War II broke out, he was near completion of that degree in London on "the dullest problem imaginable: the viscosity [flow resistance] of water."[14] During the war, he was diverted from his studies and worked for the Admiralty on the design of mines. After the conflict ended, he found that these weapons were still influencing his career:"By good fortune a land mine had blown up the apparatus I had so laborously constructed at University College, so after the war I was not obligated to go back to measuring the viscosity of water."[15]

Crick meandered for a time and then resumed graduate study at the Cavendish in an area new to him: the X-ray analysis of proteins. His presence there could hardly be overlooked. Watson has described his impact:

> Most people thought he talked too much. . . . He talked louder and faster than anyone else and, when he laughed, his location within the Cavendish was obvious. Almost everyone enjoyed these manic moments, especially when we had time to listen attentively and to tell him bluntly when we lost the train of his agrument.[16]

Despite their later gibes, it is obvious that a superb intellectual companionship developed between the two after they met. According to Horace Freeland Judson:

> There has to be an extraordinary interaction between two people, before they can do what they did. Jim and Francis talked in half sentences, they understood each other almost without words.[17]

Key ingredients in their communication were peer equality and blunt candor, without hostility, when they evaluated each other's ideas.

Crick first learned of Watson's existence from his wife when he returned from work one day: "Max (Perutz) was here with a young American he wanted you to meet and—you know what—he had no hair!" That is, he had a crew cut. In the film version, we can watch Watson's hair grow longer and longer as the plot develops until at the end, he is absolutely shaggy. Watson and Crick were soon put together in an office by Perutz and Kendrew "so that you can talk to each other without disturbing the rest of us." In Crick's words:

> Jim and I hit it off immediately, partly because our interests were astonishingly similar and partly, I suspect, because a certain youthful arrogance, a ruthlessness, and an impatience with sloppy thinking came naturally to both of us. Jim was distinctly more outspoken than I was, but our thought processes were fairly similar.[18]

The success of their joint effort came from this collaboration and also because they had the wits to seize a problem of paramount importance at the right time. In Crick's words,

> The major credit that Jim and I deserve . . . is for selecting the right problem and sticking to it. It's true that by blundering around we stumbled on gold, but the fact remains that we were looking for gold.[19]

Their contemporaries did not get that point. Watson once commented on his fellow geneticists:

> You would have thought that with all their talk about genes they should worry about what they were. Yet almost none of them seemed to take seriously the evidence that genes were made of DNA. This fact was unnecessarily chemical.[20]

Watson and Crick had grasped this idea and also that the time was ripe to nail down the connection. Soon after they met, they turned their attention away from the supposed area of their interest, X-ray studies of proteins, and set out to build a model of DNA.

In selecting a location to symbolize Mendel's efforts, I chose the pea patch, and, similarly, the fly room for Morgan. Why, then, choose the Eagle Pub as the title for the chapter that features Watson and Crick? The reason is simple. They performed no experiments in achieving their feat; their principal research tool was communication.

They spoke to each other, of course, endless hours of discussion, speculation, and debate. Crick found his thesis topic dreary and so was easily diverted. Watson was also making slow progress with his official protein problem, and so they found much time to share with one another. According to Watson, "I had nothing else to do. I was totally underemployed. And I was the one who got it." Max Perutz said of Watson: "He found it partly because he never made the mistake of confusing hard work with hard thinking; he always refused to substitute the one for the other."[21]

Watson and Crick could talk in their office, of course, but this was not always the most congenial place. On most days, they had lunch at the Eagle. I later retraced their steps. To get there, you just turn right as you leave the Cavendish. Free School Lane, never very wide, becomes a narrow pedestrian path, bordered by an old college wall on the left. You soon enter into Bennett Street and must dodge the traffic to cross it. On the other side are hinged wooden doors bearing a sign: "The Eagle. Home Cooked Food. Cold Buffet. Family Room. Fine Ales and Wines. No Parking." The doors admit you to a small cobbled courtyard. Within it stands a two-storied building with Tudor decor, a porch on the second floor, timbers, and the like. Its somewhat shabby interior was closed for renovation when I visited in 1990, but I can dimly recall having lunch there with chemist Colin Reese one day in 1960. Regrettably, no idea of significance hit us at the time.

After lunching at the Eagle, Watson and Crick would walk in the "Backs," the college gardens that lined the river, or go punting on the river. On occasion, Crick's wife would prepare dinner in their home, dubbed the Green Door. That particular door opened from the street. Behind it was a narrow staircase that ran past the lavatory to a tiny two-and-a-half-room attic apartment. The bathtub was in the kitchen, covered by a board. "It was often necessary to move a miscellaneous collection of saucepans and dishes if one wanted to have a bath." At a later stage, the Cricks could afford a larger residence—a house that they dubbed the Golden Helix.

Their professional–social life was not limited to each other, but included their colleagues from the same Medical Research Council unit and from other laboratories. Science was discussed, though sometimes in brief fragments, at a continuing series of teas, college dinners, wine tastings, conventional parties, and costume balls. Occasionally, the festivities would move to an International Congress, such as one in Copenhagen that Watson described:

> From the moment the several hundred delegates arrived, a profusion of free champagne, partly provided by American dollars, was available to loosen international barriers. Each night for a week there were receptions, dinners, and midnight trips to waterfront bars. It was my first experience with the high life, associated in my mind with decaying European aristocracy.[22]

Through this proliferation of contacts, Watson and Crick could keep informed of the most recent developments related to their interests. Their style can be contrasted with that of Thomas Hunt Morgan. He commented in his presidential address to an International Genetics

Congress that one could best speed research by working industriously, selecting favorable material, making intelligent hypotheses, "and, lastly, by not holding Genetics Congresses too often."

After knowing each other for only a short time, Watson and Crick decided to build a model of DNA together. Their strategy was simple: They would attempt to build it as a helix, à la Pauling. Any other solution would take too long; they might get bogged down like Perutz. Of course, they could not do this with their wits alone, even if the Cavendish workshop furnished the necessary rods, clamps, and plates. Fortunately, they were in an ideal place to gather the needed information. The chemical structure of DNA had been determined in the same town. Despite this advantage, mistaken chemical ideas interfered with their model building almost to the end. At the last moment they were rescued by a colleague who in an offhand conversation set them straight.

Another important clue that helped them was found in the work of Erwin Chargaff. He had analyzed the ratios of A, C, G, and T in many DNA samples with great care. They were not equal to one another as Levene had assumed. However, another curious rule applied: The amount of A present always equalled that of T, and G equalled C. Chargaff published the work, but did not attempt to explain it. Although his data were in print, Watson and Crick had not fully digested it. Chargaff was visiting England at that time, and John Kendrew tried to help by arranging for all of them to meet at lunch (or perhaps it was dinner, the details do not agree in the various histories). The separate accounts of Watson and Chargaff of this meeting, while not contradictory, are yet memorable.

The encounter according to Chargaff in his autobiography *Heraclitean Fire* went as follows:

> When I first met F. H. C. Crick and J. D. Watson in Cambridge . . . they seemed to me an ill-matched pair. . . . The first impression was far from favorable; and it was not improved by the many farcical elements that enlived the ensuing conversation, if that is the correct description of what was in part a staccato harangue. . . . In any event, I seem to have missed the shiver of recognition of a historical moment; a change in the rhythm of the heartbeats of biology.
>
> The impression; one, thirty-five years old; the looks of a faded racing tout, . . . an incessant prattle with occasional

nuggets glittering in the turbid stream of prose. The other, quite undeveloped at twenty-three, a grin, more sly than sheepish; saying little, nothing of consequence.

So far as I could make out, they wanted, unencumbered by any knowledge of the chemistry involved, to fit DNA into a helix. It was clear to me that I was faced with a novelty: enormous ambition and aggressiveness, coupled with an almost complete ignorance of, and a contempt for chemistry. . . . In any event, there they were, speculating, pondering, angling for information.

I told them all I knew. If they had heard before about the pairing rules, they concealed it. But as they did not seem to know much about anything, I was not unduly surprised. . . .[23]

What I jotted down before I left Cambridge was: "Two pitchmen in search of a helix."[24]

The encounter according to Watson went this way:

Chargaff, as one of the world's experts on DNA, was at first not amused by dark horses trying to win the race. Only when John reassured him that I was not a typical American did he realize that he was about to listen to a nut. Seeing me quickly reinforced his intuition. Immediately he derided my hair and accent, for since I came from Chicago I had no right to act otherwise. Blandly telling him that I kept my hair long to avoid confusion with American airforce personnel proved my mental instability.

The high point in Chargaff's scorn came when he led Francis into admitting that he did not remember the chemical differences among the four bases. . . . Francis' subsequent retort that he could always look them up got nowhere in persuading Chargaff that we knew where we were going or how to get there.[25]

Despite, or because of all this flak, Watson and Crick did absorb Chargaff's rules and used his data as a central element in their final model. The most important information that they needed, however, was not so readily available or offered so openly to them (with or without scorn). It was the X-ray diffraction pattern of DNA.

The best X-ray work on DNA was being performed in London in the laboratory of Maurice Wilkins, a personal friend of Crick. Watson's smoldering enthusiasm for DNA structure had flared up when he observed some preliminary photos shown by Wilkins at a meeting in Naples. Wilkins, however, was not satisfied with this result. To obtain better data, Wilkins hired the moody and fiercely independent scientist Rosalind Franklin, who unfortunately wished to pursue the investigation, model building and all, without assistance from him.[26] She took pleasure in dismissing his suggestions and their working relationship deteriorated. One entry in her notebook reads: "It is with great regret that we have to announce the death, on Friday 18th July, 1952, of D.N.A. helix (crystalline)."

Historian Horace Freeland Judson summarized their difficulty as follows:

> The conflict between Maurice Wilkins and Rosalind Franklin, though it seems petty at first and is certainly distasteful, ranks as one of the great personal quarrels in the history of science. It was marked, in the way of such quarrels, by a bitterness beyond the personal, for as it went on, Wilkins, Franklin—and Watson observing—realized very well that at issue was the most important discovery then at large in biology.[27]

Franklin, working with one assistant, succeeded in obtaining the best X-ray patterns of DNA to that date. The work was exhausting. Geometrical crystals give the best X-ray data, but DNA could not be prepared in that form. The film *The Race for the Double Helix* shows Franklin inserting a rod into a viscous solution of DNA and carefully pulling out long fibrous strands, which served as a substitute for the crystals. She then undertook a tedious mathematical analysis of the results. Having isolated herself from the communication network and lacking effective collaborators, she took one wrong turn after another. Crick had noted the value of the frank criticism that he and Watson offered to one another:"It is extremely important not to be trapped by one's mistaken ideas." Franklin, working on her own, missed the key insights.

Watson and Crick felt that they had all of the inspiration they needed. The data, alas, were not their own. Scientific protocol as practiced in England required that Wilkins direct any model-building efforts that used data collected in his laboratory.

Frustrated by Franklin, Wilkins reluctantly agreed to cooperate with Watson and Crick's efforts with fuller collaboration to take place after Franklin had been persuaded to depart. In Watson's words:

> Francis seized the occasion [a party at Trinity College] to ask Maurice whether he would mind if we started to play around with DNA models. When Maurice's slow answer emerged as no, he wouldn't mind, my pulse rate returned to normal. For even if the answer had been yes, our model building would have gone ahead.[28]

Eventually, in January 1953, Wilkins showed Watson one of Franklin's best X-ray photos. Perutz had turned a research report by Franklin over to them. The report was not confidential, but neither was it widely circulated. It would not have come into their hands had they not been part of an inner network. Upon studying the report, Crick saw the key features of the shape of DNA that had eluded Franklin. It was helical after all, but with two chains rather than one. Furthermore, the chains ran in opposite directions.

These developments left Wilkins with a residue of bitterness that persisted years later and were revealed in interviews with Judson.[29] Was he upset when Watson and Crick attacked DNA on their own?

> Upset, I was bloody annoyed . . . Rosalind Franklin put me onto the wrong track. . . . DNA, you know, is Midas' gold. Everybody who touches it goes mad . . . Jim Watson was exactly what he says he was: his eye on success and recognition. I knew them both very well. If you read the book [Watson's]—"I'm Jim, I'm smart, most of the time Francis is smart too, the rest are bloody clots." If you want to make a critical analysis, she was going at it wrong.[30]

With the key data in hand, Watson and Crick were able to complete their model in a few weeks. One cloud hung over the effort—the knowledge that Linus Pauling was preparing a model of his own. This represented their worst nightmare, their most feared competition. Watson had commented, "Pauling could not be the greatest of all chemists without realizing that DNA was the most golden of all molecules."[31]

When an advance copy of Pauling's publication arrived through their pipeline, there was a moment of anxiety, then relief. Even the gods can err. He had produced a three-chain model with obvious implausibilities. "The blooper was too unbelievable to keep silent for more than a few minutes." Their own work momentarily forgotten, they went to the Eagle to drink a toast in whiskey, rather than the usual sherry.

During that critical period, Watson was playing tennis fiercely for two to three hours each afternoon. He would take dinner or sherry at a local boarding house, and he spent most evenings at the films. Watson recalls how he joined the undergraduates in booing the censor when a Hedy Lamarr film was shown with its well-publicized nude swimming scene cut. He was certainly not confusing hard thinking with hard work.

At the end, one key feature had to be discovered to complete the model. The bases, A, C, G, and T, were interacting with each other, but how? Bases are flat by nature. Suitable metal plates were on order from the Cavendish workshop that could be used to model the ways in which the bases could connect to one another. In the interim, cardboard cutouts would have to do.

For a time, Watson simply played at placing his pieces of cardboard together in different ways, an activity that resembled that of a child in a rainy-day tabletop game. Finally, on the morning of February 28, 1953, he noticed that A and T could be put together to make a pair and that G and C would fit together in the same way, with exactly the same geometry. The combinations thus explained Chargaff's rules. Crick later described the event:

> In a sense Jim's discovery was luck, but most discoveries have an element of luck in them. The more important point is that Jim was looking for something significant and *immediately recognized the significance of the correct base pairs when he hit upon them by chance.*[32]

Watson's recollection, given in a 1984 *Omni* interview was

> I think I found the base pairs because I was the one who was doing it. You see, I had cut out the cardboard models and was playing around with them while Francis was working on his Ph.D. in the same room. When Francis saw the base pairs and their symmetry, he gave up working on his thesis.[33]

Crick then helped build this new feature into their model to create an entire satisfactory structure. Horace Freeland Judson has described their discovery most eloquently:

> That morning, Watson and Crick knew, although still in mind only, the entire structure: it had emerged from the shadows of billions of years, absolute and simple, and was seen and understood for the first time.

It was "a melody for the eye of the intellect, with not a note wasted."[34] Leaving their model with its newly fashioned base pairs behind, the human pair then set out for the Eagle to announce their discovery of the secret of life.

It is fair to ask why their model, or the more complete version that was subsequently constructed, had such immediate impact on them, those who came in the following weeks to view it, and those who later read about it in the journals. Many three-dimensional structures that have emerged through X-ray work have been rather uninformative: Years of additional study were needed to understand why the molecule functioned as it did. Watson and Crick had feared an outcome of this type: The DNA structure might turn out "very dull" and tell nothing of how a gene functioned. This was not the case for DNA. Genes reproduce themselves, and the model suggested how they do it.

When a bacterium or human cell divides, each daughter cell inherits a complete copy of its genes. An extra copy was prepared somewhere along the line. Biologists had debated for years about the way that the copy might be made. The copying could be direct, as in a modern office machine, or it could work through a two-stage, less direct process. For example, a mold (or template) could be prepared from a sculpture, and that mold would then be used to cast a new statue. The Watson–Crick model immediately showed that reproduction could be understood at the level of molecules and that the two-step template mechanism was used. This is shown in Scheme 1.

CTCAGGAGTCAGATGCAC
GTGCATCTGACTCCTGAG

Scheme 1

I have written a double series of eighteen letters, a bit longer than the length of the first model, using A, C, G, and T. The three-dimensional aspect has been ignored; the paper can represent the phosphate-and-sugar backbone that carries the letters. The letters in normal position represents a chain of DNA, while the upside-down ones represent a second chain that, as Crick deduced, runs in the opposite direction. Each A in the right-side-up chain sits opposite, or is *paired* to, a T on the upside-down one and vice versa. Each G on one chain is paired to a C on the other in the same way. The amount of A must equal T, and G equal C, as Chargaff found experimentally.

As the measurements and geometry of the base pairs were the same in the three-dimensional model, any possible sequence of letters could be accommodated in it. For reproduction, or replication, it was simply necessary to separate the two chains. Each chain could then be used as a template to make a new opposite chain. It was only necessary to place a new T opposite an existing A and so on.

A short manuscript giving the details of the model was typed by Watson's sister Elizabeth and mailed to the journal *Nature* on April 2, 1953. It was published on April 25 alongside two separate articles containing the X-ray data of Wilkins and of Franklin, each with *their* co-workers. The paper by Watson and Crick is considered to be a masterpiece of scientific understatement. It starts as follows: "We wish to suggest a structure for the salt of deoxyribose nucleic acid (DNA). This structure has novel features which are of biological interest." After the chemical description has been given, they simply state: "It has not escaped our notice that the specific pairing we have postulated immediately suggests a possibly copying mechanism for the genetic material."[35] They make no further comment on this topic.

Why was further genetic discussion not included? According to Crick: "Jim was against it. He suffered from periodic fears that the structure might be wrong and that he had made an ass of himself." Crick had insisted that at least one sentence be put in for priority purposes. A few weeks later, Watson had changed his mind, having observed how strongly the Kings College X-ray evidence supported them.

A second paper was published by them in *Nature* on May 30. This gave details of the replication scheme and mentioned another feature of the model that was of the utmost importance: "Any sequence of the pairs of bases can fit into the structure . . . it therefore seems likely that the precise sequence of the bases is the code which carries the genetical information."[36] If anyone had missed the point earlier, it was now clear that genes were written in text, and the four bases were the alphabet of the text. The sugar–phosphate backbone made up the tape on which the text was written.

Francis Crick wrote in a March 19, 1953, letter to his son (from an earlier marriage) Michael:

> Our structure is very beautiful . . . it is like a code. If you are given one set of letters you can write down the others. Now we believe that the D.N.A. *is* a code. That is, the order of bases (the letters) makes one gene different from another gene (just as one page of print is different from another).[37]

Crick later commented:

> The geneticist R.A. Fisher once told me that what we had to explain was why genes were arranged like beads on a string. I don't think it ever occurred to him that the genes made up the string.[38]

As Crick noted, the beauty of the model really affected many viewers: This "intrinsic beauty of the DNA double helix" helped make the impact. "It is the molecule that has the style, quite as much as the scientists." Likewise, Horace Freeland Judson wrote, "The structure of DNA is flawlessly beautiful."[39]

Crick tells of a dinner at a biophysics club in Cambridge, where wine was offered in liberal quantities. Watson, speaking bleary-eyed through a haze of alcohol, attempted to sum up their model for the guests, but was at a loss for words: "All he could manage to say was, 'It's so beautiful, you see, so beautiful!' But then, of course, it was." In the film version of their discovery, Watson's view is echoed as he stands in the presence of the model, struck with admiration.

Atoms and molecules themselves cannot be seen with the naked eye or an ordinary microscope as they are smaller than the wavelength of the light that our eyes can detect. Even if they were visible, we would not understand what we saw. Atoms consist of tiny dense points of matter separated by much larger clouds of electrons that have no absolute boundary. To relate to atoms visually, we must build models, and the materials that we select determine the beauty of the model. One common choice is the use of round balls to represent individual atoms. When we model DNA in this way, we get an irregular solid column notched by grooves that is more chaotic than beautiful.

Watson and Crick built a skeletal model of DNA, one which emphasized the connections between atoms rather than the atoms themselves. Such a model allows the underlying form and relationships to be grasped. The double spiral of the backbone is pleasing, and the proportions seem about right for good reason. In a recent paper,[40] it was noted that the height of one unit of the helix is related to its width by a proportion known since antiquity as the *golden ratio*. This ratio is reflected in the construction of the Parthenon and other ancient works of art. (Perhaps by coincidence, *The Race for the Double Helix* opens with Watson lurking in the shadows of the Parthenon, hoping to strike up an acquaintanceship with fellow tourist Maurice Wilkins.)

Considerations of beauty have also entered scientific disputes over the DNA structure. From time to time alternative models for DNA have been put forward that would also fit the X-ray data. A "side-by-side" model proposed in the late 1970s matched the data by having one chain come partway around the other and then reverse itself, avoiding any helix. The appearance of the model, even a skeletal one, was not pleasing, and it was dubbed the "warped zipper model." Some scientists rejected it intuitively, on aesthetic grounds, before better scientific reasons were found to disprove it.

Only a limited number of scientists could come to Cambridge to view the model. For a number of others, it was the initial article in *Nature* that awoke them to the new era. Horace Freeland Judson wrote: "To those readers who were close to the questions, and had not already heard the news, the letter must have gone off as a series of depth charges in a calm sea."[41] None of their reactions, however, was as striking as a later quote by painter Salvador Dali: "And now the announcement of Watson and Crick about DNA. This for me is the real proof of the existence of God."[42]

Even so, the number of scientists who took notice was limited. As the editors of *Nature* commented later, "There was only a small band of believers and a positive effort had to be made to convince the biological community both of its correctness and profound relevance."[43] Watson and Crick carried this out by writing additional papers, participating in meetings, and exposing themselves to the public media. Within six months of the appearance of the 1953 papers, for example, Watson was photographed with Richard Burton in *Vogue*. Crick summarized the result: "All in all, it seems that we got a very fair hearing, better than Avery and certainly a lot better than Mendel."

These tactics, taken with the style in which the work was conducted, led to a certain backlash. Watson remembered that

> in those months [after discovery] the structure was referred to by the Cambridge biochemists as the WC structure [a British abbreviation for water closet or toilet]. . . . Yes. And I figured, well, we've made the greatest discovery of the century, so we might as well stay quiet until those idiots shut up. People were saying we stole their data and so on. We got so much publicity that we didn't have to seek it.[44]

Two years later a companion of Maurice Wilkins, meeting Watson by chance in the Alps, greeted him, "How's Honest Jim?" and walked past. Crick ran into a negative reaction that might have been anticipated:

> Chargaff, when I visited him in 1953–54, told me (with his customary insight) that while our first paper in *Nature* was interesting, our second paper on the genetic implications was no good at all.

Chargaff has summarized his reactions in his own writings:

> I remember vividly my first impression when I saw the two notes that appeared in *Nature* 21 years ago. The tone was unusual: somehow oracular and imperious, almost decalogous [in the style of the 10 commandments]. Difficulties . . . were brushed aside, in the Mr. Fix-it spirit that was later to become so evident in our scientific literature. . . . I do not know whether in 1865, when Kekule put forward the structural model of benzene, which was to revolutionise organic chemistry, neckties appeared forthwith embroidered with the pretty hexagons. I should rather doubt it, since at the time mass cretinisation had not yet begun and advertising still was a home industry. At any rate, the publicity carnival that ensued upon the unveiling of the DNA model was probably unique in the history of science. . . . That in our day such pygmies throw such giant shadows only shows how late in the day it has become.[45]

I must note that there is one other time of day when people (not only pygmies) can throw giant shadows. That is at dawn.

CAMBRIDGE, ENGLAND, 1953

Fred Sanger missed most of the excitement, as he preferred to spend his time at his laboratory, which was in the basement of the biochemistry building, somewhere near the place where they kept the experimental rats. He was then in his midthirties, a man of modest stature and very ordinary appearance. He was a solid, practical person, who by appearance or conversational style might pass as a carpenter or gardener rather than a scientist. His father had been a medical doctor, and as a boy he had been influenced by an older brother who collected skulls and searched for birds' nests.

For the past thirteen years he had been involved in biochemical work in Cambridge, first as a graduate student and then as a research fellow. He was not academically brilliant, but other circumstances were favorable. He put it simply: "I would probably not have been able to attend Cambridge University if my parents had not been fairly rich."

His start in graduate research, under the direction of protein chemist Bill Perry, was inauspicious:

> Bill's chief interest at this time was in making edible protein from grass. . . . He presented me with a large bucket of frozen grass extract, and suggested that I investigate it. Unfortunately, before it had thawed, Bill had left the lab for a new job.[46]

Sanger persevered with a new advisor, and after he earned his Ph.D., he stayed on at Cambridge. He joined a research group working on the protein hormone, insulin, and from the late 1940s on devoted himself to determining its sequence.

Proteins, like nucleic acids, are made of building blocks hooked together in a row. In the case of proteins, the components are called *amino acids*. Max Perutz, at the Cavendish, was engaged in a heroic effort to determine the three-dimensional structure of a large protein, hemoglobin, using X rays. A more practical approach would be to determine the amino acid sequence (their order of connection) first and use that information as a guide for the X-ray work. No methods had even been developed to sequence proteins until Sanger started his project. Fortunately, his selection of a target was more practical than that of Perutz. Insulin has fifty-one amino acids, hemoglobin, almost six hundred.

It was a job for chemistry, not physics. By chemical demolition, or by the use of enzymes that could cut up other proteins, insulin was broken into fragments. Once separated, the individual fragments were treated as separate, simpler problems, a "divide-and-conquer" strategy. Nature had given Sanger a head start, as insulin already was divided naturally into two parts, or chains. The parts were hooked together by a different kind of connection that was easily broken.

A key to this work was the ability to separate a mixture of fragments from one another, and Sanger seemed to have a particular talent for separations. He particularly favored schemes in which the fragments were spread out as spots on a sheet of paper, giving a "fingerprint." Thus, by the divide-and-conquer strategy, a protein of sequence *abcdefghijklmnopqrst* could be split into *abcde*, *fghij*, *klmno*, and *pqrst*. The parts would be separated on paper and worked on separately. (If necessary, *abcde* could be further split to *abc* and *de*.) Finally, the four fragments would have to be lined up in the right order. This was usually done by the "overlap" method. A different splitting method would be tried for the protein. Let's say that one was found that gave *defgh* after separation. The existence of this overlapping piece established that *abcde* came before *fghij* and that together they read *abcdefghij*.

After years of effort, Sanger, with one or two collaborators, had by 1953 worked out the sequence of the two separate chains of insulin. Now he only had to figure out how they were connected to finish the job. This would take two more years. Immersed in proteins as he was, he paid little attention to the DNA action elsewhere in town. He did notice Watson and Crick, however, as they walked by his building. He described the scene in our interview:

> They used to pass by Biochemistry deep in conversation, battering away at each other, rather a crazy couple. They used to walk backwards and forwards, getting so excited about something. . . . I knew Francis better than I knew Jim—I didn't know him very much. He was rather a bit of a queer fish. I knew Francis and was already impressed with him. . . . I didn't get as excited as I should have got [about their model]. I wasn't sure whether they had got the answer or were [just] enthusiastic. . . . After a month or two it became clear that this was correct.[47]

Sanger finally went over to look at the model. "We were interested in proteins and didn't know very much about nucleic acids," he said. This would change.

POSTSCRIPT: CAMBRIDGE, ENGLAND, 1956

Watson and Crick were united again at the Cavendish, even if temporarily. Crick had returned to take a permanent position with the Medical Research Council, but Watson was to join the Harvard faculty in the autumn. They had much to discuss, but took time out to give advice and encouragement to a colleague, chemist Vernon Ingram.

Ingram had come to Cambridge to prepare chemical variants of hemoglobin for Perutz's X-ray studies. Like Perutz, he had been forced to flee central Europe in the 1930s to escape from the tyranny of Adolf Hitler. A Ph.D. from the University of London and postdoctoral work at the Rockefeller Institute and Yale had helped prepare Ingram for this work. The facility was perhaps less prepared than he was: "The Cavendish idea of a biochemical laboratory was a room with a bench and a sink and a Bunsen Burner and really very little else," according to Ingram.[48]

Meanwhile, a sample of a natural variant of hemoglobin had come into Perutz's possession, the form that is present in people with sickle-cell anemia. The disease had attracted increasing medical attention in the years following Herrick's publication, and a few years earlier, in 1949, two very significant papers had been published in *Science*.

From a study of patients and their parents, an American geneticist, James V. Neel, and his co-workers had studied family histories and decided that the disease behaved just as Mendel's wrinkled pea trait. Recessive inheritance was involved: two altered copies of a gene were needed for the disease to be expressed. Linus Pauling and his colleagues had examined the hemoglobin of sickle-cell patients and found that it differed chemically from normal hemoglobin. The information carried in the genes led to the construction of a faulty protein. In the Pauling paper, sickle-cell anemia was termed "a molecular disease."

Ingram set out to learn what change in the protein led to sickle-cell anemia. He was very aware of the work that Fred Sanger had done with insulin a few blocks away, and he used Sanger's methods to break sickle-cell hemoglobin into parts and separate the pieces. To his surprise, only one fragment that he obtained differed from those of normal hemoglobin. Furthermore, the difference seemed to be due to a single amino acid. According to Francis Crick, Ingram found this hard to accept. Perhaps there were other alterations that he had missed? "Jim and I were brasher then and refused to believe this. 'Try it again, Vernon,' we said, 'you'll find there's just a single change' and so it turned out to be."[49]

The scientific world also found the idea hard to accept, at least initially. When Ingram's observations concerning the changed fragment were presented at the 1956 summer meeting of the British Association for the Advancement of Science, he found that "only about six people attended, none of whom asked questions or displayed any reaction at all."[50] The tradition begun by Mendel and Herrick continued. Subsequently, matters changed. The realization sank in that the text of Watson and Crick was not just an abstraction. Alterations in it could have lethal effects.

During this period of his return, Watson did not neglect the parties of Cambridge. At one of them, he met and befriended another intellectually gifted and ambitious American scientist. Walter Gilbert was then twenty-three (Watson's age when he first came to Cambridge; Watson was now a ripe twenty-seven) with interests in theoretical physics rather than biology. Like Watson, Gilbert was heading back to Harvard. He had been an undergraduate there and would return, first as a postdoctoral fellow and then as a faculty member. Their friendship would continue.

POSTSCRIPT: CAMBRIDGE, MASSACHUSETTS, 1958

At the time when the Watson–Crick model first appeared, I had been enrolled in freshman chemistry at the City College of New York. I listened then to lectures on the smelting processes used for steel and the chemistry of obscure elements. I was bored to tears. It was far removed from the spectacular demonstrations of high school chemistry when the teacher would toss a shiny solid into water and it would burst into flames, or he would mix two colorless liquids and get a bright wine-red color as a result. As I had more perseverence than inspiration at that time, I stuck with my chemistry major. I avoided majoring in biochemistry because of a discouraging incident. When I went to collect my first biochemical preparation, a starch powder that

I had lovingly prepared and left in my lab cupboard for a week to dry, I found that it had mysteriously transmuted into rat pellets.

Now, five years later, my first encounter with the Watson–Crick ideas came in an evening seminar in Harvard. I was there studying organic chemistry for my Ph.D. When I heard their model explained, it hit me that this *mattered*, unlike most of the other information that I had absorbed in chemistry. I resolved to start research with DNA as soon as I had finished my doctoral project. Their idea *did* have power.

5

1975–1977: "JUST A LAB"

ASILOMAR CONFERENCE CENTER,
PACIFIC GROVE, CALIFORNIA,
FEBRUARY 1975

"We can't even measure the fucking risks," Jim Watson exploded. This remark, although delivered *sotto voce* (in a low tone not intended to be overheard), was still within hearing range of a *Rolling Stone* reporter and others who were attending the International Conference on Recombinant DNA Molecules. Watson was not listed on the official program for the four-day meeting, but by exclamations like the above one he was able "to exert a powerful presence . . . and not always in a terribly popular fashion," wrote the reporter, Michael Rogers.[1]

Watson's style has been well described by historian Horace Freeland Judson:

> Watson has always trailed after him a pungent mixture of frankness and reticence. One of the nervous springs of his intelligence is that he cannot help saying out loud the impulsive reactions—to a piece of science, to an instance of bureaucracy, to an intellectual foible of an associate—that other people usually bite back. It makes him uneaseful

> company and the subject of endless anecdote. . . . Watson is different: impatient, skeptical, forever discontent, swallowing ends of words and ends of thoughts, with a face at 50 only a little less gaunt than in the Cambridge photographs from 1953, with ice-blue, protuberant, even slightly wild eyes and fully modeled lips that retreat in a twitchy preoccupied smile.[2]

The *Rolling Stone* reporter added the following impressions:

> Watson . . . seems almost to cultivate the persona of absent-minded professor: tall, pale, thin, shirt collar turned up, wispy brown hair tugged so constantly that it stands out from his head in total disarray. He speaks with a regular punctuation of grimaces and, in the midst of any given sentence, his gaze can wander off into space; a consummate 2000-yard stare.[3]

Photographs from that period showed one other difference from his rebellious youth look of the early 1950s. The tension existing between alternative crew-cut and long-hair styles had been resolved by the considerable thinning out of his hair.

The conference itself had an importance that transcended the comments of the particular individuals in it. The presence of *Rolling Stone*, a rock music publication not usually noted for its coverage of molecular biology, underscored the significance of the event. It met on almost the exact twenty-second anniversary of Watson and Crick's rush to the Eagle to proclaim their discovery of the secret of life. Eight years after that discovery, in 1961, when Watson, Crick, and Maurice Wilkins were awarded the Nobel Prize in physiology or medicine for the DNA structure, a spokesperson for the awarding institute noted that their contribution had "no immediate practical application." Rosalind Franklin was not included in the award. She might have been left out in any event, because the Nobel Prize in any category can only be divided among a maximum of three people in a given year. This question never came up. She was eliminated for a much sadder reason: The Nobel Prize cannot be awarded posthumously. She had died at age thirty-eight of cancer four years earlier. Perutz and Kendrew were present at the same ceremony, however, making up a considerable

Cavendish party. They received the Nobel Prize in chemistry that year for their X-ray studies of oxygen-carrying proteins.

By 1975, DNA had clearly come of age. The first commercial applications were several years away, but were already anticipated by a few visionaries. The perceived hazards of DNA research were much more visible, particularly those relating to a new, but rapidly expanding specialty called *recombinant DNA*. Perhaps the most eloquent of those who were alarmed was Erwin Chargaff. Chargaff wrote in a letter to *Science*:

> You can stop splitting the atom; you can stop visiting the moon; you can stop making aerosols; you may even decide not to kill entire populations by the use of a few bombs. But you cannot recall a new form of life. . . . Once you have constructed a viable *E. coli* [bacteria] cell [carrying the DNA of a higher life form], it will survive you and your children and your children's children.[4]

This attitude was expressed more bluntly in that same year by Alfred E. Vellucci, then the mayor of Cambridge, Massachusetts (a town that includes within its borders Harvard University and the Massachusetts Institute of Technology), while proposing a three-month moratorium on DNA research in that community: "We want to be damned sure the people of Cambridge won't be affected by anything that crawls out of that laboratory."

Such oratory demanded a response, and one of the most outspoken in playing down the possible hazards of recombinant DNA was Watson: At one point he testified immediately after Chargaff in hearings held by the New York State Attorney General:

> I never know what to say when I follow Professor Chargaff but I shall try to speak calmly. What started out as an attempt by the scientific community to appear responsible takes on increasingly the appearance of a black comedy. . . . Recombinant DNA is the most overblown thing since [President Kennedy] created the fallout shelter debacle. . . . I was Kennedy's adviser on biological warfare. I knew all we had at Fort Detrick, and if I can reveal a secret about what we had, what we had was nothing. The marginal danger of

> this thing is a joke compared to [even what we had at Fort Detrick].[5]

The Asilomar (Pacific Grove, California) Conference was convened by scientists active in recombinant DNA research who were trying to avoid a bitter debate on the hazards. They failed. The site itself was idyllic: tasteful low buildings set among wind-swept Monterrey pines and other evergreens at the edge of the Pacific. After four days of confusing arguments, a consensus emerged on appropriate regulations, which were subsequently adopted by the U.S. National Institutes of Health. According to *Science*, in 1975, the conference was "generally praised for its preaching of caution, a decision seen as provident and reasonably self-denying." *Rolling Stone*, no bastion of conservatism, agreed. Their published account summed it up:

> The molecular biologists at Asilomar were the first modern researchers to assume voluntarily some measure of social responsibility for their work. . . . In a sense, then, the biologic sciences have only now suffered their first loss of innocence: out of Mendel's monastery garden, so to speak, and into J.D. Watson's stainless steel laboratory. . . . And perhaps that was the final, foggy significance of Asilomar: a promise that the scientists who deal with the most fundamental stuff of life will not sequester themselves beneath Chicago stadiums or within blockhouses in the New Mexico desert—that their work, at least as significant as the most subtle of sub-nuclear manipulations, will be done with public scrutiny.[6]

That conclusion did not go down readily with everyone concerned. Chargaff had his own epitaph for the event:

> Knowing that the desire to improve mankind has led to some of the most horrible atrocities recorded by history, it was with a feeling of deep melancholy that I read about the peculiar conference that took place recently. . . . At this Council of Asilomar there congregated the molecular bishops and church fathers from all over the world to condemn the heresies of which they themselves had been the first

> perpetrators. This was probably the first time in history that the incendiaries formed their own fire brigade.[7]

Others, Nobel Prize winner George Wald, for example, voiced similar opinions. Sixteen years later, we can tally the score. An immense amount of medical progress has been possible because of recombinant DNA research, but no casualty list has yet appeared.

A rash of advances in DNA technology had brought about this eruption of concern. The sheer size of this molecule, with its millions of units, had made it much more difficult to handle than protein or RNA. With these new techniques, the difficulties had been swept away. DNA could be treated in many ways just as if it were a text written in English.

All of the molecules that I have just mentioned, DNA, RNA, and protein, share one important feature with English. These substances are made of subunits connected in a line, just as English is written by placing a selection of letters, numbers, and punctuation marks one after the other in a row. DNA and RNA use only four letters, however, while proteins use twenty. The letters in all of these biochemical substances run continuously without spacing or punctuation.

When Fred Sanger set out in the 1940s to develop methods for obtaining protein sequences, he had at his disposal a good supply of pure insulin. The number of characters to be determined was fifty-one. This number in English would make a short sentence. (For example, the sentence just before this one has exactly fifty-one characters, counting the spaces and the final period.) A number of RNA molecules in nature have less than one hundred letters. They tend to occur together as a complex mixture, but the mixture was separated in the 1960s. The sequence of letters was then worked out for many of them.

The smallest DNA molecules known at that time, however, were those present in small viruses, which were some thousands of units long. This number of English characters would fill up a couple of pages of this book. The single chromosome present in the best-known bacteria has a letter content several times the length of this book, while the smallest human chromosome, as we noted already, approximates a mammoth library dictionary. The problems in dealing with such substances were simplified by the discovery of versatile new tools: text cutters, text splicers, and text-matching methods.

The names that I have just used differ from the technical ones selected by scientists. As I feel that technical jargon has the same effect on English prose that a field of broken glass has on a race run barefoot,

I will try to replace some of the terms with simple English equivalents in this book. When I do so, I will italicize the technical term and place it in parentheses the first time that it comes up.

The ability of the text cutters (*restriction enzymes*) can easily be grasped: These proteins recognize a particular sequence of letters in a text and cut it at that point. In their natural setting, they function as protective weapons used by bacteria to obliterate invading viruses. As a comparison, imagine an English text cutter that recognized the word *years* and cut the text after the *y* in that word. For example, the following text from the Holy Bible, King James Version (Genesis 5), would be cut after the slashed lines:

> And Adam lived an hundred and thirty y/ears, and begat a son in his own likeness, and after his image; and called his name Seth:
>
> And the days of Adam after he had begotten Seth were eight hundred y/ears: and he begat sons and daughters:
>
> And all the days that Adam lived were ninc hundred and thirty y/ears and he died.

The cutter would slash the text into four fragments, and each could be examined separately. The third piece, for example would read as follows: "ears: and he began sons and daughters: And all the days that Adam lived were nine hundred and thirty y." The same text would be cut differently by an alternative enzyme. One that recognized the word *hundred*, for example, would also make three cuts, but in different places.

The work of text cutters can be undone by text splicers (*DNA ligases*). Under appropriate conditions, they could recombine a fragment ending in *y* with another starting with *ears* to give text rejoined at the word *years*. The mixture of four pieces resulting from the cutting of the Adam text above could be rejoined to reform the original one, of course. However, this would be only one of many possible outcomes. Neither the splicer nor the fragments of text have any memory. All combinations would be produced, a result that caused the name *shotgun experiment* to be applied to such procedures. For example, the first and last fragments might be rejoined to give

> And Adam lived an hundred and thirty y/ears and he died.

Text cutters are indifferent to the source of the text. They would treat the following excerpt from "Explanation of the Tax Reform Act of 1986" of the U.S. Internal Revenue Service exactly as they had the Bible, cutting it at the position indicated by the slash.

> The tables below show the individual tax rates for each filing status for tax y/ears beginning in 1987.

Splicers, with equal indifference, will combine appropriate text passages at random, even if they were derived originally from different texts (as long as the same cutter was used to produce both). A mischievous editor could cut both the Biblical passage and the tax blurb with the "years" cutter, put both batches together, and add the splicer. One of many products would be the following:

> The tables below show the individual tax rates for each filing status for tax y/ears: and he begat sons and daughters:
> And all the days that Adam lived were nine hundred and thirty y/ears beginning in 1987.

Some of the concern about recombinant DNA came from the fear that randomly reconnected pieces of text might come together to form a hazardous set of instructions. With experience, we realize that this is extremely unlikely. If you doubt this, try a model experiment on your own. For example, recombine the set of random fragments made by shredding a cookbook and an auto repair manual, and see if any coherent but hazardous recipes emerge. Nonsense will be the overwhelmingly predominant outcome.

What use, then, could a biochemist make of these tools? He or she might get a certain sense of power from the demonstration that DNA could be joined together from apes and grapes, mice and men, duck and orange, but the joy would soon wear out unless there were more practical benefits. These benefits arise because it can be planned how things are put together. In these plans, advantage can be made of certain natural talents of DNA.

DNA is not text. With the help of some protein friends, it can also reproduce itself. For example, if you start with an ounce (or much, much less) of the DNA of a virus that infects bacteria, you can harvest a pound of that DNA later by allowing the virus to multiply in its host. If you had some other text of your own that was in short supply and you could splice it into the viral DNA, then that text would also be reproduced as the virus multiplied. You would have to locate a virus with only one cutting point at a place where its own text was of little importance. Opening the virus up at that point, you would insert the DNA text of your choice at that point with a text splicer and grow the altered virus in its host. Afterward, you could cut your passage out of the virus using the original text cutter. (Other combinations might also have resulted from the splicing, but various strategies are available to ensure that you get the one you want.)

The above process is called *cloning*. Cloning can be done by using a virus as a carrier or, as an alternative, certain small pieces of DNA (a few thousand units long) called *plasmids*. Plasmids are part of the heredity of a bacteria, but reproduce on their own separate from the main chromosome. If the main bacterial chromosome is compared to an encyclopedia volume, then a plasmid could be represented by a separately bound pamphlet. Plasmids give a bacterium such unfortunate (to us) properties as antibiotic resistance. Those alarmed by recombinant DNA research feared that in some shotgun experiment, a cancer-causing gene might inadvertently be put into a bacterium, creating a new species that could cause an epidemic of cancer.

In a cloning procedure, however, a bacterium is working as our employee. We give it a trace amount of some rare bit of DNA that might contain the genes for the smell of the rose or the horn of a unicorn, and it returns to us our investment multiplied as many times as we choose. If we wish to determine the sequence of a bit of DNA, and we are willing to put in the work, exhaustion of material will not be a problem. Once we coax a bit of DNA from nature, we will have it forever in the same way that a poem from an ancient author, once rescued from a ruin or tomb, may be given immortality by reprinting it and distributing it to many thousands of libraries and homes.

Before we multiply and preserve a scrap of DNA text, however, we may wish to know what it is. Is it the equivalent of the lost poem of a master or just an old laundry list? One additional tool, text matching (*hybridization*), may help us identify what we are looking for. This process has no existing analog in the English language. It works because DNA is written as a double message, with one chain coding the other according to the Watson–Crick base pairs, A = G and C = T. To mimic this in English, we have to introduce some innovations. Let's start with the phrase

and all the days that adam lived were nine hundred and
thirty years

We will first put it into capitals and eliminate the spaces, because DNA sequences are usually written in capitals and do not contain spaces. We then get

ANDALLTHEDAYSTHATADAMLIVEDWERENINEHUNDREDANDTHIRTYYEARS

We now need to write a second chain, running backward, matched to the first, according to a set of rules. DNA uses unlike pairing; when Watson tried like pairing, A matched to A and so on in his model, it flopped. We could invent a scheme, A = Z, B = Y, and so on or any other if we chose, but for ease of understanding, I prefer to allow like to pair with like in our model. (Just remember that it is not that way in nature.) If we follow these rules, then the second chain would hold the same message written backward.

SRAEYYTRIHTDNADERDNUHENINEREWDEVILMADATAHTSYADEHTLLADNA

The two chains are put together in Scheme 2.

SRAEYYTRIHTDNADERDNUHENINEREWDEVILMADATAHTSYADEHTLLADNA
ANDALLTHEDAYSTHATADAMLIVEDWERENINEHUNDREDANDTHIRTYYEARS

Scheme 2

As before, we have written one upside down, so that *base pairs* can be formed from the letters.

Now imagine one set is a human genetic message, the amount that you would inherit from one parent. It would be represented by twenty-three immense volumes, as we described earlier, each representing the equivalent of a huge library dictionary. You wish to find the page on which the above "Adam" passage occurs. No index exists, nor are topics grouped in any order that you can detect. The volumes are too heavy to handle, so you shred the volumes and dump the page-sized fragments haphazardly on the floor (the equivalent of using a text-cutting device on DNA). This action, which would be considered the work of a madman or vandal in an ordinary library, is redeemed by the existence of the text-matching procedure. No simple analog exists in our world for this, so we must use our imagination.

We prepare on a scrap of paper the message ALLTHEDAYSTHA-TADAMLIVED in ink that glows in the dark. Technically, this is called a *probe*. It need not be full length, but we would want it to match only one passage in the entire set. The word *and* would be useless as a probe, as it would find a match on every page. We bring this probe to the vicinity of the shredded fragments and release it. It finds its way, through some mysterious force, to the text fragment that contains the same message, illuminating it so that we can locate it by the glow. We now have the entire page that contains that phrase.

The force that makes the probe procedure work in real life is Watson–Crick base pairing. The two chains on each fragment of DNA text would be treated in advance in a way that broke their link to one another. Each would then be free to receive an outside partner: the probe. The DNA fragments would be spread out on a suitable surface, and the probe would give off radioactivity, or fluoresce, to signal where it had found a match.

By the mid-1970s, probes were being used not only to locate matching passages in shredded DNA, but to locate the approximate position of the passage in the intact volumes. Mixtures of intact chromosomes were exposed, and one would "light up" at an end or toward the middle, as appropriate, when the probe found its match.

There is a paradox in what I have just described. If you know a sentence of text (a couple of dozen characters will suffice), you can isolate the larger fragment, a page or two, that contains the passage. How do you get the first sentence to start the process? Fortunately, there were indirect methods for doing this. If you knew the amino acid sequence for a protein such as insulin, or a portion of it, the sequence of the DNA text related to it could be deduced using a set of rules called the *genetic code*. We will get to this in more detail later on. This stretch of DNA could then be prepared in the lab for use as a probe. Thus, whole sections of the human genetic text that coded for known proteins could be isolated. A formidable barrier to further knowledge existed in the 1970s, however. There was no direct way available to read the DNA text after we had located the page we wanted. People were at work, however, in two places called Cambridge, who would soon remedy this.

CAMBRIDGE, ENGLAND, 1975–77

Fred Sanger did not attend the Asilomar Conference. In fact, he traveled relatively little. One trip a year was sufficient to keep up with what was going on: "Once you've seen places, there isn't much excitement," he wrote. His first priority, as always, was his experiments.

> I have hardly done any and actively tried to avoid teaching and administrative work. . . . I do not enjoy them, whereas I find research most enjoyable and rewarding. Of the three main activities involved in scientific research, thinking, talking and doing, I much prefer the last and am probably best at it. I am all right at the thinking, but not much good at the talking.[8]

Thus, in his late fifties, when his contemporaries either directed large research groups (without directly doing the work) or had moved "up" to fully administrative positions, Sanger still labored with his own hands at the bench, still immersed in separations and sequences. This was his preference. He was much happier working with his hands than reading or writing papers.

There was a certain advantage in a life spent at laboratory work, of course. He could concentrate his energy on one problem at a time. After years of direct observation, he had a good feeling for what might and might not work. When a bottleneck appeared, he could reach back in his experience for procedures that had failed in a different situation, but might do the job now. Sometimes, he might even combine elements from two different procedures to solve a problem.

He was no longer working in the same lab, however, or on the same topic. In 1962, the Medical Research Council unit had moved out of the Cavendish, away from the ancient, history-drenched center of Cambridge. It was now housed in a new building in the functional, modern, charismaless Addenbrooke's Hospital complex more than two miles south of town. Some other scientists who had not been in the Cavendish, but were involved with the same type of problem, were also moved there. This included Fred Sanger.

Initially, this move gave him the benefit of more space. "Soon," he noted, "the laboratory was as overcrowded as most. In fact, the work seemed to go better in the overcrowded conditions."[9]

The combined facility also housed Francis Crick, who had returned to Cambridge in 1956 after a period of exile in Brooklyn. Their proximity had important consequences. "Previously, I had not had much interest in nucleic acids," Sanger wrote. "However, with people like Francis Crick around, it was difficult to ignore nucleic acids or to fail to realize the importance of sequencing them."[10] And so, in the 1960s, Sanger turned his attention to RNA.

RNA is, of course, a close cousin of DNA in its chemistry, but much shorter in length. Like its cousin, it carries information in the form of four bases, three of which, A, C, and G, are the same as the ones in DNA. The fourth, U (for uracil), is a close relative of T. RNA in our cells varies in length from about eighty units in length to several thousand, depending on its job. It occurs as only one chain, however, rather than two and so folds into shapes different than the double helix.

The main job of RNA in the cell can be expressed concisely, if in poor English, as follows: "DNA makes RNA makes protein." This idea was put forward in better grammar by Francis Crick in the late 1950s as a working hypothesis, but was dubbed the "central dogma" of molecular biology soon thereafter, because Crick did not understand the meaning of the word *dogma.* It is central, however, because it catches the most basic process involved in the workings of our cells.

A liver cell, for example, requires oxygen. A transportation system is needed, however, to move oxygen from the outside of our bodies, where it is abundant, to the liver, where it is not. The blood protein, hemoglobin, is designed for this purpose. It requires an additional nonprotein device for its work, but we will ignore that complication for now.

Hemoglobin can do its job because it has the right shape, and that shape is dictated by the sequence of amino acids in the protein chain. That sequence represents information in the same way that the assembly instructions for a bookcase does: Connect part A to part B, hook in part C, and so on, and eventually you get your bookcase. Ultimately, these instructions are stored in DNA and passed on from generation to generation as a treasured family secret. Yet DNA is not a protein and does not resemble a protein. The central dogma explains how the hemoglobin instructions, written in bases in DNA, ultimately causes hemoglobin to be made.

The region of DNA (there is more than one, but again we can skip that complication for now) that holds the assembly instructions for hemoglobin is the hemoglobin gene. When the circumstances are right for hemoglobin production (another complicated topic), the needed information is copied from one chain of DNA into a special RNA molecule whose function is to carry the message. As RNA is written in essentially the same language of four bases as is DNA, this process can be compared to the photocopying of a page or so from the huge volume that represents a chromosome.

This information, after some editing, ultimately reaches one of the cell's protein factories (*ribosomes*), intricate structures that are themselves made of RNA and protein. In this factory, the DNA information, carried in its edited RNA copy (*messenger RNA*), is translated into protein language and used to make the desired protein.

In the 1960s, the first RNA molecules were isolated in pure form from microorganisms, and Fred Sanger, stimulated by Crick, turned his interest that way. Some sequencing had been done on RNA and a Nobel Prize awarded, but Sanger thought that he could improve matters, and he did so. He stayed with his divide-and-conquer method that had worked for proteins and, using two-dimensional separations on sheets of modified paper, worked out elegant separations for RNA fragments.

Fred Sanger started his research career by working alone, and the insulin work was done with the aid of a few collaborators who came and left. By the 1960s, larger numbers of investigators were coming to his laboratory, many from abroad, to learn his methods, and he had to cope with them. He is not the type of person to assume active leadership of a group. When frictions developed among the workers, he did not know how to deal with it. "So I largely 'acted ignorant' and took no notice of it," Sanger wrote later.[11] He decided afterward that this had been the best policy. Those involved, on being treated as if nothing was happening, "were shamed into recognizing the stupidity of their behavior."[12]

He did realize, however, that he wanted a more permanent assistant to work alongside him and help with the work, someone with whom he could exchange ideas and resources freely, in an atmosphere of trust. "One of the most important things is to have someone you like and can get on well with." These qualifications were met, in the early 1960s, when he hired Bart Barrell. Barrell was a teenager who lacked a college degree, but had plenty of enthusiasm and showed a talent for sequencing work (if Calvin Bridges and Alfred Sturtevant come to your mind, that would not be surprising). He did so well that ultimately he was put on his own to direct the actual sequencing, while Sanger concentrated on the development of new methods.

Barrell was replaced as Sanger's assistant in 1967 by another quiet, congenial individual, Alan Coulson. As Coulson describes it: "I was very green in terms of research. I really learned everything from Fred. I couldn't have had a better teacher."[13]

By the late 1960s, Sanger and his co-workers were able to read the sequence of an RNA of 105 letters, the largest yet. The Sanger RNA methods worked so well that they were used by other groups to attack the sequences of RNA molecules that had thousands of units. These included RNAs in our cells that are part of the protein factory and external disease-causing substances, the RNA viruses. These viruses, of which the human immunodeficiency virus of the acquired immunodeficiency syndrome (AIDS) is a notorious example, use RNA as their genetic material. At appropriate times, their RNAs form a double helix analogous to that formed by DNA.

In April 1976, the complete sequence of MS2, an RNA virus with 3569 bases, was published in *Nature* by a research group headed by the Belgian, W. Fiers. He had eleven co-authors on the paper, and the effort had lasted a number of years. The labor involved in the divide-and-conquer method, even with the elegant techniques that had been developed, was enormous.

Fred Sanger was no longer interested in RNA by that time. He was hunting bigger game. It had always been his style to move on to greater challenges. With a permanent Medical Research Council research appointment, he did not have to "produce a regular output of publishable material." He could afford to try out "wild ideas" and attack problems that were "way out." "I like the idea of doing something that nobody else is doing," he wrote, "rather than racing to be the first to complete a project."[14] The supreme target of course was DNA, which made up the genes of all creatures except the RNA viruses.

Other workers had thought of this problem, of course. When I came to Cambridge to work with Lord Todd in 1959, I was asked to develop a method that would reveal the identity of the base on one end of a DNA chain. If it had been successful according to our hopes (it was not), it would at most have allowed us to read a few letters at one end. Ten years later, the longest DNA sequence that had been read (using other methods) was a dozen bases. When I spoke to Erwin Chargaff at about that time (he had been one of those involved in the problem), he expressed pessimism that anything substantial would be accomplished in his lifetime. Here was a field clearly in need of rescue by Fred Sanger.

He recognized at an early stage that the divide-and-conquer method, which had worked so well with protein and RNA, would not do the job for DNA. It was too slow and laborious to make much headway with the vast amounts of information present in DNA. An entirely new approach was needed, one capable of giving paragraphs-full of text in a single experiment. Somehow, out of the depths of his experience from decades of work at his laboratory bench, a new idea emerged: the read-off method. In his estimate: "This new approach to DNA sequencing was the best I have ever had, being original and ultimately successful." First, however, it had to be made to work.

The first approach of Sanger and Coulson to their new strategy was the "plus-and-minus" method. Using this, they attacked the virus phi-X-174. By the time that the Asilomar Conference was meeting, they had used it to determine the text of stretches of viral DNA up to 195 letters in length. By its repeated use, it was obvious that the sequence of the whole virus, more than five thousand bases long, could be read. Yet it did not satisfy them. The separation of fragments was not good enough, and errors crept in. An improved read-off method was needed. Further incentive was provided when another scientist from the laboratory attended a conference in the United States that summer of 1975. A research group there had also hit upon a read-off method of their own, which they had not published yet. He brought this news back to England.

Sanger later commented, in his own style of quiet understatement: "I cannot pretend that I was overjoyed by the appearance of a competitive method" that appeared to work better than his present one. Although he would have continued to seek improvements, the appearance of competition was a factor in spurring him on to do his best. He succeeded with a new innovation, called the *dideoxy* or *chain termination method.* "I suppose," he added, "the dideoxy method can be regarded as the climax of my research career."[15]

His new method did not spring up overnight: There was work to be done, but that was something that they took for granted. Four key chemicals were needed. They were subunits of DNA, each bearing one of the bases A, C, G, T, but altered in an unnatural manner. When one of these dideoxy analogs was inserted in DNA text that was being copied, as in reproduction, it jammed the copying process. It can be compared to a faulty key on a typewriter, which when struck, stalls the machinery so that no further text can be typed.

One of these analogs had been available: enough to test the ideas in principle, but not to read any actual sequences. Their test ran beautifully. Now the other three substances were needed. A drug company had promised earlier that they were going to prepare them. "I was waiting around for that." After a year, however, they changed their minds. "That was bad luck, really." Nor was anyone else willing to undertake that chore for them. In Sanger's words:

> Alan R. Coulson and I had not done any nucleotide [DNA subunit] chemistry before and were unable to persuade anyone else to make them, so we had to settle down to making them ourselves . . . we were really struggling in the dark. It was mainly Alan and I doing it. Others helped with expertise. The chemists there let us use their fume cupboards [a special compartment in a laboratory equipped for the rapid clearance of toxic odors]. That took pretty well a year. The yields [amounts of chemicals produced] were extremely poor. We just had enough to do the four reactions.[16]

The above account would be unsurprising if it represented the adventures of a graduate student embarking on his first project or an elderly chemist puttering around in the lab as a hobby. What we have rather is the account of a Nobel Prize winner in chemistry (in 1958 for his insulin sequence), one who was extraordinarily gifted in experimentation, en route to a second Nobel Prize (this double-Nobel distinction has been achieved only by Marie Curie, Linus Pauling, and physicist John Bardeen, in addition to Sanger). One might guess that rank does *not* have its privileges after all. A more accurate explanation would be that the procedure just reflected the very special character of Fred Sanger. His own comment was that "it was difficult, but it was a good experience. It was rather enjoyable doing chemistry." Alan Coulson shared his attitude: "It was quite fun, actually. The solutions were rather a funny color but they did do what they were supposed to do."[17]

With the four dideoxys available, the new method could be tried, and it worked exceptionally well. Three hundred bases worth of sequence could be read in an experiment requiring a day. By applying it repeatedly, it was possible to determine the entire sequence of phi-X-174, completing and correcting the work done earlier by the plus-and-minus method. As we shall see, later workers have complained that the repeated performance of this procedure is tedious. Sanger commented: "To me it was always a delight. At one stage in the work I would take the autoradiographs home with me and look forward to reading them in the peace of the evening."[18] Tedium is in the eye of the beholder.

The basis of the dideoxy method can readily be understood, using the English language rather than DNA for our analogy. Suppose that the message that we wish to uncover is the following: he had hot tea at the old hotel. Let us first mimic DNA by omitting spaces and writing in capitals, as if we were using A, C, G, and T:

HEHADHOTTEAATTHEOLDHOTEL

The message will be read while copying it from another chain that has the complementary message. If we follow the English version of the pairing rules that we developed earlier, the other chain would read the way we have indicated in Scheme 3*a*.

(*a*)

------------------HƎHⱯᗡHOꓕꓕƎⱯⱯꓕꓕHƎOꓶᗡHOꓕƎꓶ-----------------

(*b*)

------------------HƎHⱯᗡHOꓕꓕƎⱯⱯꓕꓕHƎOꓶᗡHOꓕƎꓶ-----------------

(*c*)

------------------HƎHⱯᗡHOꓕꓕƎⱯⱯꓕꓕHƎOꓶᗡHOꓕƎꓶ-----------------
------------------HEHADHOTTEAATTHEOLDHOTEL-----------------

Scheme 3

Of course, we would not know the letters I have shown above *in advance*. If we did, there would be no need to run the experiment. I have listed them, using hindsight, so that we can follow what is going on. The dashes that flank the English letters are meant to indicate additional unknown text.

For the experiment, we would also need a supply of the appropriate single units, each containing one letter. For DNA, the bases used are A, C, G, and T, but for the English message we would need seven letters: A, D, E, H, L, O, and T. In the dideoxy procedure, a helping enzyme and certain chemicals are also needed to make the system work. A final requirement would be a primer, a short stretch of DNA from the second chain, which does not include the portion we wish to learn. The starting situation would now look as in Scheme 3*b*.

If everything were added and the conditions appropriate, a second DNA chain would be constructed alongside the first, following the Watson–Crick pairing rules. We would have extended the second chain to give the situation shown in Scheme 3*c*. As we did not know the sequence of the original chain, however, we would not have learned the sequence of the new one. Nothing would have been gained.

Suppose now that we had "spiked" the mixture by adding a percent or two of the unnatural dideoxy version of one letter, A, for example. Remember also that a chemical experiment, such as copying a DNA chain, involves very large numbers of identical chains, not just one. Whenever a particular chain has to add an *A*, a game of russian roulette is played. If a normal *A* is selected, the game continues. The choice of an unnatural *A* ends the game at once. In the English analogy that we set up above, all of the following messages would be typed:

--------------------HEHA′
--------------------HEHADHOTTEA′
--------------------HEHADHOTTEAA′
--------------------HEHADHOTTEAATTHEOLDHOTEL-------------------

A small fraction of the chains would put in the unnatural *A* (signified by A′) in the middle of the word *had* and stop at that point. Others would continue to the end of *tea* and then stop. Yet others would stop one letter further on, while the remainder would complete the message and continue into the text I have not specified, again stopping in part after every *A*.

There would be no point to this exercise unless we could tell what had happened. Sanger developed a separation of the mixture based on length of the new message added. The method depended on the passage of electrical current through a sheet, or gel, of viscous, clear, plasticlike material (*polyacrylamide gel electrophoresis*). I have created an imaginary separation of this type for our English message in Scheme 4.

We see seven columns, one for each of the letters in the message (in actual DNA work there would be four). The three bands in the A column, from reading from bottom to top, represent the fragments ending in HEHA′, HEHADHOTTEA′, and HEHADHOTTEAA′. Again, this is written with hindsight. What we have really learned is that the letter *A* occurs in positions 4, 11, and 12 of the message. Another experiment would also be run in which a few percent of the faulty dideoxy form of a different letter was put into the mixture. We would then learn the locations of that letter. The same would be done for the remaining letters, with the separations run side by side, as in Scheme 4, to give a pattern that is called a *sequencing ladder*. By examining such a ladder, we can read off the sequence of the message by noting the order in which letters appear from the bottom to the top. We have deciphered twenty-four letters in this exercise, but in

	A	D	E	H	L	O	T
	---	---	---	---	---	---	---
24					---		
23			---				
22							---
21						---	
20				---			
19		---					
18					---		
17						---	
16			---				
15				---			
14							---
13							---
12	---						
11	---						
10			---				
9							---
8							---
7						---	
6				---			
5		---					
4	---						
3				---			
2			---				
1				---			

Scheme 4

actual practice, hundreds of letters of DNA text can be learned in one such procedure.

Using this method, Sanger and his co-workers read off chunks of phi-X-174 text three hundred bases at a time. These chunks were obtained by slicing up the virus with text cutters. The chunks were aligned in proper order using the old divide-and-conquer method until the job was done.

The full sequence of the virus was published in *Nature* on February 24, 1977,[19] two years after Asilomar. According to the *New York Times*, it "created a sensation." It represented the first publication of a DNA genome (complete genetic text for an organism). It also had some surprises, one of which was the existence, in two places, of one gene *within* another using the same text. The same letters, read differently, gave two different biologically meaningful messages. This would hardly be possible in English, but can be managed with DNA text, which uses only four letters and has no spaces.

The dideoxy method itself was published later that year in the *Proceedings of the National Academy of Sciences of the U.S.A.*[20] It became a classic. At this writing, it is the fifteenth most cited paper in scientific history, and it has not yet peaked. The rate of citations per year is still increasing and with good reason. Sanger had not discovered the nature of the gene; that was known. He had made it possible, in principle, to read all of the text that governs our heredity, as individuals and as a species, and to do the same for all of the other creatures of biology. The amount of work ahead was enormous.

In 1990, I set out on a pilgrimage to visit the place where Sanger and Coulson had worked out their method. I have already mentioned the hospital complex in Cambridge. The square, modern Medical Research Council building displays many windows, set in a background of crossed light beams alternating with red and grey brick. It is not ugly, but lacks any special character.

Dr. Sanger greeted me in his former office and showed me the two-man lab in which he and Alan Coulson had worked. It was a smallish area, about eight by twenty feet, with older wooden bench tops and some newer, less expensive Formica ones. The area "hadn't changed much" from the time that he worked there. The laboratory window looked out over the hospital complex with a view of the moors in the background. When he had first arrived there, over a quarter of a century ago, more of the moors and less buildings could be seen. I looked around his work area for some special physical feature that would make it more memorable and found none. We both agreed; it was "just a lab."

He was dressed casually in a plaid shirt. He slouched back in his chair as we chatted, a rather short man with blue-grey eyes, largish ears, and greying hair. Sanger answered questions in a soft-spoken and thoughtful way, with little flair. He appeared to be just a scientist, but what a scientist he was!

In 1980, he had won his second Nobel Prize, sharing it with the developer of the alternative read-off method (we will meet him shortly) and Paul Berg, a pioneer of cloning. He went on to participate in the sequencing of human mitochondrial DNA, an extra bit of our heredity that we will encounter later on. This substance had 16,569 base pairs of DNA. In 1982 he and his colleagues published the DNA sequence of lambda virus, 48,502 base pairs long. "Lambda was sequenced mostly by Alan Coulson and me," he commented. In the following year, 1983, he retired abruptly at age sixty-five. Why? In his words:

> A lot of people do get fed up with research, but I can't say I ever got fed up with it, really, until I was 65. I think that I was ready to retire then. The DNA work had really reached a peak but I don't think I could get much further with the setup I had. The last year, working in the lab, I was trying things that didn't work.[21]

He had encountered such periods before. He has written of the depressing "lean years" that followed his success with insulin and how he overcame his negative feelings:

> The best antidote is to keep looking ahead. When an experiment is a complete failure, it is best not to spend too much time worrying about it but rather to get on with planning and become involved in the next one. This is always exciting and you soon forget your troubles.[22]

In 1983, after his second Nobel Prize, his feelings had changed:

> The aging process was not improving my performance in the laboratory. If I had gone on working I would have found it frustrating and have felt guilty at occupying space that could have been available to a younger person.[23]

I suspect that there are few younger scientists, though, who would match his performance, even now, if he chose to return to the laboratory.

Fred Sanger lives peacefully in the countryside today with his wife. His two sons and a daughter have grown up and chosen not to follow in his steps in science. His daughter has a Ph.D. in medieval English. He keeps busy at home with gardening and carpentry and stays abreast of progress in his old area by visiting his former lab where Coulson, Barrell, and other former colleagues are still very much at work. On the day that I was there, he took some time out to visit them.

At the end, he volunteered to drive me to the Cambridge railroad station, an offer that I gladly accepted. He used a small, red, cluttered old Peugeot station wagon. He said that they had a better car, but that it lacked a legal parking sticker for the lab.

Just outside the building, a young woman, perhaps a technician, carrying a container of some chemical solvent stopped us and asked whether either of us knew the way to the delivery area. Sanger instructed her. I doubt that the young lady realized that she had been directed by a double-Nobel Prize winner.

I have wondered why, despite his achievements and many prizes, he is so little known outside his field. I found one clue in a 1979 book review by scientist Thomas Jukes of *The Eighth Day of Creation*, Horace Freeland Judson's history of modern molecular biology. In this book, conversations with Watson, Perutz, and many other scientific figures of the era are described in detail, but none with Sanger. His insulin work is referred to with admiration, but no mention is made of his nucleic acid methods. Jukes speculated: "Apparently, Fred spends too much time in the lab, and not enough talking to historians."[21]

CAMBRIDGE, MASSACHUSETTS, SPRING, 1975

Walter Gilbert peered at the gel separation of DNA fragments in his Harvard laboratories, which he at one time had shared with Jim Watson. His career in theoretical physics had been set aside a decade ago; he was now interested in the intimate details of the way in which a type of protein, called a *repressor*, embraced a particular stretch of DNA.

His shift in interest was directly due to Watson. Gilbert's wife, Celia, had worked as a technician in Watson's lab in the late 1950s. The friendship between Watson and Gilbert that had begun at Cambridge, England, deepened and they saw each other four or five times a week. The Gilberts invited Watson, who was still a bachelor, to dinner frequently, "almost mothering him." They spent a lot of time talking about science.

By 1960, Gilbert felt that physics was going very slowly. Watson invited him to come over to the biochemistry laboratory where more exciting things were happening. At that time the hunt for messenger RNA, the intermediate that carries a DNA message to the protein factory, was a hot topic. Gilbert followed Watson and his colleague around the lab for a while, watching their procedures. After a time he joined them and finally started to run his own experiments. Ultimately, he wound down his efforts in physics and became a professor of biochemistry at Harvard.

Was this unusual shift, from theoretical physicist to experimental biologist, difficult? Others have made this transition, but few with such rapid success. According to Gilbert:

> I learned just by doing. In some ways I found the transition quite easy. The critical thing that you learn in taking a Ph.D. degree is how to think for yourself. Once you've learned that, you can carry it over from field to field.[25]

In terms of specialized knowledge, "what you actually need to know about a particular field, you can learn in a short time." There was no need for him to get a second Ph.D. in biology. He was already an experienced scientist.

Before any readers leap to follow his example, we should consider his personal qualifications, as described by writer Stephen Hall in the book *Invisible Frontiers*[26]: "Testaments to his scientific style and brilliance come in almost embarrassing abundance." Hall cited, among others, biologist Philip Sharp of MIT who commented: "If I had a problem and one person to take it to, I would take it to Wally. I consider him to be the brightest person I ever met." Some of his colleagues also felt that a measure of arrogance and aggression came along with this intense curiosity and imagination, Hall reported.

Most of the theoreticians that I have known had no practical ability at all; they would have been utterly lost in a laboratory. Gilbert was made of other stuff. He had taken a good amount of chemistry as a Harvard undergraduate. Before that, it had been a childhood interest, along with astronomy (he ground his own telescope mirrors). Hall reported that at age twelve, Gilbert had suffered an explosion that cut his wrists. As he rode to the hospital with his mother, he made his first comment on the incident: "I know what I did wrong."

By the mid-1970s, Gilbert's lab at Harvard was known for its high standards, work ethic, and intensity. He set the pace himself: "I love new things, new ideas, new facts. It's very nice to have old things but a week or so later they're all old hat, and you want something new."[27] His students, as reported by Hall, picked up this theme:

> The work was always urgent in Wally's lab. It was too boring not to have results, so everything went as fast as possible.
>
> Graduate students would drift into the lab around noon or shortly after, and often work into the wee hours of the morning or on through into the next day.
>
> It was so much more than work. It was *life*!

A subculture of such intensity needed its own chronicle, and the newspaper *Midnight Hustler* was published irregularly by his students. It reported progress and tidbits of gossip from the lab and made fun of such items as Wally's purple and orange turtlenecks.

It was in this environment that Gilbert lifted one particular separation gel for inspection early in 1975. It had been suggested during a lunch-time discussion with a visiting Russian scientist, Andrei Mirzabekov. A chemical treatment with which Mirzabekov was familiar could break DNA into fragments, striking at adenine and guanine. DNA would shatter everywhere except at the place where it was hugged protectively by the repressor. The pattern produced by electrical separation of the resulting mess would reveal which part of DNA the repressor held. (This procedure today is called *footprinting*.)

When Gilbert examined the separation, he saw immediately that the experiment had worked. However, there was much more in it than he had expected. "The pattern was so sharp . . . it was so easy to correlate the positions." He was in fact inspecting a gel quite similar to the one that Fred Sanger was developing in the other Cambridge, but with only two of the four bases represented. It needed "not a terrible stretch of the imagination to look at that and realize that we could identify all the guanines and adenines of a DNA sequence."[28]

Although the other two bases could be determined indirectly by locating the adenines and guanines paired to them on the other chain, it would be both more elegant and efficient if the cytosines and thymines could be located directly. Gilbert and his principal laboratory associate, Allan Maxam, worked out a cleavage reaction for that purpose, perfecting their full method over the next year. The success of the procedure, reported at a conference during that time, in turn spurred Fred Sanger on to develop his own dideoxy method.

The two of them received the Nobel Prize together with Paul Berg in 1980. The two procedures, although discovered independently and functioning through different steps, ultimately gave the same results.

The order of letters in DNA could be read off several hundred letters at a time. Two deeply skilled men, with vastly different personalities and scientific styles, had come together in achieving the same thing.[29,30]

Sanger had followed a purposeful, almost lifelong, quest. Gilbert stated in his Nobel Prize lecture: "I came to the chemical DNA sequencing by accident."[31] In his interview with me, he added that the discovery was yet part of "a logical creative research program."[32] He was aware of the need for a good DNA sequencing method. Maxam and he had needed to decipher a string of twenty bases earlier in the decade, and it had taken them two years of intense labor. The first gel result had surprised him "and the things that surprise you most are the things that are most interesting." Gilbert was no stranger to serendipity. Accidental encounters had, in fact, turned his career in the proper direction. In his Nobel Prize lecture he acknowledged debts to many, but owed "the greatest to Jim Watson who stimulated my interest in molecular biology."[33]

SAN FRANCISCO, CALIFORNIA, JANUARY 1976

Two men were discussing recombinant DNA and business in Churchill's, a bar near the San Francisco campus of the University of California. One of them, faculty member Herbert Boyer, had helped develop the gene-splicing techniques by which foreign DNA could be inserted into bacteria. For a year, he had considered the possibility that the bacteria could be coaxed to manufacture the proteins coded by such DNA. The other gentleman was Robert Swanson, a twenty-eight-year-old venture capitalist who felt that there were vast commercial possibilities in such a process.

The two men found that they were thinking along the same lines and agreed to put up $500 each as the initial capital to start a new company. Incorporated that April, the new firm was called Genentech. This investment was "derided by scientific colleagues and businessmen alike as premature and doomed to failure," Stephen Hall reported in *Invisible Frontiers*. When the company went public on Wall Street in October 1980, less than five years later, Boyer and Swanson found their shares to be worth $66 million each. Their vision was to lead to the manufacture of such vital human components as insulin, factor VIIIC, and human growth hormone, in bacteria and cultured cells.

WASHINGTON, D.C., 1977

The Commission for the Control of Huntington's Disease and Its Consequences made its report to Congress. More than a century after the first publication by George Huntington, that malady had been recognized to be the cause of suffering for tens of thousands of Americans and greater numbers abroad. It was not just limited to Long Island as Huntington had suggested.

The commission had been set up a year earlier, after lobbying by foundations and those afflicted, "to develop a comprehensive national plan" for combatting the disease. It succeeded, and much credit was given to the efforts of a young New York clinical psychologist, Nancy Wexler. In the words of the report: "Her deep personal commitment and her intelligence graced with wit, are a force which can make anything happen."

COLD SPRING HARBOR, LONG ISLAND, NEW YORK, 1977

During the period, 1968–77, Jim Watson had divided his time between Harvard, where he taught and ran a research lab, and the Cold Spring Harbor Laboratory, where he was now head administrator. (The two institutions at the site had finally joined together in 1962. See Chapter 3.) In 1977, he left Harvard to direct Cold Spring Harbor full time. His preferred role in science was one very different than that of Mendel, Morgan, and Sanger, who were happiest in the lab. Lecturing was also not Watson's strongest point. I had one direct experience of his style.

I attended a series of his lectures in a course at Harvard in the late 1950s and always came early so that I could get a seat in the first row. He appeared to me to be a frail young man whose voice, hardly more than a whisper, could not carry further back. His style had not changed by the 1970s, according to Horace Freeland Judson. He wore a rumpled suit and

> spoke to the shoes of the people in the front row. His voice fell away so that many were leaning forward to hear. . . . A bit later, a voice from the back yelled "louder." Watson responded "Yes I'll try. I know I mumble. The only way I don't mumble is if people in the back will be rude. And I

don't mind. Call 'louder' and I'll speak louder. For at least five minutes."[34]

Watson the administrator was made of a different fabric. For this opinion, we can cite none other than Erwin Chargaff:

> Crick and Watson are very different from each other. Watson is now an able, effective administrator of science. In that respect he represents the American entrepreneurial type very well. Crick is something else—brighter than Watson, but he talks a lot, and so he talks a lot of nonsense.[35]

Some of Watson's attitudes had certainly changed since the 1950s. He stated in an interview in *Omni*:

> I am not interested in people who won't work ten hours a day and be there on Saturdays and Sundays, because somebody else will be. I don't know of any successful scientist who doesn't work harder than or as hard as virtually everybody else in his field. The thought that there are some very clever people who survive without working very hard is the sort of illusion you sometimes get in school. . . . You just don't see that in science.[36]

He was not unhappy in no longer doing science: "The thing that makes one happy is just the appearance of new science, so if you see new science appearing, then I'm happy." Large amounts of good new science did appear at Cold Spring Harbor and both its academic reputation and its endowment soared under Watson's guidance. In 1990 it was ranked by *The Scientist* as the best of the independent research institutes in the United States:

> Cold Spring Harbor, under the directorship of DNA co-discoverer and Nobel laureate James D. Watson, stands head and shoulders above all others. . . . By all accounts, among Watson's many talents is one for finding the best researchers.[37]

"Jim made it all happen," Philip Sharp of MIT said to *Science*. "What's there now, the style and substance of Cold Spring Harbor, is the creation of his energy over the past 20 years."[38]

Watson's personal life had also changed. In 1968, he married Elizabeth Lewis, a nineteen-year-old laboratory assistant: this marriage has lasted.

It might appear from the above items that I am wrapping up his biography on a happy note, but we are far from finished with him in this book. While en route to a meeting of the National Cancer Advisory board in the 1970s, Watson commented: "Washington's a pretty minor aspect of my life. . . . I see myself as an insignificant outsider."[39] This, above all, would change.

While Crick's involvement with genetics research greatly diminished, Watson grew more intensely involved in directing the forefront of the field. At about the same time that Watson moved full time to Cold Spring Harbor, Crick shifted from Cambridge to the Salk Institute in San Diego. This facility nestles on cliffs above the Pacific, near a site where brave individuals leap off the edge and glide, supported by wings strapped to their body. Below them lies a popular, if now illegal, nudist beach.

Crick appreciated his new location: "I feel at home in Southern California. I like the prosperity and the relaxed way of life." He also found a new center of interest: the study of the brain and the nature of consciousness. I recall a popular magazine article that described his theory of the nature of dreams.[40] He felt that they were not of deep Freudian significance, but simply a mechanism by which the brain cleans out purposeless trash gathered during the day.

In pursuing this new area, he had to separate himself from DNA research (except for occasional forays into the origin of life and other speculative topics), for such research has not touched on the nature of consciousness. At least, not yet.

PART II
TODAY

6

THE PROJECT

SAN DIEGO, CALIFORNIA, OCTOBER 1989

"I always stand for peace, but if there will be war, I will fight it," the speaker concluded. This was not the president or secretary of state before a joint session of Congress, but rather James Watson addressing a group of scientists at a meeting called Genome I. It was not obvious to me in the audience exactly who he proposed to make "war" on or what the war would consist of (though later I learned that it involved withholding data from Japan), but it was clear that this was only a sideshow. The principal message of his address was this: The Human Genome Project was on. The entire human genetic message would be read over a period of fifteen years. A starting date was needed ("We want to put that date as far off as possible," he joked), and he designated the start of the next fiscal year, October 1, 1990, for that purpose. Two hundred million dollars a year were to be appropriated by Congress, 3 billion dollars over fifteen years, for the Genome Project, and he, James Watson, as director of the newly created National Center for Human Genome Research, National Institutes of Health (NIH), would have the most to say of any individual about the disposition of the lion's share of the funds.

A year later, at the Genome II San Diego meeting in October 1990, the next fiscal year had not yet started as Congress and the president had not agreed upon a budget. Further, the government seemed likely to renege on a portion of the funds promised for that year. Watson's response was that "I won't use any slides. Government money is very short these days." The participants agreed, however, that the October 1, 1990, start date for the project should be maintained. The quest was on.

The meetings, of course, had no official status or power. They were convened and given their formidable titles by *Science* to make other scientists, the press, and the public fully aware of the situation that had crystallized in a number of earlier scientific and legislative meetings. In addition, one of the key events that caused the project to happen was the publication in *Science* on March 7, 1986, of an editorial by Nobel laureate Renatto Delbucco.[1] Focusing his attention on cancer, he argued that for full understanding and control of this complex disease, we must identify all of the genes related to it. Although this effort could be done "by a piecemeal approach," it would be far more useful to have the entire human genome in hand. Broadening his perspective, he pointed out that many areas of biology would benefit, and called for "a national effort . . . comparable to the effort that led to the conquest of space." He pointed out that "the sequence of DNA is the reality of our species."

This editorial gave wide circulation to an idea that had come up separately at a conference held in Alta, Utah, in 1984 organized by the Department of Energy and another at Santa Cruz, California, in 1985. The Santa Cruz meeting had been called by biochemist Robert Sinsheimer, chancellor of the University of California at Santa Cruz, in the specific hope of establishing a sequencing project there. According to Watson, "Renatto Delbucco came away excited from that gathering" and later at Cold Spring Harbor "he spoke glowingly of the prospects for cancer research if we knew the sequence of our own DNA."[2]

The Department of Energy then took up the idea and championed it. Their interest in human genetics had grown out of public concerns about the genetic effects of radiation. Further, they had experience in the management of large projects. If necessary, the Department of Energy would carry out a genome project on its own.

It was the NIH, of course, that had the primary responsibility for support of research involving human health and biology. It was obvious to many scientists that it should take the lead in any project of this sort. The agency was reluctant for a number of reasons. Historically, it had administered research by awarding many small competing grants rather than organizing megaprojects. After some foot-dragging, it accepted the challenge and established a new office for the purpose.

A noteworthy scientist was needed to head the NIH effort, one who was skilled as an administrator and who could visibly represent the field to other scientists and the public. Watson's qualifications were obvious. So he again found himself in two jobs and could no longer consider himself "an insignificant outsider" in Washington affairs.

Committees of the National Research Council and Office of Technology Assessment of the U.S. Congress were convened and in their reports endorsed the projects. Congress agreed, and some start-up and organizational monies were appropriated to the NIH and Department of Energy for fiscal years 1988 to 1990. If Congress honored the commitment beyond that, then the project was definitely on its way. To those who had championed the idea, the rate of progress toward its acceptance, with its accompaniment of debate and criticism, may have seemed painfully slow. If we place it in historical perspective, however, the rush of events is astonishing.

Often with great human undertakings, the goal has been grasped long before the means became available. The moon has always been visible to humankind, and a trip there has been a subject for fantasy over the ages. A century before the Apollo project, Jules Verne wrote of a voyage that resembled the Apollo 8 mission to an amazing extent. Explorers were launched by cannon from a site in Florida, then circumnavigated the moon and returned safely to earth. In the 1935 British film *Things to Come*, a launch was portrayed in a vessel whose supporting machinery resembled remarkably the apparatus later used at Cape Canaveral. The vision was in place; when the detailed technical advances took place during and after World War II, only the resolve was needed to make it reality.

In the area of genetics, the situation was reversed. There was no realization throughout history that the properties of our bodies were governed by a text. When this idea emerged in 1953, it seemed challenge enough to those involved to work out how this mechanism operated in principle. How could a set of characters embedded in DNA actually control what our bodies did? Erwin Chargaff had commented in 1968: "A detailed determination of the sequence of a DNA molecule is beyond our present means, nor is it likely to occur in the near future . . . We may, therefore, leave the task of reading the complete nucleotide sequence of a DNA to the 21st century which will, however, have other worries."[3] Only when Sanger and Gilbert developed their read-off methods in the mid-1970s did scientists realize that any DNA message, covering any function, could now be read. They simply had to work out a way for each specific case of locating the relevant DNA passages within the larger set of volumes that held it.

Once this was done, the text could be mastered, and the meaning extracted. Even though only a few hundred letters could be read in a single procedure, the process could be run over and over until much larger sections had been put together. What was found was often surprising and, in some cases, not immediately understood. However, knowledge, brand new and important knowledge, was there for the taking.

Sequences began to accumulate, with viruses and bacteria being obvious targets. Their genetics had been studied thoroughly by older methods, and sequence data could readily be fit into the overall picture. More and more, however, biologists turned to the challenge of higher organisms. Data on humans began to flow in as well, particularly on sequences related to known disease conditions.

In 1970, the total number of DNA characters that had been read in all biological systems was in the dozens; suddenly in 1977, it leaped into the thousands, and by the 1980s, into the millions, then tens of millions. Journals were no longer sufficient to house this material, computer data banks were needed. Genbank was established by the NIH in Los Alamos, New Mexico.

Los Alamos, of course, had already been associated with another U.S. government project of enormous magnitude: the development of the atomic and hydrogen bombs in the 1940s. The genome effort, unlike its predecessor, required no secrecy and would hopefully save human lives, rather than obliterate them. Another data bank was established in Heidelberg, Germany, by the European Molecular Biology Organization. By 1990, the combined data banks held nearly 50 million characters of DNA data, of which more than a tenth was human.

Just a few years earlier, possession of the total human text, the 3 billion characters that we get from each parent, had seemed utterly out of reach. For example, C.W. Schmidt and W. Jelanik had written in *Science* in 1982 that "even with the recent advances in DNA sequencing technology, it is unlikely that the base sequence of more than a few percent of such a complex DNA will ever be determined."[4] Others had been less pessimistic. An unsigned editorial in *Nature*, on May 8, 1980, had noted:

> Mount Everest, as we all know, was first climbed "because it was there." The Everest of those scientists who are devoted to the sequencing of DNA is the complete human genome, a peak that cannot yet be conquered. Nevertheless, it seems not out of the question, even now, to attempt to sequence an entire chromosome (or a substantial fragment) from Man or *Drosophila*.

With some foresight, they added

> a tenth of a human chromosome would not take as long [as five years] because the thought of devoting a lifetime to the task would force an order of magnitude increase in the speed of sequencing![5]

A few short years later, Renatto Delbucco could look at the accumulation of data and, writing in the same journal, deliver a message that said in essence: Let's go for it all. Suddenly, the dream had caught up with the means.

For every human aspiration, there will always be someone who looks at it and says "Why bother?" The question is valid. I was very excited over the moon landing and would still like to see more human exploration of space. But twenty years later, I cannot see where it has led us or how it has practically affected our lives. With the Genome Project, though, we are on very different ground. We are investigating our very selves, the fabric of our lives.

Jim Watson has commented, "Human beings are the most exciting thing that we can find out about."[6] A more poetic expression of these sentiments was the one expressed by Alexander Pope in 1733:

> *Know then thyself, presume not God to scan;*
> *The proper study of mankind is Man.*

For those who are unmoved by poetic appeals, however, more practical demonstrations can be put forward. Much of the initial work on the human genetic text has been driven more by an interest in malfunction than in function, an understandable desire to relieve human misery. The glimpses gained through these efforts, however, have fueled a curiosity that will not be satisfied until we have all of it, the entire plan of the human. These feelings can best be conveyed if we plunge in and see, literally, what we have learned of blood and flesh through reading some choice fragments of text.

7

THE RED OF THE BLOOD

We breath, if we wish to continue to live. My chest moves slowly up and down as I write this, and normally I would take no notice. This behavior comes with our bodies and is taken for granted. Any interruption occurs at our peril. If only our face were covered with water, while the remainder of our body were exposed to oxygen, we would still die by drowning. The closure of our windpipe by pressure from a rope or obstruction from a swallowed object would also lead to death. Thus are we dependent on our air supply.

One amendment must be made: For more than two centuries we have recognized that it is the gas oxygen, a minority component of air, that sustains us. We need oxygen to combine with the food we eat to release energy. If a glass is placed over a candle, it soon consumes the oxygen component in the closed space and extinguishes, although air remains. The same would happen to you or me.

Every minute I require about a half-pint (250 milliliters) of oxygen and much more when I run or play racquetball. The five quarts per minute that I consume when exercising must be delivered to cells all over my body. Simple diffusion through my skin would not do the job (although it works for much smaller, one-celled creatures). My body plan includes a pump, my heart, to move an oxygen-carrying fluid, the blood, past all my cells. Oxygen, unfortunately, is not very soluble in water. If blood had no special mechanism to carry oxygen, my heart would have to pump several hundred gallons every minute as I played racquetball, and I would need a fifty-pound pump rather than my one-pound heart.

The need for a specialized oxygen carrier came up early in the evolution of large creatures. Some substance was needed that could soak up oxygen reversibly, absorbing it where it was plentiful, as in the lungs, and releasing it where it was scarce and needed, as in the smallest blood vessels. Now oxygen is willing to combine with many substances; unfortunately, the products are quite stable—the oxygen will not come out again easily. A reversible oxygen binder requires a more elaborate construction.

We cannot be sure what form the earliest biological solution took, but the modern update is available for our inspection today. It is hemoglobin, the red product in the blood, a substance that we have met several times in our tale. Felix Hoppe-Seyler, Miescher's mentor, isolated it in 1865 (a year we remember for other reasons as well) and deduced that it bound oxygen. Max Perutz carried out a more than twenty-year study in Cambridge earlier in this century to learn its three-dimensional structure, and Vernon Ingram, also in Cambridge, showed that a change of one amino acid in hemoglobin led to sickle-cell anemia.

Hemoglobin increases the oxygen-carrying capacity of blood fifty-fold. Twenty-five million million (25,000,000,000,000) red cells carry it in our blood, and each cell is packed with hemoglobin to one-third its weight. They have been stripped down for this purpose: The nucleus, with its reproductive machinery, has been scrapped and the protein factories, as well. A red cell functions until it is worn down. A cell makes seventy-five thousand trips from lungs to small blood vessels over 120 days, and then it is replaced. A fresh supply is made in bone marrow cells to replace the discards, as the transport of oxygen to the cells must be maintained. The ancients erred in one way: Blood does not transmit heredity. The red cells do not even carry any genes at all, as they have lost their DNA. They were right in another way, however; blood is the essence, the soul of life. The essence of that soul lies in the red protein, hemoglobin.

Proteins are elaborate molecules when compared to the sugar and salt on our table or the natural gas and ammonia elsewhere in our homes. Hemoglobin is complex even for a protein. Four separate units, each qualified to be a protein on its own, embrace and work together in harmony. Each unit could carry oxygen by itself, but their harmony as a quartet is superb. Working together, each member of the quartet picks up more oxygen in the lungs and releases more to the cells than it would if it operated alone.

The quartet itself is made of two pairs of identical twins. One type is called *alpha-globin*, the other, *beta-globin*. Each of the four globin proteins carries the same treasure in a pouch: a ring made of a different substance called *heme*. The heme ring in turn bears a jewel, an atom of iron. The heme and iron combination lies at the heart of the machinery: It carries the oxygen and, in doing so, provides the red color. It could not do this in isolation, however; the larger and much more complex globin unit must surround it for it to work.

To live, each of us must be proficient in the art of globin synthesis. The formula can vary somewhat, as does, say, a recipe for chili, but it must work. We are furnished two recipes for each of the globins, one provided by each of our parents. They in turn obtained them from their parents, and so on, in a chain that goes back hundreds of millions of years. The recipes are genes, written in our DNA. In the mid-1970s, when humans had first learned how to read the ancient scrolls of DNA, the globin recipes were among the first to be studied. They deserved the honor. In selecting a portion of text for our own inspection in this book, we shall also follow this tradition. A further choice must be made of either alpha or beta. Ignoring alphabetical order, we select beta, as this choice leads us to the cause of sickle-cell anemia. We are now prepared to examine a portion of the human blueprint.

We could find the plan for beta-globin in almost any of the 10 trillion cells in our body. One exception, ironically, would be the very cells that carry it in large amounts: our red blood cells. As we noted before, these cells have lost their DNA. They have the item, but not the plan for it.

As we have learned, the DNA plans within cells are kept within chromosomes, which we have likened to gigantic volumes of text. If we examine the volumes, we find that they come (with one exception, the sex chromosomes) in pairs of equal size. The content of each member of a pair is roughly the same, like two independent translations of a foreign work. If we wish, we could distinguish them by placing blue and pink colors on the bindings to indicate that one comes from the father and one from the mother. This is not entirely fanciful. At least in some instances involving a limited number of genes, our cells can distinguish the maternal and paternal chromosomes in a process called *genomic imprinting*.

We want to examine the passage that deals with beta-globin, so we reach for one of the two middle-sized volumes marked No. 11 and haul it off the shelf. I can cite the volume number in an instant, but it took decades of work by scientists to reveal it. Chromosomes are elusive entities, visible under the microscope only at certain times when cells are dividing. They were discovered in the late nineteenth century, but the correct count for humans was not obtained until the 1950s. They did not come with attached numbers; they had to be photographed and the images arranged painstakingly in size order to establish identities for them. The largest chromosome, now estimated to contain some 270 million base pairs of DNA, was called No. 1, the next largest became No. 2, and so on, with one exception. Chromosome 21, estimated at 46 million base pairs, is smaller than No. 22 with its 52 million. The sex chromosomes were excluded from the numbering scheme and named X and Y. Chromosome X is midsized, whereas Y is one of the smallest ones.

Chromosomes have additional features, which sometimes strain my efforts to describe them as books. Many proteins, as well as DNA, are present in a chromosome (we may consider them to be the backbone and glue used to construct a book). The chromosome ends (*telomeres*) have a special structure. They contain DNA, protein, and RNA as well and may be thought of as covers that also contain some text. Somewhere along the length of a chromosome, possibly but not necessarily in the middle, is a pinched-in spot, which we will call the chromosome hinge (*centromere*). During cell division, it is used as an attachment point for fibers that physically haul the chromosome across the cell. The thought of a carrying handle attached to the backbone of the book comes to mind, but is not appropriate, as text runs through the hinge, as it does through the remainder of the chromosome.

In the late 1960s, geneticists learned to stain chromosomes so that each took on a characteristic zebralike appearance of dark and light alternating bands. These stripes helped identify each chromosome and were used to further divide each one into subsections to get additional points of reference. Each chromosome is thus separated by the hinge into unequal halves (the shorter called *p* and the longer *q*) and then into general regions and specific bands, according to the pattern of stripes. We can do the same for our volumes in terms of distinct sections of pages of darker and lighter color (as may be seen in certain catalogs of mail-order houses). To locate the gene for beta-globin, we now turn to section 11p15.5, a band adjacent to the end of the short arm of chromosome 11.

Once again, I have casually tossed out information that was gathered through the heroic and lengthy efforts of many scientists: They first isolated the RNA message for beta-globin from immature red blood cells in abundance. This RNA matches parts of the beta-globin gene in its base sequence and could be used as a probe for the gene. Thus, the pages of DNA that contained the gene could be obtained by shredding human DNA with a text cutter and searching with a text-matching procedure. In a parallel effort, individual human chromosomes, separated from their fellows, were matched with the probe to identify the band where the gene lay.

How may one type of human chromosome be separated from the others? The Department of Energy has now developed machines for that purpose, but the early beta-globin workers used a different ploy.

Our ancestors had thought that unrelated species could breed to give strange hybrids: centaurs would arise from humans and horses. Such combinations proved impossible when animals and humans were employed, but they can now be carried out using smaller units of life. When human and rodent cells are brought together in the presence of a certain virus, they merge to produce a hybrid that is truly "of mice and men" (my apologies to John Steinbeck). A double set of human and mouse chromosomes will not cohabit long in peace within one cell, however. An elimination contest is held, chromosome by chromosome, with mouse almost always the winner. Cell lines can be obtained with only a single human chromosome (which could be No. 11 or any other) surviving among the rodent versions. With such combinations available, probes could then be used to place the human beta-globin gene at its position on the appropriate human chromosome.

Some neighbors of beta-globin on the short arm of chromosome 11 have also been identified. They included the genes for insulin and a hormone produced in the parathyroid cell. Notably absent was the gene for alpha-globin, its close partner in hemoglobin. Logic would suggest that they be produced in proximity, but the distribution of our genes among the various chromosomes is not organized according to any filing system that we have been able to understand so far.

Thus, hemoglobin and a growing number of human genes have been assigned to particular parts of chromosomes. In 1990, the number of genes placed in this way was almost two thousand. The job is far from done, as the total number of genes in our full library has been estimated at fifty to one hundred thousand. Some scientists have continued the search for additional genes, while others have wanted deeper understanding of the genes already identified. We shall join the latter group for a time and take a closer look at the beta-globin area.

When biologists have laid their hands on a particular stretch of genetic text, they can then use established techniques to identify adjacent areas and expand the region that has been sequenced. By such efforts, more than seventy thousand characters of continuous text have been deciphered in the region that includes the bega-globin gene. We shall tour this area, starting at the point where that gene begins. Scheme 5 provides an excerpt from the actual human text.

TAGACCTCACCCTGTGGAGCCACACCCTAG
GGTTGGCCAATCTACTCCCAGGAGCAGGGA
GGGCAGGAGCCAGGGCTGGGCATAAAAGTC
AGGGCAGAGCCATCTATTGCTTACATTTGC
TTCTGACACAACTGTGTTCACTAGCAACCT
CAAACAGACACCATGGTGCACCTGACTCCT
GAGGAGAAGTCTGCCGTTACTGCCCTGTGG
GGCAAGGTGAACGTGGATGAAGTTGGTGGT
GAGGCCCTGGGCAGGTTGGTATCAAGGTTA
CAAGACAGGTTTAAGGAGACCAATAGAAAC
TGGGCATGTGGAGACAGAGAAGACTCTTGG
GTTTCTGATAGGCACTGACTCTCTCTGCCT
ATTGGTCTATTTCCCACCCTTAGGCTGCTG
GTGGTCTACCTTTGGACCCAGAGGTTCTTT
GAGTCCTTTGGGGATCTGTCCACTCCTGAT
GCTGTTATGGGCAACCCTAAGGTGAAGGCT
CATGGCAAGAAAGTGCTCGGTGCCTTTAGT
GATGGCCTGGCTCACCTGGACAACCTCAAG
GGCACCTTTGCCACACTGAGTGAGCTGCAC
TGTGACAAGCTGCACGTGGATCCTGAGAAC
TTCAGGGTGAGTCTATGGGACCCTTGATGT

Scheme 5

We see in front of us an almost impenetrable thicket, 630 characters long, of the letters A, C, G, and T. With nine times that number, we could describe the heredity of a small virus, while more than seventy-five hundred such texts would cover the genome of the bacterium, *Escherichia coli. Nature* lamented, back in 1980, that "there is the problem, faced by journals, of printing screeds of ATCG permutations, less appetising to those who have not acquired the taste, than even the stock exchange page of a newspaper."[1] (*Webster's New World Dictionary* defines a screed as a long, tiresome piece of writing). A bit later, a review in the same journal concluded that "*Nucleotide Sequences 1984* must rate as the most boring biotechnology reference book of 1984."[2]

Fortunately, genetic sequences in the future will be stored primarily in computers, which generally do not complain of boredom. Further, the stock market pages, although dull to some, have proved gripping enough to others to provoke celebration or suicide. Genetic texts will have the same power. Let us take another look at Scheme 5 and pay attention first to distortions that I introduced in presenting the text.

In order to fit the letters onto one page, I arranged them in a tidy box, 30 characters wide and 21 lines deep. This was done strictly for convenience, and other arrangements could have been chosen. Similarly, the book you are holding has its text subdivided into rows of lines, which are collected into pages, for your convenience in reading it. The particular point that lines are broken usually has no relation to the narrative, and the book could have been printed on one continuous tape without any loss of information. If this were done, it would stretch more than three-quarters of a mile. You would have to fold it up in some way to carry it about, and there would be problems in finding a particular place and in reading it. The genetic text in each chromosome is printed as a single tape, as far as we know, and is folded up to one ten-thousandth of its outstretched length within that structure. Margins and pages do not exist.

A feature that adds to the monotony of Scheme 5 is the presence of only four characters with no spaces or punctuation marks. Chinese or hieroglyphics might be equally unintelligible, but would be more pleasing to the eye because those characters have greater visual interest. Of course, the same amount of information can be carried any written language, as long as two or more characters are used. The fewer the number of characters available, the greater the amount of them that must be printed to express the same message. In International Morse Code, for example, only the dot, dash, and space are used. The five-letter English word *hello* can still be expressed by using a longer text:

.... . .—.. .—.. — — —

H E L L O

Similarly, I am typing this text right now on the English keyboard of my computer, but the information will be stored in a language with only two characters, represented by the choice of positive or negative magnetic polarization at a particular site on the disc that holds it. I can still recover the information and view it in English when I choose.

The selection of four (rather than two, sixteen, or whatever) characters for heredity, as represented by the four different chemical bases attached to the backbone of DNA, was made early in the evolution of life for unknown reasons. Their representation by A, C, G, and T is an historical accident, the product of routine assignment of names by Albrecht Kossel and other chemists in the nineteenth century. They are derived from *aden*, a Greek word for gland; *cyto-*, a prefix signifying cell adapted from Greek *kytos* (hollow); *guano*, Spanish for sea bird manure, originally an Inca word for dung; and *thymus*, a gland near our throats whose name in turn comes from Greek *thymos*, spirit. Francis Crick once said: "I cannot help wondering whether some day an enthusiastic scientist will christen his newborn twins Adenine and Thymine!"[3] To my knowledge, that day has not yet come, although the Adenine Press has joined the lists of book publishers.

Any other four characters could, of course, be chosen to represent the information in DNA. From time to time an alternate system is suggested that would lead to more efficient storage or analysis of the patterns. The suggested replacements usually are abstract and made of lines or boxes. I hope that they are not adopted. If we remember that the ultimate storage of the characters will be magnetically within computers or on discs, I would prefer that the representation viewed by humans be kept as human as possible. The four letters are familiar to readers of many languages, and upon scanning forbidding-looking forests of text, we can still allow the occasional CAT, TAG, and ACT to keep us in touch with normal existence.

A different approach was taken by Susumu Ohno and by some other scientists. They transcribed genetic text into music, assigning one or two tones to each genetic letter. Ohno reported that "Chopin's *Funeral March* is very much like the coding sequence for tyrosine kinase [an enzyme], the heart of many cancer-causing genes."[4] Is nature trying to tell us something? Does the sequence of the gene for insulin, a hormone of carbohydrate metabolism, conceal the tune for the old barbershop quartet favorite, "Sweet Adeline," now changed to "Sweet Adenine"? I suspect that Ohno was just having fun. I am not as sure about the motives of one Hiroshi Nakamura who published in *Acta Astronomica*[5] the suggestion that the genetic sequence of virus SV40 contained a message sent by extraterrestrials. He attempted to interpret the letter sequence in terms of a starmap with ambiguous results.

Let us return to our effort to relate the gene fragment in Scheme 5 to biology. It needs to be deciphered just as any other unknown language. Sydney Brenner, yet another accomplished biologist at the Medical Research Council laboratories in Cambridge, has described genetic text as "a very ancient script—the oldest there is, in fact." Noting that the Watson–Crick structure was published one year after the decipherment by Michael Ventris of Linear B, an ancient text from the isle of Crete, Brenner dubbed genetic text "Linear G." He concluded that "the decipherment of Linear G—the generation of organisms from their genetic scripts—is *the* fundamental problem in biology."[6]

Michael Ventris had one advantage over genetic scientists, however. Linear B is a single language, whereas genetic script carries several.

Consider the problem, for example, that a Chinese scholar who had seen no Western language might encounter with the following text:

GAGATOBEORXORJSHFGBRLSIFGLAHWNOTTOBEISHKABIBBLEIBBLECESTLAVIEETAOIN
SHRDLUETAOINSHRDLUABBLEIBBILQUEESELBURRONOITSEUQEHTSITAHTRAUCHENVER
BOTENIBBLE

This mélange contains messages in four languages, English, French, Spanish, and German. The English message "to be or not to be" is interrupted by a group of syllables typed at random on my typewriter. The subsequent "that is the question" appears further on, but backward. The phrase "etaoin shrdlu" is nonsense, but has significance in typesetting. When a typesetter wishes to mark a flawed line for removal, he fills it out with this text, which can be produced easily by running a finger down the left-hand vertical rows of a linotype keyboard. I have included other phrases of nonsense, sometimes using varying spelling: ibble versus ibbil. All of the peculiarities I have put in have their equivalents in genetic text.

A gene has been understood for many years to be that length of DNA script that describes one enzyme or other protein. I will give the name *protein code* to the language that relates DNA sequence to protein sequence (the common phrase *genetic code* is more ambiguous). We now recognize that this language fills only a few percent of our genetic text. Another language that I will call *control* is also vital. The instructions for making hemoglobin, for example, are present in most cells in our body, but we do not want our eyeballs or skin cells to be making hemoglobin. We need instructions that dictate *where* in our bodies and *when* it is appropriate to make each substance.

I will give the name *structure code* to another language that is written in our DNA. In addition to controlling events elsewhere in the cell, chromosomes must take on particular shapes and structures to function in their life cycles. Local sequences of DNA act as flags to mark particular places that are important for this purpose. At such locations, the DNA may take on special shapes not anticipated by Watson and Crick, yet appropriate for their particular jobs. Unique text sequences mark the chromosome ends and hinge, for example. This language is not yet understood. Further codes and purposes may exist in DNA that we have not as yet grasped, but will after we have read enough of the text.

An additional complication exists in genetic language that does not occur, to my knowledge, in conventional human ones. The characters in Scheme 5 are written as if they were a single line of text. Yet Watson and Crick showed that DNA is a double helix, with two texts running alongside one another. Because geneticists can write the second text, once they know the first, they often save space by showing only one of them. The line of text that matches the last line in Scheme 5, for example, is

ACATCAAGGGTCCCATAGACTCACCCTGAA

I prepared this by interchanging A with T, G with C, according to the pairing rules of Watson and Crick, but I also wrote it backward, as the strands run in different directions. Thus, the final GAA above matches the starting TTC in the last line in Scheme 5.

When geneticists write only one strand, they use a rule to select the strand. They are guided by the process in which DNA is copied to RNA in accordance with the "DNA makes RNA makes protein" dogma. Only one of the two paired strands of DNA in a gene is normally read and copied. (*Felt* would be a better term than *read*. The copying enzymes make physical contact with the text; in this way the language resembles Braille.) The RNA copy that is made matches the text of the DNA strand that was not read (with U replacing T). This unread sequence, the one that matches RNA, is the one that is published.

Of course, when a new and uninterpreted stretch of DNA appears, a geneticist cannot know in advance which strand is which. Both must be examined for meaning. To confound matters, a protein code message may appear on one strand in a given area and on the other strand in an adjacent area. To mimic this, I wrote one message in backward in the above mélange.

Finally, the disturbing possibility exists that much DNA may have no function at all. Such areas are dubbed unflatteringly "junk," "nonsense," or "garbage." Simple repeating phrases, such as IBBLIBBLE or GAGA in the mélange analog, are always candidates for this distinction. However, we cannot be sure. They may represent important messages in structure or an unrecognized language. It may also be important that certain regions of DNA be kept apart by a certain distance with the actual text of no importance. I would term such separating DNA *filler*. Even after such exceptions are removed, we will undoubtedly remain with substantial hard-core areas of true nonsense.

The reason for the existence of stretches of nonsense in DNA is not well understood. The suggestion has been made that some sequences maintain themselves and even spread despite cellular efforts to remove them. They have been termed *selfish DNA*. With these possibilities in mind, we can return to our actual scroll of genetic text. We will consult a more user-friendly version, however, which I have placed in Scheme 6.

1 TAGACCTCACCCTGTGGAGCCACACCCTAG
2 GGTTGGCCAATCTACTCCCAGGAGCAGGGA
3 GGGCAGGAGCCAGGGCTGGGCATAAAAGTC
4 AGGGCAGAGCCATCTATTGCTTACATTTGC
5 TTCTGACACAACTGTGTTCACTAGCAACCT
6 CAAACAGACACC**ATGGTGCATCTGACTCCT**
7 **GAGGAGAAGTCTGCCGTTACTGCCCTGTGG**
8 **GGCAAGGTGAACGTGGATGAAGTTGGTGGT**
9 **GAGGCCCTGGCAG**GTTGGTATCAAGGTTA
10 CAAGACAGGTTTAAGGAGACCAATAGAAAC
11 TGGGCATGTGGAGACAGAGAAGACTCTTGG
12 GTTTCTGATAGGCACTGACTCTCTCTGCCT
13 ATTGGTCTATTTCCCACCCTTAG**GCTGCTG**
14 **GTGGTCTACCTTTGGACCCAGAGGTTCTTT**
15 **GAGTCCTTTGGGGATCTGTCCACTCCTGAT**
16 **GCTGTTATGGGCAACCCTAAGGTGAAGGCT**
17 **CATGGCAAGAAAGTGCTCGGTGCCTTTAGT**
18 **GATGGCCTGGCTCACCTGGACAACCTCAAG**
19 **GGCACCTTTGCCACACTGAGTGAGCTGCAC**
20 **TGTGACAAGCTGCACGTGGATCCTGAGAAC**
21 **TTCAGG**GTGAGTCTATGGGACCCTTGATGT

Scheme 6

In this scheme, protein-coding areas have been highlighted in boldface, important control areas indicated with a single underline, and other noteworthy locations with a double underline.

How can a DNA text tell a cell how to make beta-globin? This substance, like other proteins, is a complex three-dimensional machine. It embraces the vital oxygen-carrying heme and also rubs against its other partners in the hemoglobin quartet in order to deliver oxygen more efficiently. On the side it helps with additional chores: carrying carbon dioxide and excess acid wastes away from the cells and back toward the lungs. The exact shape in space of beta-globin allows it to do these jobs.

A lot of text would be needed, in any language, to describe this shape, but fortunately DNA does not need to do this. The beta-globin gene simply specifies 146 amino acids that, when connected in the order named, fold up on their own to give the desired form. Other proteins of our body may have as few as 50 amino acids, like insulin, or more than a thousand. The number and order of amino acids for each are stored in the appropriate genes. If the recipe is correct, then the product will fold up in the appropriate shape according to the laws of physics and chemistry. The job of the gene is to list the amino acids in the proper order. We will take a look at the way that the beta-globin gene carries out its job.

Find the ATG on line 6 of Scheme 6. These three letters mark the beginning of a protein-coding area (*exon*). From this point, each group of three letters signifies one amino acid in the beta-globin recipe. The GTG that follows indicates the amino acid valine, which starts the beta-globin chain. Continuing, the next five groups of three letters (*codons*) indicate the amino acids histidine (CAT), leucine (CTG), threonine (ACT), proline (CCT), and glutamic acid (GAG) in that order.

Why are they read in groups of three, the so-called triplet code? If only one letter signified an amino acid, then, mathematically A, C, G, and T would only be able to code for four amino acids. Two letters could accommodate sixteen, while three could handle up to sixty-four amino acids. This is enough to take care of the set of twenty that are used naturally, with room for special start and stop signals and a lot of excess space. This space is filled by allowing alternative spellings to signify the same amino acid, in the way that "color" and "colour" have the same meaning in English. Thus, the double-underlined CAC at the end of line 19 in Scheme 6 also means histidine, just as CAT did on line 6. Most spelling differences that signal the same amino acid occur in the third position of the triplet of letters.

If we follow along the text, we find that the boldface letters continue until about the middle of line 9 and then stop. The amount of continuous text to that point was enough to list twenty-nine amino acids and part of a thirtieth. At this point, we come to an interruption (*intron*). We experience the same effect when reading a magazine article and the text breaks with the note "continued on page 12" or when a commercial intrudes when we are watching television. Such interruptions occur often in the genes of higher organisms, but hardly ever in bacteria.

The discovery of the interruptions in our genes stunned the scientific community when they were first announced in the late 1970s. Nothing in the teachings of Mendel, Morgan, or Watson and Crick had suggested that our genes would be so untidy. The start of each interruption is marked by the letters TG, which represent control language. Further letters are also involved in the language switchover, otherwise the switch would have taken place immediately on line 6. The interruption continues until the underlined AG near the end of line 13 (another control marker), and then protein code returns. There is little understanding of the function of the actual text of the interruption, except to mark the changeover points. I am reminded about the joke concerning the man who was asked about the value of his marriage. He responded that he now had a partner with whom he could discuss problems that he had not even had when he was single.

The second protein-coding area starts in line 13 and continues into line 21, where a second interruption begins. This area covers the code for amino acids 31 through 106, a central region that makes up almost half of beta-globin. This part of the protein is the one that holds the heme and makes the important contacts with the partner molecules in hemoglobin. The double-underlined CAC at the end of line 19 specifies the amino acid that actually contacts the oxygen-carrying iron.

This central area of beta-globin code, bounded by its two interruptions, acts almost as if it were a separate unit of gene function. Observations of this type made in beta-globin and other proteins have led to the idea that the positions of the interruptions are logical, even if their text makes little sense. They subdivide a gene into large subunits. In some cases these subunits can be combined in different ways to make different proteins. Imagine that the magazine article said "turn to page 6 or page 12," and you got different endings depending on your choice.

Interruptions are removed, and any choices of the above sort are made within the RNA message after it has departed from the gene, but before it reaches the protein factory. If the interruptions were not removed, they would be treated as normal protein code at the factory, and a very faulty product would result.

Walter Gilbert and others have suggested that the division of genes into smaller units that could be combined in various ways was also a mechanism that sped evolution. The history of the interruptions may be cloudy, but their existence can hardly be ignored.

I have not specified the beta-globin gene in Scheme 5 or 6 beyond the second interruption, but I can describe what happens further on in the gene. The second interruption continues on for more than eight hundred letters before the third and final protein-coding area begins. At the end of this section, after the specification of amino acid 146, comes the triplet TAA. This is one of three that have the special meaning: stop, end of protein. The amino acid recipe has been given, and the protein code stops, but other meaningful stretches of text are present. Some of them occurred just before the first protein code region began, and others follow the end of the last one. These areas contain control messages needed by the RNA in its voyage to the protein factory. In other words, they have credentials issued by DNA: "I am a legitimate message sent by DNA. Make protein according to the recipe I enclose."

How big is the beta-globin gene? Before the discovery of interruptions, the size of the protein-coding area might have served as a measure. Perhaps a more useful definition would equate it with the area copied into messenger RNA. This region starts at the double-underlined A in line 4 in Scheme 6 and terminates about two hundred fifty residues beyond the end of the protein code. On this basis, the beta-globin gene measures 1720 base pairs, each the equivalent of a letter of text. This length and the possession of only two interruptions afford it the classification as a small gene. The largest gene encountered thus far produces a protein dystrophin, whose malfunction is a cause of the disease muscular dystrophy. Dystrophin has more than fifty interruptions and extends over an area greater than 2 million base pairs.

Many areas that lie outside the actual boundaries of a gene by the above definition are still important to its proper function. The top three lines in Scheme 6 contain several underlined sequences in control language. These phrases are called *promoters*. They are essential if the gene is to function at full capacity. In test tube experiments where one or another of the underlined letters was changed artificially, the activity of the gene in producing beta-globin message decreased by seventy to ninety percent.

The promoter sequences are not unique to beta-globin, but are present in many other genes and carry characteristic names. The sequence on line 3 often contains the letters TATA, it is called the *TATA box* (here it has become the ATA box). The CCAAT-containing sequence on line 2 has been dubbed the *CAAT box*. Line 1 contains two slightly different copies of another control sequence that lacks a colorful name. The group shown does not exhaust the control elements. Additional control elements lie outside the text shown in Scheme 6. They are located both further "upstream" (earlier in the text) or "downstream" (later in the text) from the gene and bear colorful names such as *enhancers* and *silencers*. In the case of beta-globin production, they help deal with contingencies that can arise in adult life. I might need to step up my beta-globin production, for example, if I lost blood because of a wound.

The gene that we have toured belongs to an adult with normal, healthy beta-globin production. Yours and mine need not be the same. That does not mean that we are ill, necessarily, only different. As I have no identical twin, my genes are different from every one of the other billions of human beings on this planet. As we are part of the same species, however, these differences are limited. I have seen estimates of the difference that run from one-hundredth of a percent to as much as one percent. They can be found in many places, for example, in the beta-globin gene.

If we return to Scheme 6, we note that the first two amino acids after the ATG start on line 6 are coded by GTG and CAT. The second amino acid, histidine, coded by CAT here, is coded by CAC at some other places in the protein as I pointed out earlier. Either triplet can be used in our genes to the same effect. Returning to line 6, I will now confess that the article that I copied the sequence from actually listed CAC there. I put in CAT so that I could make the above point about coding. But I did not invent the sequence either. The use of CAT there was listed as an *alternative*.

Both forms exist abundantly in the human population with CAC in the majority. You or I may have either. In fact, it is possible to have both, as each of us normally has two different copies of chromosome 11 (and all the other numbered ones). I will call small variations of this type *spelling differences* (*polymorphisms*), including cases where the two forms differ by the loss or gain of a few letters, as well as simple substitutions of one letter for another. Larger changes will be called *text differences*.

Why do two forms occur at all? Most scientists believe that there was only one usage at an earlier date, but at some point during history a mutation occurred. That is, some error led to a chemical change being introduced. About twenty-five years ago, a student and I discovered that the letter C can change spontaneously to U (a close relative of T). If this should occur to a C in DNA, a mutation would result. Our cells are well aware of this natural hazard and possess a repair system that corrects this misspelling whenever it is introduced. Occasionally, however, a typographical error of this sort slips by and becomes part of the permanent record. In this particular case, the mutation is a silent one. That is, it causes no change in the amino acid recipe for beta-globin and, therefore, has no practical effect. Its presence can be detected easily though by using a text cutter that will recognize GTGCAC but not GTGCAT.

When the beta-globin gene area is sliced with such a text cutter, and the resulting fragments are located with a suitable probe, two results are possible. The text region containing the start of the first protein-coding area will appear as one piece if GTGCAT is present (no cut will be made), but as two shorter fragments if the text reads GTGCAC (the cutter will have sliced it in two). The two possibilities can be distinguished by this text fragment length difference. (Geneticists have selected the much less comprehensible jawbreaker: *restriction fragment length polymorphism*. As this phrase is unwieldy, they abbreviate it RFLP, then pronounce it "rifflips" at meetings. We finally get a slangy phrase that rolls readily off the lips, but gives no clue at all of its meaning to an outsider. I will stick to the phrase I chose above.)

The spelling change described above is harmless, but as almost all users of language have learned, some typos can be much more perverse. I leave it to the reader to dream up English sentences that can be ruined by the substitution of *lice* or *mice* for *rice* or *nice*, or the removal of the prefix *un*. Error mechanisms exist to permit every letter interconversion in our genetic text, or the loss or gain of letters. Whole blocks can be moved about, and segments can be duplicated. In a protein-coding area, harm is possible whenever a spelling or text change alters the amino acid recipe for a protein.

Beta-globin takes part in an oxygen delivery system. With 146 parts, it is less complex than a delivery van; perahps a bicycle would afford a better comparison. How many ways are there to break a bicycle? Some design changes would matter little: the color of the seat or presence of handgrips on the handlebars. An inconvenient seat height or loss of the bell would be a nuisance, but the vehicle could still function. Break the chain, however, or disable the turning mechanism, and you are done.

Many changes in beta-globin are cosmetic. In fact, if you know what you are doing, you can redesign, then structure extensively and still have an oxygen carrier that works, just as delivery bicycles can differ in design. Alpha-globin has five fewer amino acids than beta, and only 64 of its 141 amino acids correspond to beta in position and identity. Yet both globins have roughly the same shape, and both work as oxygen carriers.

It is very different to take a working mechanism and introduce changes at random. Consider, for example, the change of CAC at the end of line 19, Scheme 6, to TAC. A change in the third position of a coding triplet is often silent, but in the first position, you mean business, and, in this case, bad business. This position codes an amino acid that binds directly to the oxygen-carrying iron, a crucial function. The coding alteration leads to the replacement of histidine by tyrosine, another amino acid that is willing to bind to iron. This recipe change, however, completely alters the appetite of the iron.

This key atom now changes its preference from oxygen to water, just as a diner may shift her preference from red wine to mineral water when the menu changes. One commodity that we have no need to transport in our blood is water. It is as if our delivery bicycle has its luggage container stuck shut. The vehicle can still move about normally, but nothing can be carried.

The loss of a single hemoglobin molecule, of course, is of zero consequence to our health. In fact, millions of my own red cells, each crammed full with hemoglobins, are being trashed each second as I write this. A genetic change is not a vehicle malfunction, however. It represents a design error. Every vehicle made to the plan has the same error, and there is no possibility of a recall. Fortunately, we each order our beta-globin supplies from two sources, which are the genes on our two separate copies of chromosome 11. If one provides a defective product, the other may still function.

Individuals with the above genetic substitution of TAC for CAC or others that produce the same effect have a bluish skin and blood that is chocolate brown due to the presence of an abnormal hemoglobin, called *hemoglobin M*. In northern Japan, the condition has been known for centuries and is called *black mouth*. Fortunately, the presence of one normal gene for the affected globin unit protects them from serious disabilities. No one has been studied yet in whom both copies of a beta-globin gene are disabled in this way. The condition is lethal. Text changes involving one letter in 3 billion can kill. Poet George Herbert described the situation eloquently:

> *For want of a nail, the shoe is lost.*
> *For want of a shoe, the horse is lost.*
> *For want of a horse, the rider is lost.*

DECIPHERING THE SICKLE

Hemoglobin M is a classic case of a damaging mutation. If the text change occurs in a gene, all of the globin produced by it is defective. It cannot function at all. The change that we examined earlier was harmless. These two possibilities represent the extremes that can occur in nature with many fascinating circumstances falling in between. Mutations that affect the amino acid recipe of a protein can sometimes have unexpected consequences.

Let us return to Scheme 6 and look at line 7 of the text. The first triplet, GAG, codes for the amino acid glutamic acid. A one-letter change of the A to T would alter the text to GTG, which codes for a different amino acid, valine. What difference would this make in the recipe?

To answer this question, we will invent an imaginary biochemist, expert in genetics and protein chemistry and who further has studied all of the X-ray and other data on the function of normal hemoglobin. She would not have any training in human diseases, however, or have encountered this particular change before.

Our biochemist might observe first that the two amino acids concerned had very different properties. Yet the location of the change was such that it would ordinarily matter little. The alteration lay on the surface of the hemoglobin molecule, normally an unimportant location. It had nothing to do with the oxygen-binding area or the parts where the quartet of alpha- and beta-globins made contact with one another. She would conclude that the altered beta-globin would bind oxygen normally and cooperate with alpha globins to form hemoglobin that functioned well. If pressed, she might speculate that an individual who had the mutation would be inconvenienced very little by it. The first two predictions would be on target, but the third would be dreadfully wrong. For the change we have listed is the very one involved in sickle-cell anemia!

The case described by Dr. James Herrick in 1910 represented a mild form of the disease. More severe versions are marked by crises with agonizing pain in the back, limbs, or abdomen. Joints can be swollen and painful, and damage can accumulate in the bones, heart, lungs, kidney, spleen, and other organs. The patient becomes more vulnerable to infections, which together with the failure of vital organs can lead to death. In many cases, the victim does not survive early childhood.

Awareness of the above symptoms has persisted in many African tribes for centuries, along with the knowledge that the disease runs in families. Each tribe has its own name for it. As in the case of hemoglobin M, two altered genes are necessary for the disease to display itself. The presence of one sickle gene and one normal one produces an adult who is in good condition. He is said to have sickle-cell trait. One demonstration of the capabiliity of those with sickle-cell trait came from test results on black National Football League players. Of 579 tested, sickle-cell trait was found in 39, a ratio similar to American blacks in general, where sickle-cell trait has been estimated to occur in seven to ten percent of the population.

Recently the *New York Times*[7] estimated that 2.5 million out of 30 million American blacks carry sickle-cell trait. The disease itself has sixty thousand sufferers in this country, or about one in five hundred. On a worldwide basis, its incidence is highest in regions of West Central

Africa, including Nigeria and Zaire; twenty-five percent of the population can carry sickle-cell trait in those areas. It is the most common known genetic disease that afflicts black populations. The disease can also be found in countries bordering the Mediterranean, including Spain, Algeria, Morocco, Southern Italy, and Greece. Further pockets occur in Saudi Arabia, India, and Southeast Asia.

A one-letter misspelling in the genetic text can have vast consequences. In this case the change is one that superficially looks harmless, but is not so. We now understand why. The change in shape of the surface of a beta-globin molecule alters its contour so that two beta units can stick together. This stickiness only occurs when neither is carrying oxygen. The slight adjustment in shape that occurs when oxygen is taken on ruins the fit.

Of course, two beta-globins are packed together with two alphas to make the hemoglobin quartet. If each beta also sticks to a partner on another quartet, then many hemoglobins can be linked together in chains. Further, the chains pack nicely alongside each other forming long fibers. The fibers are rigid enough to stiffen and deform the red cells that contain them, forming sickles and other distorted shapes.

The normal doubly indented disc shape of red cells affords them both a large surface for oxygen exchange and the flexibility needed to squeeze through the smallest blood vessels. Sickle cells have lost that flexibility and can clog up those vessels. The disruption of circulation creates organ damage, and the cascade of effects caused by the disease results.

If these unfortunate events are to happen at all, they must take place rapidly. Hemoglobin molecules give up their oxygen in the tiny capillaries and then are vulnerable to sickling. If the change does not occur quickly, however, the cells move into larger vessels, less easily clogged, and then into the oxygen-rich lungs, where sickling is reversed. If sickling is to take place, many hemoglobins must come together to form a fiber, so the concentration of sickle hemoglobin is crucial. Sickling time has been measured at two seconds for those with two sickle-type genes, and seventy seconds for those with one. As a consequence, those with sickle-cell trait may have one percent sickling under conditions in which those with the disease have fifty percent. One individual feels no consequences, the other can suffer immensely.

Why has an unfortunate hereditary condition of this type spread to such an extent? We shall see shortly that the disease probably had just a few origins, tens of thousands of years ago, and has spread through population migration. Ordinarily, such a condition would disappear over the millenia due to the lesser survival of those afflicted. In this case, it has persisted because a compensating advantage exists.

Those areas where the disease has long endured are also afflicted by malaria. The malarial parasites, in one stage of their life cycle, take up residence within red blood cells. They are not pleased, any more than you would be, when their residence suddenly takes on an unfamiliar and distorted shape. The causes are complex, but the ultimate effect is simple—they are killed. This protection from the most lethal form of malaria is not limited to those with sickle-cell disease, but extends to those with the trait. The parasites unwittingly alter the acidity within red cells of the latter individuals in a direction that favors sickling. The amount of sickling can rise from two percent to as high as forty percent upon infection. Thus, over the ages, a balance is struck between those killed by sickle-cell disease and those saved from malaria.

Not too many die of malaria, of course, in New York, Detroit, or Los Angeles. Even in the tropics, there are less painful alternatives available for control of that disease. A cure for sickle-cell anemia would be very welcome, but no effective one has been devised so far. Before the mechanism was known, doctors could do little but treat the symptoms or build up the physical strength of the patient through good nutrition and rest. Now that the mechanism is known in much detail, and the question frequently comes up: Why no cure yet? We are very much in the situation that occurs when a child is trapped in a collapsed building. We may know that he or she is in there, but not know where. Finally a cry is heard deep within the wreckage. One problem has been solved but another one remains: The victim must be dug out.

We are at that stage with sickle-cell anemia. We need to dig, and that can take time. Efforts have been made to find a targeted remedy, for example, to devise a drug that would change the shape of sickle beta-globin and destroy the fit that leads to stickiness. Beta-globin is a protein, however, and much of our bodies are made of protein. Any protein-changing substance that we apply to our body or any substantial part of it will have an immense number of effects, not all of them good. We could get very lucky, but a more specific solution would be very desirable.

ANOTHER HOPE: GENE THERAPY

Some clever remedy may emerge from the efforts of those involved in medicinal chemistry, but the ultimate solution would be to find a way to rewrite our genetic messages. A general solution of that type would apply not only to sickle-cell anemia, but to other diseases as well. The effort to correct a genetic defect by altering the DNA text of a patient's body cells is called *gene therapy*. In September 1990, the first government-approved experiment of this type was run using a patient who suffered from a different disease.

A four-year-old girl received a transfusion of some of her own blood cells in a hospital in Bethesda, Maryland. Her blood, specifically the T cells (part of the immune system), had been treated in a test tube with a virus that had been equipped with a human gene for the enzyme adenosine deaminase. The girl's own DNA lacked an effective text to specify that enzyme. As a result, her immune system malfunctioned, leaving her at severe risk for infections and early cancer. She had received weekly injections of the enzyme itself. Hopefully, the altered version of her own cells that she had just been given would survive and function for some time providing an internal supply of the enzyme. If so, she might have to come in only twice a year for additional treatments.[8] After three months, the early results of the treatment looked very promising to the research team.

The disease from which the girl suffered, adenosine deaminase deficiency (ADA), is rare, with only a few dozen sufferers known around the world. Further, the treatment was not permanent in its effects, as the altered T cells would die out sooner or later, and they would have to be replenished. The case drew wide media attention anyway as the first of its kind. Its ultimate success will have to be determined, but much more expertise will be needed before it could be applied to a disease such as sickle-cell anemia.

A scenario for such a treatment can be pictured, though. Bone marrow cells would be removed surgically from the patient and altered by viral or other treatment outside her body. This represents the tricky part. We would not only want to introduce a normal gene for beta-globin, but ideally to silence the sickle globin one as well. The less the production of the sickle globin, the less the chance that sickling might still occur in adverse circumstances. The altered cells would then be reimplanted with the hope that they would establish themselves and proliferate.

Cure scenarios still seem some years away. What use can be made of the genetic information on sickle-cell anemia right now? A very hot topic comes up: prenatal diagnosis and abortion. Various ethical and religious groups have considered this issue and come to well-reasoned, but opposite conclusions. This book concerns science, and I have no need here to add fuel to an already well-supplied fire. I will make the following simplifying assumption: that each group has the freedom to follow their own moral dictates. For those who have considered the ethics of abortion and decided that it is moral, genetic knowledge has helped a lot.

Before the genetic text was read (about 1976–78), diagnosis required fetal blood from the placenta or umbilical cord. This procedure had some risk, as much as five to ten percent mortality for the fetus in some cases. Further, it could not be carried out until week 20 of pregnancy, close to the practical and legal limit for most abortions. Now that we know the exact text change for sickle-cell anemia, almost any cell will do. They all carry the same full genetic message. Amniocentesis and other procedures that have less risk and can be applied earlier may be used. The appropriate passage can readily be separated from the full genetic text and read.

The diagnosis or abortion option has not been used as often for sickle-cell anemia as it has for other conditions. The severity of the disease can vary. In some cases, it is lethal, in others, almost normal lives can be led with occasional crises. The effects are generally less severe in Arabia than they are in Africa. Other factors, not yet understood, affect the course of the disease. One medical review summarized the situation: "The extreme clinical variability of sickle cell anemia significantly diminishes the potential interest in its prenatal diagnosis."[9]

Still, it is nice to have the choice. In a 1984 *Newsweek* article, one couple at risk, Zandra and Linwood Lockett, was described. Each had sickle-cell trait. Their child would draw one gene from each of them at random. Its chance was fifty–fifty in the case of the first parent. If that draw went badly, there was again a fifty–fifty chance on the second gene. The odds of a doubly bad result and sickle-cell anemia were thus one in four. Zandra, a nurse, had seen enough of the disease to deter her: "I didn't want to go through that with a child of my own." The new analyses permitted amniocentesis, and the result was favorable. "I was hysterical," she said, "I just sat at my desk and cried for joy."[10]

I can add nothing to her words.

BEYOND BETA

The beta-globin gene is part of a much larger area of over seventy thousand base pairs whose text has been read. This section lies for the most part upstream of beta-globin and is well worth exploring, but we will need a new metaphor to describe our journey. Until now, when we considered the bases one letter at a time, text reading sufficed. Now we will travel more quickly, pausing only occasionally to scan the script. Unfortunately, the technical literature offers only a confusion of terms for this voyage.

The terms *upstream* and *downstream* that we met earlier imply river travel, perhaps by canoe. Another phrase that is used is *chromosome walking*. This term describes a method in which one fragment is used as a probe to identify neighboring text that it partly overlaps. The newly discovered fragment is then used as a probe in turn to locate the next bit further on and so on. This term *walking* suggests a hike. A related procedure that skips over some areas and travels more rapidly is called *chromosome jumping*. This description could mislead the casual reader, who might think that the process involved leaping from one chromosome to another. Even if we limit the concept to travel along a single chromosome, "jumping" does not reflect the way people travel. Humans may jump once to get across a gap, but do not move continuously that way—that style belongs to grasshoppers and kangaroos.

We want to travel as tourists, moving through territory that has already been explored. The linear construction of our genetic text further limits us to travel along a line, as on an unbranched railroad. This illustration will set our metaphor. With apologies to travel writer Paul Theroux, we shall set out on a trip up the old globin local (chromosome 11 section) with our stopping points, the genes, represented by towns.

We shall lead our tour through the beta-globin gene region of an immature red blood cell of an adult, myself, for example. The cell has not yet lost its DNA and is actively making beta-globin. If we selected a skin or liver cell instead, our journey would be dreary. Everything we encountered would be shut down and in storage.

We want to see action, so we begin the trip in "downtown" beta-globin, the metropolitan depot of our line, where the RNA assembly line is in full operation. As we travel up the line, we leave the RNA factory and pass by a series of suburban control areas, which gradually thin out. Our rail line enters open countryside. In an actual railroad in Europe, the distance between destinations would be marked in kilometers, or km (thousands of meters), with one kilometer equal to about five-eighths of a mile. Here, in geneland, our progress is marked in kilobases, or kb, with one kb equal to a thousand base pairs of DNA.

After our locomotive had moved some 7 kb from beta-globin center (in a leftward direction in our genetic text, as written in Scheme 6), we would start to pass through control areas again, and at some point we would encounter a sign that said, "Welcome to delta-globin." The town that we now entered would look strangely familiar.

I have an experience of this sort at times when my wife and I visit her parents, who live in a large urban housing development. It is planned with alternating tall buildings and town houses arranged around a series of plazas, all built to the same design. On occasion, we have taken a wrong turn and entered a square that looks very much like the one where her parents live, yet it is *different*. The bricks vary a bit in color and damage can be seen in different places. So it is with delta-globin.

The plan for delta-globin describes an oxygen carrier that functions much like beta-globin. It holds heme in the same way, and a delta-globin pair will link up with a pair of alphas to form hemoglobin just as beta does. Each protein has 146 amino acids, and the interruptions in the protein code occur at exactly the same place. The DNA sequences of the protein-coding areas show thirty-one spelling differences, twenty of which are silent (they do not change an amino acid in the protein). If the exact amino acids were compared to one another, ten differences in the 146 amino acids could be detected. The amino acid changes do not harm its function; delta works.

When we turn our attention away from the town plan of delta-globin and look instead at its level of economic activity, we do see a vast difference. Beta was booming, whereas delta is severely depressed. In fact, the plant is run by a skeleton crew, and only a trickle of RNA is produced. Delta is a competitor of beta, but has lost the contest, with only a one percent share of the market. The problem lies not with the product, which works quite well, but with the management. In genetic terms, the control areas are messed up. For example, the CAT box sequence, CAAT, has become CAAC, in delta. A single letter change in a control area, as in a protein-coding area can have a profound effect.

Rather than linger in this depressing place, we will board our locomotive and continue on up the line. Unfortunately, worse sights lie ahead.

As we chug perhaps 3 kb beyond delta, we find a settlement that might best be described as a trailer park. It occupies some three hundred base pairs of DNA text. Although trailers are mobile in principle and must move at least once to get from the point where they are made to the place where they settle down, once there they may become permanent. This particular settlement on the beta-globin line was firmly in place, but doing little. Nothing obvious was being manufactured;

certainly not RNA for any globin. A sign carried a cryptic message: "South delta camp; a member of the Alu family." (The name Alu was taken from a text cutter that recognizes this family. The cutters in turn draw their names from the initials of microorganisms that produce them.)

Let us hold our puzzlement for a bit and continue up the line for another kb. We would encounter another such camp, with its overall plan laid out in reverse (in actuality, the relevant text lies on the opposite strand of DNA, an item that does not fit tidily into our metaphor). The sign here might read: "Reverse south delta camp, another member of the Alu family." If either of the signs had cared to boast, it might have added: "Almost 1 million members, genomewide. The most prevalent of all human genome families." Stunning, but true. When noted human families are discussed, it is not the Flintstones or Simpsons, or even the royal Windsors of England who should head the list, but the Alus. They are five and a half to six percent of all of us, named not for some hero, but merely for the text cutter that is useful in detecting them.

A group of such prominence must have had an extended and noteworthy history. The Alus do have their written record; it is in our genes, but so far not much of it has been interpreted. From the parts that are understood, however, we have enough for the following tale.

A FABLE: THE LIFE AND TIMES OF THE ALUS. Once upon a time neither human nor monkey nor familiar mammal walked the earth. Reptiles ruled, but beings existed who would evolve into mammals someday. Our ancient ancestors employed an RNA whose close descendent, called *7SL RNA*, serves us today. This type of RNA works in partnership with proteins to help other proteins cross membrane barriers in our cells.

One day, a 7SL RNA was abducted by a virus. The virus itself was of the type that used RNA as the genetic material. While in captivity, the 7SL RNA learned many of the habits of its captor, including key ones: how to have a copy of itself written in DNA text (reversing the usual process in which RNA sequences are copied *from* DNA) and how to smuggle this copy back into the genome. Once there, it would resemble the normal DNA gene for 7SL RNA, but it would not be surrounded by the normal controls of that gene, and it might lack other items such as interruptions.

In its new existence, this 7SL DNA had no duty to make 7SL RNA (the normal DNA gene still carried that burden). Occasionally, over the generations, it produced a baby version of itself, which departed to take up residence elsewhere in the genome. For the most part, it only had to survive and await the adventures that evolution brought it as millions of years went by. Thus, it may have mourned when 150 base pairs were lost from its middle, rejoiced when another copy of its shortened self was created alongside it, and mourned again when the left half of its new double self lost another twenty-five base pairs. The last two events came when the primates had separated from other mammals in the course of evolution. Only primate Alus bear these particular scars.

Events of the above types did not affect the Alu's ability to proliferate within our DNA. Today, in each of us, nine hundred thousand Alu sequences lie spread throughout our chromosomes. A full-length Alu will run for about 380 letters of DNA, although versions that have been shortened by additional accidents are also abundant. They occur on the average about every three thousand letters of normal DNA, with the exception of some places like the chromosome hinge area, which somehow have managed to avoid them.

Further, they are a rich source of individuality. As nonworking genome areas, Alus can accept spelling changes with impunity and therefore vary greatly. An Alu in one location in my genes will vary from Alus elsewhere in my DNA. Further, my "south delta" Alu is likely to differ a bit from the same Alu on my other chromosome and from your "south delta" Alus as well. Alu sequences will differ much much more than protein-coding areas generally and therefore will provide a good way to tell one person from another. We will get back to the use of DNA in establishing human identity later in this book.

THE VOYAGE CONTINUES

The above fable may or may not be true in all particulars, but catches the essence of the Alu story. Our genes are not a tidy set of instructions, as we may have imagined in the time of Morgan, but are rambling, redundant, and, in many cases, full of unnecessary passages. (One observer has said that our DNA text more resembles human conversation than written human prose.)[11] Yet arguments persist that some of the Alus may have some subtle role in control or some other cellular function. We cannot doubt that some of the genome is junk, however, or if we did, the next encampment on the railroad line would cure us.

We travel 5 kb or so past the Alu pair, and we again approach the area of a gene. A sign announces: "Approaching eta-globin." Beyond that is another one, however:"Warning! Abandoned gene (*pseudo-gene*). Do not stop."

Heeding the notice, we do not slow our locomotive, but notice that we are passing a gene area that resembles beta and delta, but is utterly in disrepair. Enough damage has been inflicted so that eta-globin is utterly nonfunctional, producing nothing of value. As we commented earlier on genetic diseases, there are many ways to ruin a gene. This area differs from the gene for defective hemoglobin M, for example, in that various kinds of injury had been inflicted on it, while only one letter was changed in hemoglobin M.

One effective way to destroy the meaning of a gene is simply to remove a letter. Genes are read in protein-coding areas in units of three. If one letter is taken out, all triplets to the right suffer "frame-shifting" and lose their original meaning. This comes clear with an illustration. Take the following English three-letter word sentence:

THE BIG CAT ATE THE FAT RAT

Remove the *B* of *BIG*, and shift every other letter to the left, keeping the three-letter word structure, and you get

THE IGC ATA TET HEF ATR AT

Another effective gene-killing device is simply to convert an amino acid-coding triplet to a stop message early in the text. Alternatively one could ruin the "start here" signal. Once a gene is spoiled, of course (and provided that some other gene carries on the function to ensure survival of the being and its descendents), there is no constraint that operates to avoid further damage. Eta-globin started, presumably, as a cousin of delta and beta, but has diverged some twenty-five percent in its text since its ruination. It is not an isolated oddity in our genome, either, as abandoned genes are fairly abundant. The marvel is that they have remained in place.

Beyond the abandoned gene, our speeding train passes yet another Alu encampment and, then, some 6 kb beyond eta and 23 kb from beta center, we encounter another sign: "You are approaching the gamma twins." Yet two more genes lie ahead, 6 kb from one to the other, relatives of beta-globin, but not so closely related to it as delta is. In order of approach, the not-quite-identical twins would be named gamma-A and gamma-G to signal the only amino acid difference between the proteins described by them. One has the amino acid alanine (A) at a position where the other has glycine (G).

The gammas, like delta, produce a small amount of a fully functional globin. In this case, however, the low production is not caused by poor local management, but by occupying troops who forcefully prevent any greater level of activity.

What happened? To find out, we will invent an imaginary gene-dwelling old-timer who witnessed everything that transpired:

"In the good old days," the old-one reminisced, "this place was running full blast. Why our body *depended* on its output. There was hardly anything trickling out of the beta plant then." When was that? "Oh, back almost to the beginning of things; this body hadn't even been born yet. When our body was a fetus, just three months old, the epsilon plant up the line started to shut down and there was a real demand for our product. Our gamma-globin, together with the stuff they made on chromosome 16, made up fetal hemoglobin. This was the important oxygen carrier right through birth. Then those foreign control agents moved in and told us to shut down. The beta plant was taking over the business. Things have been fairly quiet since then. We turn out a little, but not much."

From this conversation, we've also learned that we have at least one more stop to make on the globin line. We set out again in search of epsilon-globin. Before we reach it, however, we pass another Alu, then reach another area best represented as a trailer park. A sign says "Member of the LINE-1 (or L1) family." Once again we meet a sequence that, like Alu, has spread widely in the human genome. Again, its construction suggests a past history of mobility, perhaps within a virus, and quiet survival now. We could traverse an Alu sequence quite quickly, but the LINE sequence runs on for about three thousand letters of DNA. This one is a junior member rather than a full-fledged one, which can run for more than six thousand base-pairs of DNA. Perhaps fifty thousand LINE members are spread around our genes. This is considerably less than the Alu membership, but because of their greater average length, the LINE-1 family occupies over three percent of our DNA.

Many other repeated themes decorate (or clutter) our genome, for the most part shorter than Alu in their basic unit. Whereas the Alus and LINEs usually exist in isolation, the briefer ones tend to cluster in repeats. In the shortest cases, only the repetition makes them distinctive. Take the sequence CA, for example (with a matching TG running on the other chain). This pair is one of sixteen possible two-letter combinations and should show up frequently in DNA. It could occur in a control sequence, in a protein-coding area, or as part of any other language. Appearances change when we run into the string of letters: CACACACACACACACACACACACACACACA, or CA repeated fifteen times in a row.

I will dub a run of this type a CACA sequence. We have to look at it with suspicion. It could code for protein, specifying the amino acids corresponding to CAC and ACA in strict alternation. This is possible but unlikely, as proteins seldom follow such simple-minded patterns.

Our suspicions grow when we find runs of CACA not just once or a few times, but all over the genome. In all we have perhaps fifty thousand of them, with the number of CA repeats usually ranging from twelve to thirty. Further, this number can vary from one individual to another for the CACA in a particular location. About two-thirds are pure uninterrupted CACA, whereas the remainder have one or more breaks.

Many other simple repeating themes exist in profusion, each with its own peculiarity. The TATs, for example, like to take up residence in the vicinity of an Alu, becoming, in essence, the camp followers of a parasite. A shorter member of repeat length nine would run as follows: TATTATTATTATTATTATTATTATTAT, but repeats of up to forty are known. Many other variants exist. Taken together, the Alus, LINE-1s, and repetitive phrases of various length take up some twenty-five or thirty percent of the human genome. Many appear to have no purpose, but as we shall see shortly, others may be vital elements of chromosome structure.

Continuing on our trip, we pass another shortened LINE-1 member and finally reach epsilon-globin, 43 kb up the line from our starting point in the middle of beta. This gene is completely shut down now, but again the inhabitants have memories of glory. "We were first," they can boast. During the first eight weeks of life, epsilon-globin played the role presently taken by beta in the hemoglobin molecule. After that, external controls were imposed, and production shut down.

If we continued up the line further we would traverse the control region of epsilon-globin and then enter open countryside again. Further Alus would be encountered and finally, some 50 to 60 kb from beta-globin, a new set of controls. This region affects the behavior of the entire group, or cluster, of globin genes, rather than a single one. It undoubtedly bears some responsibility for the successive start-ups and shutdowns of the various genes at different points in development. It has been aptly named as the dominant control region.

The stretch of continuous DNA that has been sequenced runs out at this point, and if we tried to ride further, we would do so in darkness. We can imagine a sign that says, "Leaving globin region," and beyond that a passport control before we entered the realm of some new gene. A different option open to us would be to return to beta-globin and ride the rails in the opposite direction down the line (downstream or forward in the text). We would have a shorter and less interesting trip that would take us past a couple of Alus and a full-length LINE-1 before we again reached the end of the sequenced area (and the other border of globin country). We have covered the largest stretch (at this writing) of continuous sequenced human text. Yet we have covered only an insignificant portion of chromosome 11, let alone the whole genome.

In historical exploration, a certain type of person would be the one who first ventured into unknown territory, noted the dangers and safe places, and made a rough map of the terrain. He would then move on to his next adventure, while others moved in to settle the land. The same pattern may well be followed in the human genome.

The first explorers will simply decipher the text, recording the DNA sequence in one human cell sample. They or others may also probe its division into different languages, noting protein-coding, control, and structure areas; interruptions; LINEs and Alus; and any unusual features of interest. A group of specialists will then move in for a longer stay. They will prepare and test the newly discovered proteins, explore their role, and learn how they fit into the complicated network of interactions that makes up human biochemistry.

Other groups, using the same sequence, will want to answer entirely different questions. They will study how this DNA script varies from one organism to another. In doing this, they will be comparing species at the molecular level and gaining insights into the history of evolution. Yet others will study the differences in the text from one human being to another. Their goal is similar: to understand human diversity, relationships, and evolution.

Our experience with the globin gene cluster can give us a taste of what is to come. DNA tracts have been compared, organism by organism, and the degrees of divergence noted. The closer the DNA of some species resembles our own, the more closely related to us it is. If we assume that the amount of text change is proportional to the passage of time (the details of this idea have been hotly debated), then a chronological history of the beta-globin gene area can be constructed. To complete the picture, information from fossil and geological studies is added. The following anecdotal beta-globin family history emerges.

THE GLOBIN FAMILY HISTORY

Once upon a time, hundreds of millions of years ago, there was only one ancestral globin. Creatures were not complicated then. This ancestor served muscle and blood alike. In the next generation, specialization set in, the muscle globins separated from the blood globins, and the latter then duplicated themselves. One branch became the alpha-globin genes, who migrated far away, while the other developed into the betas. That happened 400 million years or so ago. The beta types then split further 200 million years ago. One line gave the father of the related group, the gamma twins and epsilon, another produced the father of delta and beta. Gamma and epsilon came apart 120 million years ago, and gamma twinned only 15 million years ago, while beta split from delta 40 million years back. I would rather not discuss the other globin, eta. He came to a bad end.

This fanciful history is superimposed upon that of the species. The higher primates appeared after the delta–beta split. The common ancestor had this split, and therefore so do all primate descendents. The gorilla, orangutan, chimp, and human have virtually the same cluster structure, as they diverged late in the game. Following the separation, smaller differences appeared among the four species.

Although our beta-globin chain is the same as the chimp's, it differs from the gorilla, gibbon, and rhesus by one, three, and eight amino acids, respectively. The various globin genes have had different fates among species: In the rhesus, the damage to delta is more extensive than in humans—it has been abandoned. An article by Allan Wilson and co-workers at Berkeley, describing this history, was called "The Rise and Fall of the Delta Globin Gene."[12]

Mammals that diverged earlier can vary considerably. The goat, for example, has twelve genes in its beta-globin cluster, in three sets of four. An original set of four was duplicated, and then one group duplicated again. In this family, gamma got lost and delta was abandoned, but eta remained respectable as an embryonic gene. Biologist Nick Proudfoot has summarized the colorful scene of species divergences.

> Globin gene evolution has provided a very flexible set of genes capable of providing for the individual needs of each species for oxygen carriage at each developmental stage. Every newly created globin gene is then randomly mutated and the mutations either selected for functional globin var-

iants or are put on the genetic scrap heap as useless variants.[13]

The animal sorters deserve their sport, but I suspect that genetic comparisons among humans will prove more fascinating to most of us. Individual humans can sometimes show the types of changes that have marked species evolution: duplication or loss of genes. One case is known where three copies of the gamma-G gene was present. Misalignment in crossing over between chromosomes is one mechanism that can cause gene duplication. This process can also lead to the loss of a gene from a chromosome, a more serious event.

Loss of one copy of the beta-globin gene need not be serious, but loss of both can lead to severe anemia. In such cases, fetal globin genes may come into use to provide the needed units for hemoglobin formation. An individual who fails to make any or enough beta-globin (in contrast to one who produces damaged beta-globin, such as hemoglobin M) suffers from beta thalassemia, a disease named after the Greek word *thalassa*, the sea. This disease is common around the Mediterranean and in Southeast Asia and can be caused by the loss of genes or an error in control sequences.

Localized spelling differences are more common than wholesale gene loss. The estimates of differences among humans run from one in one hundred to one in one thousand or more. Most known sequences, though, have been obtained from protein-coding and control areas, where differences can be more harmful than elsewhere. As more sequences in other areas are learned, I suspect that the figure may move closer to one in one hundred. Further, most recorded spelling differences have been reported at sites where text cutters act, as these have been easiest to detect. Methods are proliferating, however, that will allow the easy detection of all kinds of change in DNA.

To sample what may be learned by the study of spelling differences, we will return to sickle-cell anemia. Pockets of this disease and of sickle-cell trait exist in a number of areas around the globe. Some very different alternatives may have caused this. At one extreme, the condition may have originated separately in each area, while at the other one individual alone may have suffered the mutation with all existing cases representing descendents of that person. To sort things out, we need some reference points that can distinguish one person's globin gene area from another. Such points have already been described.

Investigators who studied the human beta-globin cluster found ten or eleven locations within it where individuals varied apart from the change that caused sickle-cell anemia (more undoubtedly exist, but the above ones were easy to score). Spelling differences were followed by noting whether a text cutter cut (+) or failed to cut (−) each location. Thus one individual would classify as − − − + − + + − − + and another as + − − − − + + + + +. If ten sites were followed, then 1024 different possibilities could be found in the population in theory. However, far less were encountered, thirty-eight in one study.

Why should only a small fraction be present? Let us consider only three sites and presume that in the original ancestor of all globins, these sites were + + +. Mutations might have occurred over the eons that created + + −, + − +, and − + +. They should persist in the descendents, giving four types in the population. To get one of the absent types, say − − +, one would need a second mutation in someone who already carried one or a crossing over event (of the type studied in fruit flies by Thomas Hunt Morgan) between + − + and − + + parents. If events of this kind were very rare, then no such individual would appear. I will call this behavior *incomplete scrambling* (*linkage disequilibrium*).

One study compared spelling differences in the beta-globin area among those who carried at least one sickle-cell gene. By including parents in the study, geneticists could keep track of which chromosome was which in an individual and score two chromosomes for each person. From these comparisons, three distinct sickle-cell backgrounds emerged, which I will call Atlantic, West African, and Bantu. By "background," I mean the pattern of + and − that is obtained by testing a person's DNA at ten or eleven positions with text cutters.

In Benin, West Africa, all twenty of twenty sickle-type chromosomes examined carried the identical background (West African). In Senegal, forty-six of fifty-six shared the Atlantic pattern, one that was quite different from the West African type. Eight were of the West African type, while two represented the thousand or so other possibilities. In the Central African Republic, twenty-four of twenty-eight were of the Bantu type, two of the West African type (both in the same individual), and two others atypical. The data suggest that three separate mutations gave rise to the sickle condition in these backgrounds, with little migration or scrambling since then. Data were also taken in Algeria, where twenty chromosomes of twenty were of the type I have called West African. As malaria is not widespread in Algeria, one likely inference is that population migration from the West African area introduced the gene into Algeria. The authors suggested "these haplotypes [DNA spelling patterns on one chromosome] could provide an objective tool capable of defining the place of origin of Blacks dispersed through the Americas by the slave trade."[14] They went on to suggest that one could perform this calculation on whole communities that had appreciable amounts of sickle cell genes. (Their argument could be expanded. I will suggest later on that the origin of individuals could be learned as well. Not only blacks, but individuals of every ethnic group could be analyzed, if appropriate DNA reference points were found.)

In a different study,[15] chromosomes of both the sickle-cell and normal types from Americans and Jamaicans of African descent were analyzed. More variety was encountered, as might be expected, as these populations had descended from individuals uprooted and compelled to migrate. In 170 sickle-cell chromosomes examined, 16 different patterns were encountered, but 151 of 170 belonged to only three types: the Central African (108), Bantu (32), and Atlantic (11) ones. These proportions were similar in Americans and Jamaicans. Of the normal chromosomes, only six of forty-seven had backgrounds that fell into these three patterns, whereas the majority were divided among thirty-five other patterns that were only slightly represented in the sickle-cell chromosomes. The results suggest that individuals who did not have sickle-cell trait or disease got their globin genes from ancestors that came from other areas of Africa.

The same study was extended to individuals from Spain, Italy, and Greece who were Caucasian in appearance and had sickle-cell trait. Of eleven Mediterranean sickle-cell chromosomes examined, nine were of the West African pattern and two, the Bantu pattern. Not one of 122 normal Mediterranean chromosomes was of any of the three African types discussed. "Poorly documented population movements" from two areas of Africa can again be inferred to provide the gene ancestors. In other words, it was inferred that Mediterraneans with sickle-cell trait had an African ancestor in some previous generation.

A study in greater depth was performed in the town of Coruche in central Portugal, which has a local high concentration of the sickle-cell gene in a population of normal Portuguese appearance. Rice farming is common in the area, and malaria had existed there for centuries until quite recently. A group of 181 school children and teachers were examined, and 14 had sickle-cell trait. None had the disease, but six cases were known to doctors in the area. The background was analyzed for eleven chromosomes, and they scored as seven Bantu, three Atlantic, and one West African.

The authors concluded that these results

> reflect the extent of Portuguese naval explorations. It is concluded that the sickle cell gene in Portugal has probably been imported from Africa and been amplified in comparison with other genes characteristic for African races because of selective advantage of "sickle cell trait" in an area endemic for malaria.[16]

Historically, the principal Portuguese colonies in Africa, Angola, and Mozambique were inhabited by Bantu, whereas lesser colonies, Portuguese Guinea and the Cape Verde Islands, were in the Atlantic region of Africa. The authors continued, "These data suggest that the sickle cell gene has been introduced into Portugal by gene flow related to Portuguese naval explorations."[17]

The situation was contrasted to that in Sicily and Greece where the West African pattern was dominant, "indicating a North African origin dating back either to the Saracen raids on many areas of the Mediterranean or to the Venetian, Ottoman and Frankish occupation involving the settlement of West African slaves."[18]

If we recall that the partly explored (in terms of spelling differences) beta-globin cluster represents perhaps one-one hundred thousandth of the human genome, we can see that there is much, much more about human history to be learned from our DNA. We will return to this topic later, but for now we wish to explore a bit further on our chromosome.

Let's stick with the comparison, 1 kilobase = 1 kilometer. With this scale, the beta-globin gene cluster does have the size of a suburban railroad line. The morning commute from my home to my office in New York City would compare to a trip from the more distant gamma-globin to downtown beta-globin. If we were to continue our trip on the old globin line in that direction, we would finally reach the tip of the short arm of chromosome 11 after we had covered a distance in kb that would have taken us from New York to Rio de Janiero, if it were in km. The chromosome end is so novel, however, that it might be worth the trip.

As we approached within a few thousand bases of the tip, we would find that simple repeats made up most of the text with a particular one, TTAGGG, becoming more and more prevalent. This simple phrase, taken over and over again, would finally become the entire text. At the very end, the same repeat would stick out for two units (twelve letters), but without its companion strand. It is as if our rail line ended with a short stretch of a single rail.

This feature is not whimsical or lightly chosen. We would encounter the same arrangement at both ends of all of our chromosomes. Further, the same repeat is used by all of the other mammals and other vertebrates examined so far. To find some divergence, and a mild one at that, we would have to examine species as distantly related to us as microorganisms and higher plants.

We would expect that a recipe that has been preserved with such care must have a vital function, and this is so: DNA is copied during cell division in a way that works well for sequences in the middle, but would lead to some fraying and loss of text at the end. Unless some special device were adopted, there would be a progressive loss of text as cells divided again and again, and chromosomes would gradually wear away. The TTAGGG repeat sequence at the end of chromosomes carries a message in structure language. It causes DNA to adopt a special arrangement of its chains in space, one that is currently unknown. This arrangement is recognized by an enzyme system that keeps the end in repair, or at least it does so in our sperm and egg cells.

There is a fascinating speculation just out in print that was first confided to me by Jim Watson. It suggests that chromosome ends are involved in the aging process.[19]

According to this theory, our normal body cells lack an effective end repair system. As they divide, the TTAGGG repeat shortens and is ultimately lost. When it is gone, the chromosome becomes unstable and cannot function properly. It has grown old. By the time that you read this, the idea may be proven or proven wrong. The very possibliity that it may be true shows us, however, that we are exploring terrain that holds some of the deepest secrets of our existence.

The trip from the beta-globin gene to the end of the short arm of chromosome 11 is small compared to one that we could take if we set out in the other direction. Using our scale of 1 km to 1 kb, we would have to ride a distance greater than two times around the world to reach the chromosome hinge that separates the short and long arms.

Upon arriving, we would again encounter text written in the language of structure, but with substantial differences from the chromosome end. The area involved would be much larger, perhaps a couple of hundred thousand base pairs, and more diverse. Although the sequences present are related enough for them to be loosely grouped into one large family, much more variation exists among chromosome hinges than chromosome ends. Basic units of some 170 letters seem to be grouped into larger arrays of some three thousand characters, which in turn have some larger organization.

You may recall that the hinge serves as the attachment point for apparatus that hauls the chromosome about during cell division. The unusual DNA passages that it contains must serve that purpose in some way.

On our tour of part of chromosome 11, we have explored one cluster of five working genes and sampled in part the hinge and end areas. The entire genome may have five or ten thousand such clusters and thousands of independent genes as well. In addition, there are vast areas between the genes that are little understood. What would we gain if we extended this exploration to all the remainder of our chromosomes?

In answering this, we should first note that the globin genes represent almost familiar terrain, as the substances they produce are among the most common of human materials. The ancients wrote of the color of our blood, and the responsible protein has been explored for more than a century. Miescher discovered DNA under the supervision of the individual who deduced the oxygen-carrying function of hemoglobin. Watson and Crick discovered their DNA structure in the same building where Perutz was attempting to learn the structure of hemoglobin and where shortly thereafter Ingram would demonstrate the amino acid difference in sickle-cell anemia. At the time when the globin genes were sequenced, the various proteins produced by them were already recognized and their roles understood.

Imagine how things would differ if we broke into an unexplored area of genetic terrain, selected only by its location. We would encounter new genes of unsuspected purpose. Our lack of knowledge would not mean that they were unimportant, but rather that they governed basic human functions that were less accessible to us than is our blood.

For millenia, our ancestors understood little about blood. They did not recognize that it circulated, carried oxygen, and fought microorganisms. Clotting was a mystery. Today much is clear, but we are still ignorant of the mechanism of so many other human functions: What governs our development from one cell to a very specialized organism? Which molecules are involved in emotions, memory, pleasure, pain, sleep, and aging? Which clocks control puberty, menopause, and the rate of aging? Why do some of us resist certain infections and chemicals and others succumb? In deciphering each new sequence of DNA, we will gain vital information about ourselves. In the long run, the order in which we explore the sequences will not be crucial, but rather the end result. To know ourselves, we will want to have them all.

At the current moment, however, certain passages have more urgency than others. The critical ones are those known to be associated with human misery. These sections contain spelling changes of the type we sampled with sickle-cell anemia. I call them the terrible typos.

8

THE TERRIBLE TYPOS

BALTIMORE, MARYLAND, NOVEMBER 1989

The elevated walkways that surround Baltimore's convention center offered a better view. The lecture halls provided the excitement of a live presentation. However, it was the windowless poster area in the convention hall basement that fascinated me most of all.

Scientific poster sessions were familiar to me. I had attended many chemical meetings and looked, with poorly concealed boredom, at the interminable displays of formulas and patterns of spots, peaks, and bars. Occasionally I was even the perpetrator of such an exhibit. The display at the annual meeting of the American Society of Human Genetics, however, was jolting and reminded me of a visit to a musuem of the Holocaust. Mixed in among the inevitable abstract patterns were photographs of suffering humans, sometimes deformed and usually children. They were victims of a different kind of war: our struggle with hereditary disease.

The title above the photographs named the genetic condition that caused each disease, names that were usually obscure, but sometimes evocative: maple syrup urine disease (nerve damage, seizures, coma), hereditary startle disease (intense rigidity with unchecked falling after an unexpected touch or sound, occasionally fatal muscular rigidity), Praeder–Willi syndrome (shortness, learning problems, obesity), and

Crouzon syndrome (middle of the face compression, deformed features). The photograph of a laughing, blond, blue-eyed child provided a welcome contrast, but my delight was quickly dissolved by the description that came with it: happy puppet syndrome (Angelman's syndrome). This disease produced prolonged, but inappropriate, unprovoked smiles and outbursts of laughter in addition to jerky puppetlike movements, drooling, a protruding tongue, albinism (lack of normal coloration), and mental retardation.

There were many, many more diseases displayed in that poster area. In the majority of cases, the parents' health was normal. Similar photo collections, or drawings and descriptions, could have been assembled by interested rulers or organizations at many points in human history. For the most part, the afflictions are not new. What has changed over the last century, and particularly in the last decade or two, has been our ability to understand them and take action.

The first opening was provided by our old acquaintance, the monk Gregor Mendel. Many of these diseases obey his rules quite well. Some of them, sickle-cell anemia, for example, follow the pattern of his wrinkled pea trait in which two affected genes of the same kind are needed to develop the disease. (A one-eyed man is hardly blind.) In such a case, the unexpressed gene is termed *recessive*, whereas the normal one is dominant. Perhaps one or two out of every thousand live births suffer from some recessive disease.

In some cases, one defective gene is enough to cause a problem. (One flat tire will disable a bicycle.) Such maladies follow the behavior of Mendel's smooth peas: The poorly functioning gene is dominant. Estimates of such diseases run up to ten per thousand live births. One subcategory of this division are diseases that involve the unmatched chromosome pair, the X and Y sex-determining chromosomes. We shall give this subject its full due later on.

Victor McKusick of Johns Hopkins University has compiled a list of conditions that follow the above rules in his treatise "Mendelian Inheritance in Man." This compendium has grown rapidly in its successive editions over more than two decades until it now lists about five thousand traits.

Not all of these traits produce illness. When I opened the volume at random, the first item that I saw described blond hair: "Striking blond hair may be recessive," the entry began. "I know of families in which both parents and many other sibs of the blond child are relatively dark haired, although both parents have very blond relatives. Red hair also seems to be recessive." Immediately after this rather chatty, personal account came a formidable description of

Bloom's syndrome, bristling with technical terms and studded with three dozen references.

In Baltimore, then, I was exposed to only a small part of a vast list, which becomes much larger when we include multigene diseases that do not follow Mendel's rules. Through human history there was little to be done about genetic afflictions except to discourage certain kinds of mating. If John has a rare and harmful recessive gene, then his sister Jane has a much greater chance (one in two) of having the same gene than does an unrelated member of the population. If they were to wed, their children would be at substantial risk (one in four) of drawing the deadly pair. Without understanding the mechanism, many societies learned by observing the consequences that incest was not a good thing.

In the age of DNA sequencing, we have much more power. At some future date, if we hold our course, we shall have the whole list of one hundred thousand genes and understand the role of each. We shall also have a repair manual, which will take somewhat longer to get. At the end, though, we shall know what to do in each case to relieve or avoid the misery.

In the interim, we live in an age of heroes. These include the disease victims, martyrs to the imperfections that have come with the human condition, and their rescuers, who refuse to accept the imperfections as permanent. The contest is a new one, however, and the would-be champions must improvise their weapons, even as they rush into combat. The struggle is worth our attention, and we shall explore several battles and the issues that have come up during their course.

TOIL, TEARS, AND SWEAT

The cameras of PBS's science television series "Nova" passed through the window of an apartment complex and into a living room. A girl, perhaps three years old and with rich reddish-brown hair, lay on her tummy on a padded bench or table while her mother rhythmically and gently thumped on her back, as if beating a drum. After a time, the mother turned her daughter on her side and continued to beat. The girl, Philippa Brody, endured the treatment with her thumb in her mouth and a woebegone expression on her face. On the floor, an older brother imitated the ritual by pounding a doll, while a younger brother looked on. No game was being played. Philippa was receiving a session of the three-times-a-day physical therapy that together with heavy medication would maxmize her chances of surviving into adulthood with the disease cystic fibrosis.

The physical massage was intended to loosen the thick mucus that would otherwise clog her lungs, obstructing her breathing and providing a breeding ground for infective bacteria. Her infection was also controlled by antibiotics. In eighty-five percent of the cases of this disease (I do not know Philippa's status), the victim's pancreas also malfunctions, creating digestive difficulties and leading to malnutrition. In addition, the sweat glands also behave abnormally, giving off unusually salty perspiration.

Although the cause of these different malfunctions was only diagnosed as a discrete disease, cystic fibrosis, in this century, the abnormal sweat at least was readily detected. Historical anecdotes tell of midwives who licked the forehead of a newborn infant. If the sweat tasted excessively salty, then the baby was destined to die of lung congestion and its side effects. Death in childhood was common. With modern medical countermeasures, now "many affected persons survive into their teens and even early adult life," according to one report. A late 1990 review,[1] put the median age of survival at about twenty-six years.

It is one thing to stem the flood, however, and quite another to find and fix the leak. Despite its many manifestations, inheritance studies suggested that the disease was due to a single gene defect. Philippa's parents were healthy. Yet each was a carrier and held one copy of the causative gene. They were not unusual in this way. Cystic fibrosis has been termed the most common lethal disorder of Caucasian populations, with an occurrence of about one in every twenty-five hundred live births. The rate is much lower among blacks and Asians. To reach those levels, one in twenty-five whites must be a carrier. (About 1 in 625 marriages, 1/25 x 1/25, would then involve two carriers. In such a marriage, a child would have one chance in four of drawing two defective genes. Philippa was unlucky, while her brothers fared better.)

Given that much information, how do we move ahead? In the case of sickle-cell anemia, the breathlessness and fatigue of the patient had suggested his anemic condition to James Herrick. Examination of his blood revealed the sickled shape of the red cells. Hemoglobin, the most abundant protein of red cells, was ultimately inspected and found to differ from the normal pattern in these cells. The amino acid change involved was determined by Vernon Ingram, and finally, about seventy years after Herrick's paper, the genetic text was deciphered. This order (if not pace) of events is considered normal genetics. Scientists work from an observed body type (*phenotype*) to the genetic spelling (*genotype*).

In the case of cystic fibrosis, the defect lay within cells that were not so readily diagnosed or understood. It was necessary to isolate the gene first, deduce the protein from its text, and then work out its function. Appropriately, this approach is called *reverse genetics.*

In reverse genetics, one must locate a specific passage of text, a few pages long, in a collection of twenty-three unabridged dictionary-size megavolumes of the type we described earlier. The entries are not arranged alphabetically, no index exists, and, further, the pages are unnumbered. Certain aids are needed, for example, some reference points within the volumes. We have described such items earlier as probes, phrases of known text that can be prepared by chemical means in the laboratory, with built-in radioactivity or fluorescence. Such probes can be assigned to bands or locations on particular chromosomes ("mapped") by a text-matching procedure.

For a probe of this type to be useful in the search for a disease gene, another qualification must be met. The DNA area that it matches must contain a spelling difference that can tell one individual from another.

Let us suppose that we use the probe to pull out a page from my DNA text and the same one from yours. We test both DNAs with a text cutter or other tool that can pick up spelling differences. If our texts are the same, then this probe will not be of use to us. If they differ, then we are in business. We will have discovered a marker. If we had used a text cutter, then one of us would be scored as "+" and the other "−", as we described in the case of sickle-cell anemia. We could go on to classify anyone else who would give us a sample of blood or other source of his or her DNA.

Suppose now that we analyzed large families with histories of cystic fibrosis. We could test each member with our probe and try to learn whether the + or − trait was inherited independently of the disease or whether it was transmitted in parallel. If the latter were true, then the disease gene and our marker are close together on the same chromosome; they would be linked. The closer that they were, the better, for our data would be less likely to be confused by crossing over. The marker could now be used for diagnosis: New families could be analyzed, and carriers could be identified.

There are complications. If the mutation that causes the disease arises frequently, then it could be linked with the + trait in one family and with the − result in another. Neither + nor − on its own signals the presence of the disease gene; this must be established for each new family. To make a correlation, you need a large family with several living cystic fibrosis sufferers. Ideally, one would want three full generations with many brothers and sisters in the youngest one and, of course, plenty of cooperation.

Further, if + were linked with the disease in this family, but it was also common in the population, then many people in that family who had a + would not be carriers. It helps further if more than one marker can be followed, or a marker is used that can have more than two spellings (CA, CACA, CACACA, and CACACACA, for example).

Let's look at an example taken from a real Amish–Mennonite family group (I have invented the names of the individuals). Two different markers closely linked to the cystic fibrosis gene were available, and two different probes were needed to detect them. We know now that they flanked the cystic fibrosis gene. Each occurred as + or −, giving four possibilities in all: + +, − −, + −, and − +. We will just call them A, B, C, and D.

One couple, Duncan and Roberta, have eight children. Two of them have the disease. Both afflicted children scored as AB, that is, each had one chromosome of type A and one of type B. Both parents are healthy, but each must be a carrier. In one of them, the defective gene is linked to marker A and in the other to B, but which was which? To find out, the parents' DNA had to be tested.

Duncan had a BD constitution. His B marker was the one he passed on to his affected children, so that must be linked to the disease. Roberta had type AB. If we did not have Duncan's data, we would not know which chromosome she gave to her afflicted children. Since we do, we know the chromosome with the A marker was passed on and had the disease gene. Their six healthy children all fall into one of the three types BB, AD, and BD. The BD type has inherited two disease-free genes; those of the other classifications must be carriers (they have Roberta's A or Duncan's B).

Duncan has a brother who we will call Dan. Dan is married to Ruth, Roberta's cousin. They have five children, two with the disease. Those two are also of type AB, but so are their healthy parents. The three healthy children of Dan and Ruth typed as AA or BB. This situation is more confusing and would be ambiguous without the data from Duncan and Roberta. Of the four chromosomes in the parents, we have a good A, a bad A, a good B, and a bad B. From the family data, however, we can presume that the Duncan–Dan line provides the bad B, and the Ruth–Roberta line, the bad A. This correlation was found to be valid when data were taken from the available parents for the couples.

Family studies of this type can be used of predictive testing once a linkage is established. How does one find linked markers in the first place? That is the hard part. You need a whole collection of markers with probes to detect them scattered over the set of human chromosomes. (Much trial-and-error work is needed to assemble such a collection.) Then with your set of markers in hand and blood samples from suitable families, you attempt to establish correlations as we have described above. If your data hold up, the marker and the disease are linked. It can get messy. A marker some distance away on the same chromosome will give an imperfect correlation, with the failures due to recombination. A lot of statistics are needed. If a distant marker is found, then you would examine others known to be on the same chromosome to get a closer one. Finally, when you had come as close as you could in the text, you would start to read sequences directly to find the actual gene.

With these rules in hand, we can now appreciate the dynamics that underscored the Great Cystic Fibrosis Gene Race that occurred between the mid-1980s and August 1989 and was covered most thoroughly by Leslie Roberts in *Science*.[2,3] There were a number of runners, but we will focus on a few of the most prominent. The stakes were the usual ones: money and glory. Robert K. Dresing, the president of the Cystic Fibrosis Foundation in 1988, described the latter reward: "We're talking Nobel Prize material."[4] The money is harder to calculate, but we can estimate its dimensions. Four million pregnancies take place in the United States each year, of which perhaps four-fifths come from the high-risk group, Caucasians. If half chose to be tested, at say $100 per test, then we would have $100,000,000 per year in revenues. One Wisconsin geneticist described the prospect bluntly: "This is the gold rush. This is the Klondike. The potential market for this screening is at minimum a one billion dollar a year industry."[5]

A Massachusetts biotechnology firm, Collaborative Research, found this prospect attractive and entered the race at an early stage. It had invested heavily in probes, but guarded them "zealously, for proprietary reasons."[6] They had no direct access to suitable families, however, as did those academics who were affiliated with hospitals. Collaboration was needed, and ironically, this proved to be their weak point.

At an early stage they enlisted Lap-Chee Tsui (pronounced "Choy") of the Hospital for Sick Children in Toronto, which has a very large cystic fibrosis clinic. Together, they indentified the relevant chromosome as No. 7, using a loosely linked probe. Friction arose when they chose to withold that information at an important meeting. "We were nervous about making a mistake," commented one company executive.[7]

Although the news was not released officially, it leaked out by rumor nonetheless. Two other groups learned or deduced that the gene in question was on chromosome 7, discovered closer linkers, and almost beat Collaborative into print on the location of the gene. Perhaps the company was overconfident. Their executive officer, Orrie Friedman had once boasted: "We own chromosome 7."

One of these two additional groups was headed by Robert Williamson of St. Mary's Hospital in London. He had focused his research on this problem, committing his fifteen-man group to the hunt. Williamson, according to Leslie Roberts in *Science*, was "an ardent socialist" and felt that "no one should profit from publicly-funded work on cystic fibrosis." He had no reservations about glory, however. One scientist commented to Roberts: "Bob is pretty single-minded. He wants to solve it. He wants to be famous for cloning the gene."[8]

Another effort was headed by Ray White of the Howard Hughes Medical Institute at the University of Utah. White has had a broader spectrum of interests and has earned various awards in human genetics. He did not focus his efforts on cystic fibrosis. He had a formidable probe collection, however, and an ideal population on which to draw, the Mormons of Utah and southern Idaho. An account in *Genetic Engineering News* summarized the virtues of this group:

> This region has a tradition of large families that remain in the area. The people are long-lived, marry early and have many children, so it is often possible to maintain genetic material from three generations.[9]

The publications in which the White and Williamson groups identified chromosome 7 as the one that held the cystic fibrosis gene did not acknowledge the priority of the Collaborative–Tsui efforts in the field. One *Nature* reviewer called this "immoral but not criminal." Those who were accused naturally disagreed, White countering that the charge was "almost a joke. All interested parties knew it was on 7." Williamson characterized Collaborative as "a company interfering in normal scientific communication at an early stage."[10]

Collaborative, in the interim, attempted to regain its lead by forming a new partnership with White to replace the one with Tsui that had foundered. Their new effort led to no better result. In December 1986, seven groups agreed to pool their data at a meeting. Collaborative scientists attended, "but nobody wanted to talk to them", according to a report in *Science*.[11] Eventually the company moved to a more open position. "Let's just say we learned," commented an executive.

The next episode in the developing saga came in the spring of 1987, when Williamson announced that he had discovered a "strong candidate" gene. His competitors were discouraged, and it was rumored that some grant applications were turned down because the reviewers assumed that the job was done. A few months later, however, this candidate failed and "the envy many of his competitors felt turned to anger." At the Gordon Research Conference in this area in 1987, "some of the cystic fibrosis workers were barely civil to each other." A summary by Leslie Roberts in *Science* the next April commented that

> even within the highly competitive field of human genetics, the search for the cystic fibrosis gene stands out for the intense nature of the rivalry.

Another observer added that "this is not your average ego-driven science. This is nasty."[12]

Shall I use these events as fuel for a sermon on the greed of humanity or the unworthiness of modern scientists? However unpleasant the interpersonal events were for the participants, rapid progress was being made. Family diagnoses were being performed. Moira Brody, the mother of Philippa, for example, was able to turn to Williamson's laboratory during her third pregnancy and receive welcome assurance that the fetus was free of cystic fibrosis.

At the height of the furor, an effective collaboration was formed that did lead to the isolation of the gene in late summer of 1989.[13] Lap-Chee Tsui and Francis Collins of the University of Michigan Medical School met at the American Society of Human Genetics Conference in 1987 and agreed to pool their resources. Collins, in particular, was expert in chromosome jumping, which was needed in the search for the gene.

We mentioned earlier the technique of chromosome walking in which a section of deciphered DNA is used as a springboard for the exploration of an adjacent area. This slow but steady progress along the text can be blocked for technical reasons, however, when an area of repeating sequences is encountered. The best text equivalent of chromosome jumping that I can think of is the stapling of adjacent pages together in a book. One can then jump from known text to the point that has been stapled to it, bypassing an entire page or section. Leaps of up to one hundred thousand base pairs are possible, bypassing the obstacle. The new entry point can be used either for a thorough search in its vicinity or as a base for another jump.

Starting at a closely linked marker, the group of Collins, Tsui, and their collaborators carried out a walk-and-jump journey of some 280,000 letters through band q31 of chromosome 7, until they entered

DNA text that described the protein involved in cystic fibrosis. Their first job had been to choose the proper direction for their search. At first, they explored both choices, but one was abandoned when they ran into a known marker that was further from (less tightly linked to) the gene than their starting point.

Another problem was to recognize their goal when they got there. They were moving through unexplored terrain, and it was not easy to learn when they had entered the cystic fibrosis gene area. One tool that they employed was called a *zoo blot.* They used a rough form of text matching to see whether the new sequences they had uncovered were also present in other mammals. If so, they examined it further. It seemed likely that a protein of importance would not be limited only to humans. Each promising new stretch of DNA was also matched with the RNA messages that were produced in active sweat glands. Eventually, they entered a protein code area that looked right and decided to read the entire gene. It turned out to be a whopper.

The gene they had discovered was divided into twenty-seven separate protein-coding areas, which together coded for a product of 1480 amino acids. Its total length, interruptions and all, covered some 250,000 base pairs, five times as many as are present in the entire beta-globin cluster of genes.

The most crucial test was yet to come: the sequence of portions of the gene was determined in patients with cystic fibrosis and compared to that present in normal individuals. Of those afflicted, 145 had a specific change in the text that was not observed in the 198 normal people who were examined. Three letters were removed from the text in those patients, leading to the loss of the amino acid phenylalanine from its normal location in position 508 of the protein chain. Sixty-nine other cystic fibrosis patients, however, did have those same letters in place. Some other spelling difference was responsible for the disease in those cases.

Other workers discovered some of those differences a few months later. Changes of a single letter in the DNA region that described amino acids 549 to 559 could also trigger the disease. Many ways exist to spoil a recipe; more than sixty have been discovered thus far. One is just more common than the others. The three-letter change was found upon further study to correlate with the most severe form of the disease and to be represented much more in Northern Europe than in countries bordering on the Mediterranean.

The protein recipe itself confirmed the advance speculations of some observers. The order of amino acids was appropriate for a substance that criss-crossed a membrane, shuttling back and forth between the inside and outside of the cell. That location was suitable for a regulator that had something to say about the transportation of salts or other substances in and out of the cell. Perhaps it served as a pump for these materials. You could start to understand how a defect in the protein could lead to changes in the thickness of lung mucus and saltiness of sweat. The connection to pancreatic malfunction remained a mystery for now.

Knowledge of the protein produced no immediate drug remedy, but at least the line of attack was now clear. Scientists could then harvest the protein to learn how it works and what goes wrong in the mutants. Then they could try to devise a remedy. The many groups that raced for the gene did not have to fear unemployment. They, and some others as well, set off in that direction. As we shall see shortly, the chase has become very hot.

The discovery of the gene, especially by a group of cooperating, noncontroversial, and nonprofit-making academics, was warmly greeted in the press and at the scientific meetings that I attended that autumn. A new wave of ethical questions arose, however, some of which extend beyond cystic fibrosis to genetic testing in general. They will appear again and again as more of our genetic text is read.

Moral questions came up because direct screening for the disease was now possible. Three-quarters of the carrier population could now be detected by a direct and simple test. Neither linked markers nor family input was needed. The genes of the parent or fetus in question could be examined directly to see whether the three-letter deletion existed. If that deletion was present in one chromosome, the individual was a carrier; if it was present in both, he or she had cystic fibrosis. The difficulty arose because one-quarter of the faulty genes would not be detected by that test. No conclusive assurances concerning the disease could be given if the test for the three-letter deletion was negative.

Let's be specific. Both my wife and I are Caucasian and have perhaps one chance in twenty-five of being carriers. As we have no family history of the disease, no diagnosis could have been made until the gene was discovered. Our chances of conceiving a child with cystic fibrosis could only be estimated from its rate in the Caucasian population, or one in twenty-five hundred. The same situation had applied to the parents of Philippa Brody before they had her.

Assume that we both take the test and pass. We are not in the clear, but our odds have improved. Our chances of conceiving a child with cystic fibrosis would have fallen from one in twenty-five hundred to one in forty thousand. On the other hand, if one of us were shown to be a carrier and the other did not carry the three-letter deletion, then our chances would be one in four hundred of producing a child with cystic fibrosis. It is not certain what we would do about this intermediate level of risk.

The significant result would arise if we both tested positive. We would be forewarned that we have a one in four chance of having a child with the disease and could consider prenatal testing (with abortion as a possibility) or not having a child at all.

The debate among geneticists was conducted at several levels:

> 1. Should the test be offered at all at this stage? The results might be misinterpreted. Those who "passed" and had a child with cystic fibrosis might file a lawsuit (but failure to offer the test might also be a cause for a malpractice suit).
>
> 2. Should mass screening be conducted to reduce the incidence of the disease? Opponents felt that incomplete information would be ineffective without extensive counseling, and few resources were available for that purpose.

Fortunately, the above situation is temporary. I used the circumstances after the initial gene discovery for my illustration, because the mathematics was simple. Since then, additional cystic fibrosis cases have been examined, the list of spelling changes that can cause the disease has grown, and the chance of a false negative has diminished. Eventually almost all of the harmful spelling changes will be indentified, even if a large number of them are involved. The test may grow more expensive, but questions based on its reliability will diminish.

At what point should a test be made available to interested parties in the general population? That is a decision of values, not of science. Our tradition is this country has not been to deny health information to interested individuals when they claim that they can handle it and are willing to pay for the cost of getting it.

In the particular case of cystic fibrosis, a chance exists that advances in research will make the debates obsolete before the ink has dried on the polemics of the moralizers. By the autumn of 1990, one year after the discovery of the gene, human cells had been cured of their symptoms in the laboratory by gene therapy.

Two separate research groups had stitched a working copy of the gene into viruses, then allowed the viruses to carry the genes into cells taken from cystic fibrosis patients. The sickly cells then began to function normally with their clogged channel unplugged. One of the students involved was reported to run out of the lab saying, "I can't believe it worked." One group leader, Dr. James Wilson of the University of Michigan, commented to the *New York Times*:

> My tendency is to be very conservative, but the hazard with that is you end up underestimating how fast the field of gene therapy is expanding. At this point, it's impossible for me not to be optimistic about cystic fibrosis.[14]

Somewhat different techniques and many more precautions are needed to treat living patients in a hospital than their cells in a test tube, but there was room for optimism. Another scientist, Paul Quinton of the University of California, Riverside, caught the mood in his comments in *Nature*:

> Only a few years ago the idea of introducing normal genes into the cystic fibrosis lung to correct its fatal susceptibility to infection was science fiction. Now the accomplishments of these investigators seem to press the fiction inspiringly close to reality.[15]

Gene therapy experiments were contemplated in which a patient would inhale an aerosol that contained the normal gene within a suitable carrier. By the end of 1990, Robert Beall of the Cystic Fibrosis Foundation could hope that his organization would be "out of business" by the year 2000.

ETHICAL QUARRELS

Cystic fibrosis, of course, represents only one of thousands of genetic diseases. As genetic research progresses, tests will be developed that can predict many of them long before they can be cured. The above arguments and others will appear again and again as the Human Genome Project progresses, so perhaps the time for us to deal with them is now.

Mass screening on a voluntary basis (we have no tradition of compulsion, even for AIDS) is an old issue. The need for trained counselors is simply part of the cost. We will have to judge whether the gain in public health for each new expenditure will be as great as it might be if the money were spent on alternatives, such as enhanced screening by conventional means for cancer. This debate is really a budgetary item to be handled by elected representatives rather than an ethical one.

Much can be accomplished voluntarily by private initiatives, as illustrated in the case of Tay–Sachs disease. This affliction occurs primarily in individuals of eastern European Jewish background, at a rate of one in thirty-six hundred live births. Two defective genes must come together to produce this devastating illness in which children grow vegetative, become blind, and die at an early age. The incidence of this disease fell by about eighty percent between 1970 and 1980, as orthodox Jewish communities instituted their own screening efforts and discouraged marriages between carriers. In the case of Tay–Sachs, carrier status can be learned from a blood test, as the crucial protein involved in the disease has been identified and its activitiy can be measured.

Some controversies may simply be a cover for deeper and more bitter ones that involve sharp disagreements on questions of values. A body of thought exists, for example, that questions whether disabilities should be avoided at all. The alternative would be to accept them as part of nature. For example, Mary Johnson, editor of *The Disability Rag*, commented to the *New York Times*: "We are not really willing to confront disability. What we really want is to get fewer and fewer disabled people. That is what really worries me."[16] In a similar vein, sociologist Barbara Katz Rothman made the following remark to *New York*: "In gaining the choice to control the quality of our children, we may rapidly lose the choice not to control the quality, the choice of simply accepting them as they are."[17]

The most visible and vocal opponent of all forms of genetic manipulation has been Jeremy Rifkin. The following quote from his book *Algeny* is emblematic of his philosophy.

> In all of humanity's past experience, living things enjoyed a separate, unique and identifiable place in the order of things. There were always rabbits and robins, oaks and ostriches, and while human beings could tinker with the surface of each, they couldn't penetrate into the interior of any. Now, as we move from the age of pyrotechnology to the age of biotechnology, people are beginning to learn

> how to reorganize living things from the inside out . . . [This marks] a qualitative break with man's entire past relation to the living world.[18]

His philosophy is often an attack on Darwinian thought. Rifkin states, "Attacks on Darwin's theory . . . are going to increase in the years ahead, leaving Darwin a lifeless corpse, a distant memory of a bygone era." This position is shared by Creationists, but certainly not the scientists most closely concerned with evolution.

In Rifkin's system, the welfare of individual humans matters less than the sanctity of nature: In return for securing our physical well-being, we are forced to accept the idea of reducing the human species to a technologically designed product. Genetic engineering poses the most fundamental of questions: "Is guaranteeing our health worth trading away our humanity?" In his view, even our ultimate survival may be less important than what he sees as the integrity of the universe.

> Not once in the long history of Western civilization have we ever said no to our own future. . . . Over the centuries, we have constructed countless cosmologies to lend an air of legitimacy to our ceaseless drive for self-perpetuation at all costs. We have deceived ourselves into believing that our interests were in accord with the interests of the universe, when in fact it was only our limited needs that were being projected onto the cosmos.
>
> Can any of us imagine saying no to all the great benefits that the bioengineering of life will bring to bear? Can any of us, for that matter, entertain even for a moment the prospect of saying no to the age of biotechnology? If we cannot even entertain the question, then we already know the answer. Our future is secured. The cosmos wails.[19]

I myself and many others would reject this value choice for a simple reason: We judge human suffering to be an evil that outweighs considerations of what may or may not be natural. The cries of a child in pain are very real to us; those of the cosmos are not. Suffering is not limited to the afflicted child alone, of course, but also extends to the parents. Writer Deborah Batterman expressed her feelings on this issue quite strongly in *New York*:

> Twenty years ago when a woman gave birth to an impaired child, people would say, "Oh what a heartbreak that she has to live with that situation." . . . That's the martyrdom indoctrination. We don't have to do that now. We're not raised to be martyrs.[20]

Disagreements of this type can seldom be bridged by discussion, and in such cases, some advocates will call for coercion. For example, geneticist Margery Shaw was quoted as follows in D. Nelkin and L. Tancredi, *Dangerous Diagnostics*: "The law must control the spread of genes causing severe deleterious effects, just as disabling pathogenic bacteria are controlled."[21]

The record of the twentieth century on the use of coercion rather than the persuasion to control reproductive behavior is so dreadful that it is unlikely to be brought up seriously again in this country. Earlier in our history, miscegenation laws were passed to prevent the mixture of "blood" between races that were felt, on arbitrary grounds, to differ in quality. Several thousand individuals were sterilized in the United States in the first decades of this century as a measure to control "feeblemindedness." Finally, the Nazis carried out the ultimate escalation and murdered millions of Jews, Gypsies, and members of other nations for proclaimed genetic reasons. We do not need to explore this path again.

Even those who agree that genetic infirmities are unfortunate are still divided on the measures that may be taken to avoid them. Dr. Brian Scully, a Catholic and an infectious diseases specialist, commented to *New York*: "I don't want children to have cystic fibrosis. But to say that if you have cystic fibrosis I'm not going to have you, I think that's wrong."[22] The issue here, barely concealed, is the morality of abortion. As I mentioned earlier, I do not think that the authentic ethical differences here can be reconciled by debate, but perhaps some form of peaceful coexistence or segregation of the two camps into separate communities can be obtained.

Those who do accept abortion and choice will find themselves confronted with new moral decisions in the future as tests expand and new options appear: What grounds are sufficient to terminate a pregnancy? All who accept abortion would agree about spinal muscular atrophy, a disease that kills within nine months of bith, or Tay–Sachs disease. What about a disease like Alzheimer's that strikes later in life or a disabling condition such as deafness? What should be done about those who would, and in fact do, abort a child of undesired sex?

Genetic advances may take some of the sting out of these questions for those who are willing to accept scientific intervention in the most intimate processes of life, but a new set of issues will then arise: *Nature* in 1990[23] described a test-tube fertilization procedure that was selected by several couples. All of them had progeny at risk for lethal diseases that affected only males. For obvious reasons, they wanted females.

In this procedure, a number of eggs were removed from the ovary of each mother, fertilized by sperm from the intended father, and allowed to divide to the stage of about eight cells. One cell was removed from each embryo at that stage (this can be done without harm), and a technique that we will call *DNA amplification* (*polymerase chain reaction*, or PCR) was used to permit genetic analysis on that modest a basis. The presence of DNA sequences from the Y chromosome identified an embryo as male. Two female embryos were reimplanted in each of the women, and in two cases of five, the procedure succeeded. Both women appeared to be carrying healthy female twins.

The authors, a group from Hammersmith hospital in London, commented: "Preimplantation diagnosis will be a viable option for many families carrying genetic defects." An earlier paper had shown, in principle, that the same procedure could be used to screen for cystic fibrosis. The disadvantage inherent in this procedure at the present time is that the rate of successful implantation is low, some fifteen percent. Undoubtedly, this will improve with further research. The potential applications stun the mind. Avoidance of genetic disease will no longer involve the termination of a fetus well down the road in development. The parents will select which of several barely developed human possibilities shall be brought to fruition. "Which would you prefer, Mrs. Brown, the dark-haired girl or the fair-haired boy?"

Embryo selection represents one of the coming genetic techniques that will raise novel legal and moral headaches for society. One case has already come before a court in Tennessee in which a divorcing couple quarreled over custody of seven frozen embryos, each of only a few cells. The mother wished them kept alive for future use, while the father wanted them terminated. In the initial decision, a judge declared them "children" and awarded them to the mother. Obviously, much more will be heard on this issue.

Whether by embryo selection or abortion, we are obviously entering an era when much greater choice or control by parents over the nature of their offspring will be possible. We will tackle the full implications of this development at greater length in Chapter 19, when the largest issues of all come into focus. For now, I want to touch on another disagreement that was embedded in the quotes I selected.

Should we attempt to cure genetic disease at all, or would it be more moral to take a passive attitude and learn to accept what nature chooses to offer us, even accepting disability as part of the natural scheme of things? Once again, we have a decision based purely on values. There is no way to determine which is "correct." Personally I feel that the human condition, with its inevitable course of aging and death, has enough imperfection built in so that no additional burdens are needed to remind us of our limitations. Again, individuals can disagree on the issue, and there is no need for consensus. I would feel better, however, if those who bow to the dictates of nature were selecting their own fate rather than that of others. The division raised by this issue conceals a greater one that has not yet surfaced into active public debate: the ultimate goals of the human race. We shall have a chance to wade into a corner of that immense subject before we are done.

A separate issue involves the feelings of those who, suffering from a genetic disease, see others terminate a fetus because the same condition is present. Some victims identify with the illness and see the abortion as a rejection of them. In Marc Lappe's *Genetic Politics*, the reaction of photographer Edward Weston is described when he found he had Parkinson's disease: "He was sure the disease was genetically caused, a direct read-out of his flawed nature."[24] He reacted with anger and self-hate. When he learned that the disease was not genetic after all, he felt better and more able to accept it. In the same book, however, it was pointed out that reactions vary. For some, a genetic diagnosis is a relief; they bear no responsibility for their disability or illness.

I had never considered the question of genetic flaws with respect to myself until recently. When I interviewed Dr. Nancy Wexler (she heads the Committee on Ethics of the Human Genome Project), she turned her head directly toward me, looked into my eyes, and asked me how I would feel if I found out that I was a carrier of a deadly disease. I drew a blank. I have read that each of us is likely to carry five or ten recessive and lethal genes, which if combined with their match would produce a dead individual, but never had applied it to myself.

The word *carrier* has a dreadful ring to it with echoes of those who spread infectious disease, typhoid, or the plague. A genetic disease has less capacity to inflict mischief on others, but those of us who are curious about our own natures or careful in having children may need some reeducation to protect our self-esteem. In one Canadian study, school children who were found to be carriers of Tay–Sachs disease were shunned by their noncarrying schoolmates, which led them into anxiety and depression.[25]

Perhaps the ultimate challenge of this type was raised in the recent film *Twins*. Genetic material from six fathers and one mother was pooled and sorted for strength and purity to create a superior human, Julius (played by Arnold Schwarzenegger). The rejected portions, "all the crap," were set aside, but by some miscalculation came together to form Julius's twin, Vincent (played by Danny DeVito). Their appearances matched their origins, with Julius tall, strong, and morally upright, and Vincent, the opposite. Vincent naturally did not react well when he learned of his makeup: "I'm genetic garbage!" Fortunately, Julius managed to convince him of his innate human worth. The two were reconciled and prospered together.

I do not suggest that this fantasy be studied as a clinical example of the proper method to use when counseling those who carry genetic flaws (that is, all of us). However, it does contain a message. We already are who we are, and any genetic descriptions that simply describe us without predicting unforeseen harm need not be taken so dreadfully seriously.

9

HARD TRAVELIN'

I waited for my interview with Nancy Wexler[1] in the hallway of a medical facility that was part of George Huntington's old alma mater. The corridor that led to her office was decorated in extraordinary fashion. Diagrams of family histories, with squares representing males and circles, females, covered the wall for fifty feet on one side and fifteen on another. Filled-in squares and circles recorded diseased individuals, whereas hollow ones represented healthy people. The lines that connected them and lines with diagonal slashes through them indicated marriages, births, and deaths. Yellow, orange, green, and brown dots also decorated the wall; I asked her their meaning. They signified individuals of high priority—their blood must be sampled. Lowest priority presumably fell to those whose symbols were now hidden behind filing cabintes!

These individuals lived on a remote shore of Lake Maracaibo, Venezuela, in an impoverished fishing village where wood and tin houses were constructed on stilts that were anchored in the shallow waters of the lake. In every year since 1981, Dr. Wexler has led a scientific expedition to the area, spending weeks taking blood samples and cataloging the family relationships.

The natives of the region have not attracted this attention for any usual reason of anthropology or sociology, but because they represent one of the world's largest local concentrations of sufferers of Huntington's disease. In that settlement, over one hundred cases exist in a community of several thousand, with one thousand more suffering a substantial risk of developing the disease. "The families are large—15 or 18 children is typical—and they are beautiful for genetic studies," Nancy Wexler said in *Science*.[2] She made it clear to me that they were equally beautiful to her as humans and that she was very involved in their lives and their cause.

Pictures of Venezuelan children covered her office wall. "One of the compelling aspects is the number of children who are at risk. They are very lively, cheerful and beautiful children," Nancy told *Columbia*. "What has been very difficult for all of us is that you know that over three hundred of them will certainly die if we don't find a cure."[3] She emphasized to me that genetic research offered them the only chance they had.

The incidence of Huntington's disease is much lower on a global basis, with perhaps one sufferer for every twenty thousand people. In the United States, twenty-five thousand cases are known. The DNA samples of the Venezuelan community have provided the extensive family data needed in efforts to find markers for the disease-causing gene. In fact, the entire afflicted Venezuelan group represents one large extended family. They are all the descendents of Maria Concepcion Soto, who lived in the area 150 years ago. We still do not know how she got that gene.

A number of workers have suggested that the disease had a single European origin and then spread through migration and exploration. An ancestor of Maria Concepcion Soto probably introduced the affliction to Venezuela. At present, concentrations of sufferers are scattered from Wales to Tasmania. Many cases in the United States have been traced to two brothers and another relative who emigrated from Suffolk, England. Several of their descendents were subsequently burned as witches in New England. As in the case of cystic fibrosis, the incidence is low among blacks and Asians. The group that Nancy Wexler investigated had first been reported in the 1960s by Americo Negrette, a Venezuelan physician and poet. He had followed up rumors that individuals from a particular village consistently showed drunken behavior.

The staggering, swaying, discoordinated movements of a victim of Huntington's disease have invoked thoughts of intoxication as much as dance. Both terms were used in George Huntington's original article. What we have gained in a century plus are first-hand descriptions of the disease circulated by groups such as the Huntington's Disease Society of America:

> When I walked I staggered. . . . People actually took their kids to the other side of the street because they thought I was drunk. . . . Several times I saw the same movie twice because I hadn't remembered that I had seen it the first time [this was due to loss of memory]. Although I love my children, if I had to do it over again I would not have children and I don't recommend that others do either. The suffering is too high to bring others into it also.[4]

In most cases, this awful disease first strikes the victim when he is in his prime; the usual onset is at age thirty-five to forty-five, although much younger and older instances are known. Gradually, over a decade or two, he loses everything, body, mind, and personality. It begins with small abnormal movements, twitches, and clumsiness. "Gradually," Nancy Wexler said in *Science*, "the entire body is encompassed by adventitious movements. The trunk is writhing and the face is twisting. The full-fledged Huntington's patient is very dramatic to look at."[5] Movement difficulties are only one part of the symptoms. Mental faculties such as memory and ability to reason and think clearly also go downhill. Victims suffer paranoia and delusions and show violent fits of temper. Of those diagnosed for the disease, about a quarter attempt suicide; six percent succeed.

The interpersonal relationships of those who endure the disease also suffer. A clinical psychologist by training, Nancy Wexler has interviewed a number of the adult children of Huntington's patients. All of them had suffered strained and disagreeable relations with the ill parent before the disease was diagnosed. A number were relieved to learn that it was because of the malady that the parent was ill-tempered and not their fault. A 1980 study of fifteen wives of Huntington's victims reinforced this picture: The wife became inextricably involved in the disease and suffered continuous trauma from it.[6]

In the later stages of the disease, the sufferer loses the ability to walk and talk and requires institutionalization. Autopsies have shown shrinkage of the brain areas involved in motor control and a loss of brain cells. In sufferers, a hole was literally created within the brain.

A FAMOUS VICTIM

The most celebrated individual who succumbed to Huntington's disease was folk singer, song writer, and political organizer Woody Guthrie, who died in 1967 at age fifty-five. In his earlier days, he had wandered through the United States, sharing the struggles and sorrows of the vagabonds he met and organizing workers' movements. His mother had died of Huntington's disease at age eighteen, but he had been misinformed that only women acquired the disease. As his illness progressed, he gradually lost the abilities that had characterized his life: to walk, talk, write, play the guitar, and sing. His last wife, Marjorie, has written a moving account of his struggles at various stages:

> I began to notice that his anger seemed longer or more severe than before. . . . His beautiful handwriting was deteriorating. . . . He could no longer stand and could no longer walk or come home.[7]

Woody Guthrie wrote over one thousand folk songs including popular favorites such as "This Land Is Your Land," "Roll On, Columbia," and "Hobo's Lullaby." The title that most aptly caught his own plight, however, in a way that he did not foresee was one called "Hard Travelin'."

The hereditary conditions that we have discussed so far have made their presence known quite early and usually killed the young. If Huntington's disease had these characteristics, however, it would not be with us. For this is an example of a dominant genetic disease, one in which a single defective copy is sufficient to bring on the illness. In fact, when sufferers were found in the Venezuelan community who had two defective copies of the relevant gene, no difference in symptoms could be noted. The second flat tire on the bicycle still left it immobile. The late onset of Huntington's permits those who carry the gene to have children and pass it on before they are stricken.

The one-bad-gene-is-enough situation, together with delayed onset, alters the mathematics of inheritance and creates a different set of circumstances than we found with cystic fibrosis. We do not have unaffected carriers who harbor the gene and show no effects. Instead, we have those suffering from the disease and their relatives. The latter can be divided into two groups:

1. Those who have not received the gene. They and their children will lead normal lives.

2. Those who have received it. They may be healthy now, but can expect the symptoms to commence, sooner

> or later. If the parent of an individual has the disease, the chances of his or her falling into group 1 or 2 are fifty–fifty, exactly even, depending upon which of two copies of the gene was drawn from the afflicted parent.

Until the 1980s, there was no way for a person in this situation to tell which category he or she was in until the disease began, if it ever did. This could create a situation of extraordinary anxiety. Every misstep or dropped object might be taken as a sign of the onset of Huntington's. By the mid-1980s, however, the efforts of James Gusella at Massachusetts General Hospital, Nancy Wexler in Venezuela, and others had uncovered closely linked markers. A test was available.

The possibility had been forseen a generation earlier by J. Bell and J. B. S. Haldane in 1937. In their pioneering paper in which they reported the first human gene linkage between color blindness and hemophilia, the authors speculated:

> The present case has no prognostic application, since haemophilia can be detected before colour blindness. If however, to take a possible example, an equally close linkage were found between the genes determining blood group membership and that determining Huntington's chorea, we should be able, in many cases, to predict which children of an affected person would develop this disease, and advise on the desirability or otherwise of their marriage.[8]

The authors did not comment on the possible psychological consequences of the information. The possibility of early diagnosis created another set of agonizing circumstances. To test or not to test? Nancy Wexler commented to *Genetic Enginnering News*:

> We now have a test that can tell people they are going to die of Huntington's disease and there is nothing they can do about it. . . . The requirement for psychological counseling is monumental and critical.[9]

For some, the suspense is unendurable. They wish an answer, one way or another. Life-style choices must be made. Should they start a family? Should they begin to save money toward the heavy medical costs that would accumulate with the disease? Or perhaps they will be cleared and need no longer live under a cloud. Janice Blenkharn was one such person. Her story was described on BBC television and in Robin McKie's *The Genetic Jigsaw*. Many family members on her mother's side were afflicted, and she was concerned about her two sons. She had been unaware about her risk situation when she had her children, but felt guilty about the possibility of having passed the gene along: "Because of the guilt that I have, I feel if I do have the opportunity, I have to take the test, and put them out of the misery of being at risk."[10]

Unfortunately, the test was positive, but Janice was able to resign herself to the situation. "I have no regrets at all. I feel better. There are people to help the children, and by the time they are adult, there will be even more things to offer them. So I'm quite content with the result."

Another case was that of Karen Sweeney, which was described by Alan Newman in an article in *Johns Hopkins Magazine*. Karen's mother had been diagnosed for Huntington's at age thirty-three, and Karen was now in her late twenties. Karen's at-risk condition caused her much distress. "Whenever Karen forgets to lock a door or shivers from the cold or drops a glass they [she and her husband] think of the disease."[11] Like Janice, Karen had also wanted an answer when a test became available.

She approached her results with overwhelming anxiety: "My heart fell through the floor. Fear went through me. A week from now, to learn, am I going to live or die. Am I going to end up like my mother. All that went through my mind, 24 hours a day." At the end, she and her husband were asked again if they wanted the information. They again said yes. The result was negative; the copy of the relevent chromosome that she had inherited from her mother was the unaffected one. With ninety-five percent probability (there was a five percent chance that crossing-over took place between the marker used and the Huntington gene), she was free of the disease. Their reaction was joy, crying, screaming, jumping joy. Afterward, she said: "After 28 years of not knowing, it's like being released from prison. To have hope for the future . . . to be able to see my grandchildren."

DIFFICULT CHOICES

The new tests have raised ethical problems as well as personal ones. The ones we raised earlier dealt more with the status of the unborn. Huntington's typifies another category where adults are affected. Debates have been held on questions such as insurance and employability.

For example, consider someone who has learned from the Huntington's test that he will get the disease. He faces formidable medical expenses in the future. Should he be eligible for insurance on the same terms as any other citizen? Should he volunteer the test information or provide it if requested? These questions are troubling, but many who have spoken on them have suggested that no really new issues are involved, and I agree. I have tried to sum up this reasoning:

Insurance has been designed to share costs in a manner that is proportional to risk. The acceptance of those who are at unusual risk at normal costs would penalize the other subscribers. Smokers, fliers of private aircraft, those of advanced age, and other risk groups may be charged additional premiums. Exceptions have been made on the basis of gender, for example. These exceptions, however, were made by legislators, not insurance brokers, and imposed uniformly on all companies. Someone who conceals information to secure more favorable rates is in the same position as a Wall Street trader who profits from inside information on stocks. (Nancy Wexler agreed with this diagnosis enthusiastically, but added that one could sympathize more with the motives of the patient than the profiteer.)

If society decided, out of compassion, to shield Huntington's sufferers from medical costs, one alternative would be to establish a separate program outside of the normal insurance. The same considerations would apply to victims of many other diseases. Nancy Wexler felt that this would become more likely as tests became available that predicted other conditions, such as cancer and heart disease:

> Suddenly, a new group may become uninsurable by virtue of genetic flaws and this group may include Congressmen, captains of the insurance industry and other policy makers. Even a national president may not be immune. When enough of the right people are uninsurable, new forms of coverage may emerge.[12]

Different questions come up when someone at risk for Huntington's wants to be admitted to a training program. Should someone who may have the disease be admitted to a lengthy residency to prepare her for a career as a brain surgeon? Her service to society would be shorter than normal, and one or more disasters might occur at the onset of the illness.

Cost, rather than principle, might be the crucial factor here. No absolute right to employment despite handicap has been established in the United States. The blind are not employed as airline pilots or bus drivers, for example, although elaborate radar systems and autopilots with touch-responsive control boards might make this a possibility. On the other hand, most people agree that work should not be denied for trivial or exaggerated reasons. Individuals with sickle-cell trait were not allowed to become pilots in the 1970s, a decision that most experts now recognize as a blunder. For cases that fall in between, some balance must be struck with compassion on the one hand and cost on the other. If economic costs are involved in making a profession accessible to someone who would not normally qualify, then the funds spent will be unavailable for alternative purposes, for example, research into the disease or sharing of medical costs.

Many additional ethical questions come up in diagnosing those who are at risk. At present, the test requires the cooperation of relatives. What if they do not want to be tested? If a person and his or her parent were both at risk and neither knew, then the child's positive test would also establish the parent as positive. What if the parent wants to remain in ignorance? Should the child be tested?

Another case came up when prospective adoptive parents and the sponsoring agency wanted a child tested before adoption. The child's mother had the disease. The geneticist concerned, Michael Conneally of Indiana University Medical Center, refused. He felt that the decision to know or not to know belonged to the child, who was too young to make it on his own behalf.

For many, the danger of knowing is worse than the uncertainty of living in doubt, as a positive diagnosis carries with it a high risk of depression and even suicide. One of those who has opted for this status is Arlo Guthrie, Woody's son and a successful folksinger in his own right. Arlo commented on the television program "Nova,"

> I have learned to live every day without looking over my shoulder—without looking back. I'd be doing exactly what I'm doing whether I had Huntington's or didn't have Huntington's. The biggest genetic disorder of all is death. And everybody has that programmed into them. I'm not

> unlike everybody else who's walking around this planet. The only difference is that I might leave it sooner than other people.[13] He has four children.

Another at-risk individual expressed similiar sentiments to *Time*,[14] "Nova,"[15] and myself[16]: "Before the test you can always say 'well, it can't happen to me.' After the test, if its positive, you can't say that anymore." . . . "I'm happy now doing what I want to do. Sooner or later I'm going to know anyway. I don't think that I would be much happier or do anything differently if I knew that I was free of the disease." . . . "There would be greater distress to know that I had it than there would be happiness to know I was free of it." These quotes came from someone who has had every opportunity to be familiar with the test, the disease, and its consequences: Nancy Wexler.

She was twenty-two when she learned that her mother, Lenore, had been diagnosed for Huntington's disease. She had graduated from Radcliffe with specializations in English literature and clinical psychology and had started graduate work in the latter field. Her father, Milton, brought her and her sister together and told them the news. "Until then, it was a very, very hush-hush," despite the fact that three uncles and a grandfather were also stricken. "My father sat my sister and me down and really laid it on the line."[17] He also told them that they were going to fight the disease. A successful psychoanalyst with a strong positive nature and considerable financial resources, he was not just delivering niceties. He founded the Hereditary Disease Foundation, one of the organizations that has mobilized a national effort to combat the illness.

One of the decisions that Nancy and her sister made was not to have any children. Another decision that developed more gradually was for Nancy to base her career around the struggle with the disease. As we have seen, much of her research has centered about the Venezuelan community. Upon her father's retirement, she also replaced him as head of the Hereditary Disease Foundation. In this role, she helps distribute research funds to research groups concerned with the disease. Funds can be obtained to supplement those provided from the greater resources of the National Institutes of Health and may be used to pay the salary of a Ph.D.-level research scientist, for example.

Nancy has used her position to steer the Huntington's community away from the quarrels that marked the cystic fibrosis gene search. The key word in the Huntington's effort is collaboration. Those supported by the Hereditary Disease Foundation are requested to share data freely with one another. Communication is aided by frequent meetings called by the foundation. Once a year, a gala is held at the home of the actress Julie Andrews, who is sympathetic to the effort. Funds for such occasions are donated separately and do not deplete those raised for research purposes.

The price to be paid for extra support and social amenities comes from the egos of the investigators. When the gene is discovered, the announcement will be made and paper published on behalf of the collaboration rather than any subset of individuals. One of the participants, Francis Collins (whom we met earlier as part of the cystic fibrosis story), summarized the atmosphere: "Knowing her status, you can't look her in the eye and say 'I can't work with so and so.' Her drive and enthusiasm holds it together."

The one participant that has not cooperated has been the gene itself. It proved quite willing and encouraging in the earliest moments, but since then has turned elusive and fickle.

The first efforts, early in the 1980s, were to find linked markers and so learn which chromosome it was on. Some pessimists felt that this search by a trial-and-error strategy might prove exhausting. As James Gusella commented to *Newsweek* in 1984: "We figured we'd have to do a few hundred probes, but on the twelfth one we found it. We were lucky."[19] The first found probe was 4 million base pairs away and established its location on the short arm of chromosome 4. Closer markers were then found, but only on one side of the gene. They allowed testing to be carried out with ninety-five percent and then ninety-nine percent probability of accuracy. The gene was reported to be in band 4p16.3 and within 325,000 and then 100,000 base pairs of the very chromosome end. It was near the very end, perhaps lurking within the repeats that proliferate near that structure. An alternative possibility, suggested to me by Robert Moyzis of Los Alamos National Laboratory,[20] was that Huntington's might be caused by a defect *in* the end structure rather than by a gene defect. Finally, a group headed by Hans Lehrach in London reported that they had cloned the end of chromosome 4, capturing the last 115,000 base pairs within a synthetic chromosome made of parts donated by yeast (this type of construction is called a *YAC*, or yeast artificial chromosome).[21] Location of the gene itself seemed close at hand.

At this point, a new set of data arose that contradicted the first. The relevant gene was not near the end at all, but several million base pairs inside. Other reports confirmed this, but the original information would not go away either. Both sets of data looked valid, but contradicted each other. At present, the location some distance from the end seems to have the inside track. In Nancy's words: "Everyone seems to feel that it's down below but it's not definitely ruled out that it's up there."[22]

We have a situation that is common in science: ambiguity. Some subtle mistake has confounded one group of measurements, leading to a misinterpretation, or both are somehow correct, and the Huntington's disease-causing area can lie in two different environments. Contradictions of this sort can occur in the normal course of research; eventually they are resolved. The development is particularly frustrating in this field, because of the continued suffering of those who have the disease or who live, like Nancy Wexler and Arlo Guthrie, under the cloud of it. At some date not too far away in the future, we will be spared this particular type of frustration.

When the Genome Project has been completed, the sequence of the terminal few million base pairs of the short arm of chromosome 4, like all other human chromosome areas, will lie within a computer, for access as desired. Clever programs will also have identified the likely gene areas and perhaps drawn some preliminary conclusions about possible gene functions. To find the gene for Huntington's, one would have to examine spelling differences in the most likely areas for a number of Huntington's victims and necessary controls. A number of strategies already exist that permit a scientist to scan the text of an individual in an area where the common sequence has been determined and quickly locate changes from the general pattern. If Huntington's victims consistently show changes within a particular area that do not occur in healthy people, then that region will be flagged as the likely gene. In fact, if the search for the Huntington's disease gene by conventional methods should lag, then the short end of chromosome 4 would be a likely choice for one of the early massive sequencing efforts of the Genome Project, so that the above strategy can be implemented.

When that gene has been located, diagnosis will be possible without the need for a family history, but no immediate cure will necessarily result. Some critics of the Genome Project have pointed out that no remedy exists yet for sickle-cell anemia, although the cause has been known for some years, a situation that can create great despair and frustration. At such times we must remember that scientific progress cannot be taken for granted. It requires time and human effort.

In 1865, art, music, government, and international politics were in no worse shape than they are today, perhaps better. In science, however, Mendel was first formulating his rules, DNA had not yet been discovered, and spontaneous generation was still a live issue. Science progresses, unlike some other areas of human activity. We have moved only one generation beyond Watson and Crick, but in that time, we have already learned how many proteins work, the way they are assembled, and the genetic code that controls the recipe. We have developed ways to read genetic text on a massive basis, and when we choose, we shall be able to record our entire genetic plan.

Someone who presumes that some magic wand will be waved once a gene has been discovered and a cure found is dreadfully naive. Scientists have no more hours in their day than anyone else, and they spend many of the ones they have at work. Those who have spent time in a laboratory learn that it takes much more than the pouring of liquid from one test tube to another to move things forward.

A cure will be found for sickle-cell anemia and Huntington's as well when enough effort and intelligence has been invested. In the interim, research progess justifiably brings hope. Nancy Wexler put it memorably at a recent meeting: "Hope is the only diet for the dying." If she who is at risk has patience, we can afford to have it as well.

Not everything need be taken on faith, however. We can look at a case where modern gene technology has brought the needed product to market.

10

THE ILL-MATCHED PAIR

GREAT NECK, LONG ISLAND, NEW YORK, 1988

I was attempting to open a stuck window in my home. It did not yield easily, so I struck the frame with my right palm to loosen it. I heard the crash of shattering glass; a pane had broken. More alarming was the red fluid that was coming out of a semicircular gash in my palm. I felt no pain as I wrapped a towel around my hand to control the flow of blood. Fortunately, my wife was home. She covered the wound with gauze and tape and drove me to a nearby hospital. In the emergency room, a physician removed tiny fragments of glass with a tweezer and closed the cut with stitches. I wore a bandage for some days before having the stitches removed.

The most important emergency measures had already been completed before I reached the hospital. My body had called upon an internal technology, perfected by hundreds of millions of years of evolution, to clot my blood and prevent its further loss after the cut. There is no simple on–off switch that governs the process, but rather an elaborate set of checks and balances involving two separate, but cooperative pathways and many proteins. It would not do to have the process triggered accidentally; internal clots forming in critical blood vessels can kill.

An early step in the clotting process involves small sticky cell fragments called *platelets*. They are always present and clump together at the site of a wound to afford a partial, fragile seal. To close the opening more firmly, a network of insoluble fibers must be laid down, binding the platelets firmly. This key step requires the cooperation of a number of blood-clotting factors that stimulate one another. The entire series has been called a *cascade*.

It is a pleasure to watch any well-rehearsed cooperative group do its task, whether a theatrical company or a basketball team. My conscious mind could not watch my body executing the steps smoothly as I rode to the hospital, but I could note the final result: the bleeding stopped. Accidents over the years have lulled me into security, and I take this result for granted.

As we have learned from our household applicances, however, the more working parts a machine has, the more opportunity for one of them to fail. When we examined the disease sickle-cell anemia, we saw that the failure of an individual hemoglobin molecule would be unimportant. A genetic design flaw in which every copy of the molecule produced was defective would be much more serious. The same rules govern the clotting of our blood.

A LONG-KNOWN MALADY

Hereditary blood-clotting diseases have been known through history and have different symptoms and patterns of inheritance, according to the particular substance that is flawed. They affect perhaps one in every ten thousand to twenty thousand males, but far fewer females. We will focus on the one that has caused the most human misery, accounting for perhaps eighty percent of the total: hemophilia A. It results from a defect in the production of blood-clotting factor VIIIC.

The role of factor VIIIC in the cascade gives no clue as to its clinical importance. It appears to operate as just another member of the team. It enters the action as two protein fragments carried by a much larger protein. This complex, which is called *factor VIII* (without the C), assists another substance, factor IXa, in its role of turning loose yet another one, factor X. The unleashed, motivated factor X (now called Xa) races off in turn to awaken another factor, and so on. The entire sequence is enough to give a headache to even the most motivated medical student. The importance of factor VIIIC may come from its low concentration; it may represent the weakest link in the chain.

Whatever the explanation, the results of factor VIIIC insufficiency have been dramatic through history. (I cheat slightly here. Before this century, hemophilia A could not be distinguished from the less frequent hemophilia B, which is due to factor IX deficiency. I will presume that the cases I cite were the former.) One instance described in the Jewish holy writings, the Talmud, dates to the second century A.D. Three sons of a woman had bled to death from the small wound produced by the circumcision ritual. The distinguished scholar Rabbi Judah then excused her next son from the rite. More general rules were codified afterward describing the family circumstances under which circumcision should be waived.

Unlike sickle-cell anemia or cystic fibrosis, hemophilia has selected no ethnic group for special attention, but has appeared, often without warning, in the various branches of the human family. We have already described its notorious intrusion into the heredity of Queen Victoria of England. For a more contemporary and vivid description of the effects of the disease on an individual, we can turn to Robert and Suzanne Massie in their book *Journey*.[1]

They were forced to become familiar with the disease when their son Bobbie was diagnosed with hemophilia at an early age. Bobbie had been born with many bruises and bled much longer than normal when he was circumcised and a blood sample was taken from his toe. There had been no previous occurrence of it in Suzanne's family in nine recorded generations of family history. They learned that the severity of the disease varies and that sufferers do not normally bleed to death from the slightest scratch. If the cut is bound tightly, the other components take over, and healing (not necessarily true clotting) slowly takes place. Cuts in awkward places such as the mouth are more difficult.

Much more serious is the internal bleeding, or hemorrhages, caused by blows and injuries. The pressure from the flow of blood into internal spaces can damage organs. When the fluid is trapped in the confined space of joints, the pressure can cripple the victim. More severe cuts and injuries can kill. Through history, more than half of the sufferers died before age five, and only eleven percent survived to twenty-one.[2] Parents and relatives of hemophiliacs could go frantic in trying to protect the child. The Spanish royal family padded the trees in the park in which the afflicted prince played. As we have seen, Empress Alexandra hired sailors to watch Czarevitch Alexis constantly. The Massies could not help but note that Alexis's affliction was milder than their Bobbie's, with "strikingly fewer bleeding episodes."

Bobbie did have one important advantage over Alexis, however; by the late 1950s and early 1960s, suitable blood transfusions could be supplied to hemophiliacs in emergencies. Physicians had learned to match donors to recipients so that no destructive reaction against the donor's blood took place. The fragility of the relevant factor was also recognized, and donated blood was stored in the deep freeze. With these precautions followed, someone else's factor VIII could rush to the rescue in a life-threatening situation. Thus, when Bobbie bumped his forehead in a car and bleeding in his brain resulted, transfusions given every six hours in the hospital pulled him through. At age eight he was also able to survive a fall in a restaurant that broke his arm, again through transfusions. This resource has allowed hemophiliacs to survive into adult life.

Alas, transfusions have their cost, and as Bobbie grew older his need for blood increased. Joint problems in particular required massive amounts. The Massies became dependent on the donated blood of relatives and friends and the efforts of their church, which organized a collection drive in Bobbie's behalf. However, begging for blood was psychologically hard, and they found that tending to their son's disability had a crushing effect on their ability to lead normal lives in other ways. It was hard to have conventional friendships, and they had little patience for superficial activity. In their particular case, however, they found an ingenious way to ease their situation. Robert Massie wrote the best-seller *Nicholas and Alexandra*[3] using his personal experience with a hemophiliac son to enrich his biographical account of the tragedy of the last czar and empress of Russia.

In one way the experience of the Massies matched that of Nicholas and Alexandra and almost all others in history: Only sons, not daughters, suffered from the disease. The Massies differed from the czar and empress in that they had no previous history of hemophilia in their families. In that respect, they shared the experience of Alexandra's grandmother, Queen Victoria, and her husband, Prince Albert. Perhaps two cases of five of hemophilia arise through a new mutation in a family that has no previous history of it.

The overall pattern of hemophilia inheritance now seems quite clear, but it took many centuries of patient observation to unravel it. If a woman carries a gene for hemophilia, she will remain healthy, but be a carrier. Her offspring, male or female, have one chance in two of receiving it. If they do not, they and their descendents are clear. If a son inherits a gene for hemophilia, he will suffer from the disease. All of his daughters will be carriers, but his sons and their offspring will be free of it. A pattern of this type is caused by a sex-linked recessive gene.

Queen Victoria had nine children. Only one of the four males, Prince

Leopold, inherited the disease from her. (He could not have received it from his father, because if Albert had carried it, he would have shown the symptoms.) The other males, including the future King Edward VII, were luckier. Leopold survived only to age thirty-one, but before passing on he had one child, a daughter Alice (do not confuse her with Victoria's daughter by the same name). She was a carrier, and her son Rupert was a hemophiliac.

Of the five daughters of Victoria, we can deduce that two were carriers. One daughter, Beatrice, had two hemophiliac sons and a carrier daughter who married King Alfonso XIII of Spain. Thus, hemophilia was passed on to sons of the royal family of Spain. Another daughter, Alice, married Grand Duke Louis, ruler of the German principality of Hesse-Darmstadt. One of their sons, Frederick, suffered the disease, and two of their daughters were carriers. One of them was Alexandra, who married the czar of Russia.

In *Nicholas and Alexandra*, the advisability of this marriage is brought into question. A number of Alexandra's relatives had demonstrated signs of the illness at the time, and medical opinions that advised against marriage in such cases had been published. Geneticist J.B.S. Haldane commented in 1939: "Kings are carefully protected against disagreeable realities. . . . The hemophilia of the Tsarevich was a symptom of the divorce between royalty and reality."[4]

Robert Massie suggested that Nicholas and Alexandra most likely considered the disease to be "a matter in the hands of God" rather than a textbook illustration of the laws of genetics. Queen Victoria appears to have had the same attitude. When one of her grandsons died of the disease, she wrote: "Our poor family seems persecuted by this disease, the worst I know." In fact, the rules governing the inheritance of the gene for factor VIIIC are quite analogous to those describing the white-eyed mutants of the fruit fly that Thomas Hunt Morgan investigated during the lifetime of Czarevitch Alexis. Both were cases of sex-linked inheritance.

THE ROLE OF SEX

If we return in our minds to the library that represents the forty-six chromosomes of humans, we will find that all but two of the volumes are paired, matched in size and content. We will call them the matched chromosomes (*autosomes*). The remaining two, the sex chromosomes, make an ill-matched pair, with the X chromosome much larger in size and gene content than the Y. As their name implies, they govern sexuality.

As a male, I have one X and one Y chromosome in my cells; the presence of the Y determines my sex. My wife has two X chromosomes. The sperm cells that I produce contain only twenty-three chromosomes. They draw a copy of each of the twenty-two matched pairs and either an X or Y at random. The egg cells of my wife have one of each of the twenty-two matched chromosome types and an X. My son Michael has inherited an X from my wife and a Y from me. According to current knowledge, the selection of an X-carrying sperm or a Y-carrying sperm takes place at random. Had an X-carrying sperm won the race, I would have had a daughter instead of a son.

The Y chromosome that males possess is a rather strange affair. It ranks among the smallest ones, but it can vary in size among individuals and ethnic groups. For example, a recent genetics text noted that this arm "is substantially longer in Japanese males than in males of most other populations."[5] This variability resides in the long arm, which is comprised largely or entirely of repeating sequences (two families of longish repeats make up seventy percent of the Y chromosome).

The short arm of the Y chromosome, while not richly endowed with genes, has a more significant role to play. A region of about 2.5 million base pairs at its end is closely matched in sequence to the corresponding portion of X and can trade material with it. These portions of X and Y behave as if they were part of a matched chromosome pair. In one male who had lost this part of his Y chromosome so that no crossing-over could take place during sperm production, only immature, infertile sperm were produced.

The end of this matched area of the Y chromosome is marked by an Alu family member. Beyond the boundary is men-only terrain. No exchange with the X takes place. This area must contain the gene or genes that determine maleness. Most scientists think that a single gene may be involved. Its product would be a master substance of some type, whose presence affected the activity of many other genes. In jargon, it has been called *TDF*, the testes-determining factor. Some excitement came up in 1989 when it was thought that this gene had been found. A Y chromosome gene was located that produced a *zinc-finger protein*. This structure is a trademark of many proteins that act to control DNA. The excitement died down when a very similar gene turned up on the X chromosome, and some individuals were discovered who had lost that gene from their Y and were nonetheless males.

The field was not set back for long. In 1990, three groups reported that a new candidate gene had emerged.[6] This contender nestled a bit closer to the border of the male terrain than the earlier one and also seemed to have the right stuff. It was probably also a gene that grasped DNA, although not necessarily with zinc fingers. The researchers went on to check whether the gene occurred in other mammalian species. They investigated pigs, mice, and tigers, among others. The candidate appeared in all cases. The *New York Times* headline reported "Scientists Say Gene On Y Chromosome Makes a Man a Man."[7] *Nature*, more conservative, titled its story "What Makes a Man a Man?" Their reporter, Anne McLaren, commented: "None of this evidence establishes conclusively that SRY [the new gene, named for the *s*ex-determining region of *Y*] is TDF."[8] She had a thought of her own, however, which strengthened the argument: Yeast contains a relative of the SRY gene, called *Mc*. In Gaelic, of course, *Mc* signifies "son of." Case closed.

In contrast to the Y, the X chromosome, 166 million base pairs long, bristles with genes. We might conclude from this that women have more functioning DNA in every cell than men, but we would be wrong. Within a few days of conception, most of one X chromosome in every cell in a female's body is shut down, or inactivated. The part that pairs with Y escapes this fate, and a few other genes do so as well.[9] Nature appears to be equalizing the amount of active DNA in a typical body cell of a female and a male, with each having one fully active X chromosome and a bit more activity on the Y or the other X. Although each cell obtains no advantage, the whole woman does. The chromosome that is selected for inactivation is chosen at random. In a group of cells in an organ, some will have the mother's X chromosome present, others, the father's. The products of both X chromosomes are available for use by her body.

To appreciate this difference, let us turn our attention back to hemophilia. To find the gene for factor VIIIC, we must travel far from the area on the short arm of the X chromosome where Y-pairing takes place, to band Xq28 at the very end of the long arm. We have now arrived in the heartland, the most celebrated terrain, of human genetics.

The sex-linked genes, those on the X chromosome, have been the most studied and best documented of all human genetic passages. Long before the arrival of DNA analysis, they could be followed because of the simplicity of the hereditary situation in males. A male has only one copy of such a gene available. If it is defective, he suffers a disease or condition. In contrast, the female has two differing copies. (It does not matter if each cell has only one in working condition.) If one is defective, the other will protect her, provided that the disease is recessive, as in

TABLE 1

Some X-Chromosome-Linked Diseases

Adrenoleukodystrophy (nervous degenerative disease)
Duchenne muscular dystrophy (progressive muscle deterioration)
Fragile X syndrome (severe mental retardation)
Glucose-6-phosphate dehydrogenase deficiency (anemia due to destruction of red blood cells)
Hemophilia A (blood clotting deficiency)
Green color blindness
Ichthyosis (thick, rough skin due to an excess of the protein keratin)
Lesch–Nyhan syndrome (mental retardation, self-mutilation)
Red color blindness
X-linked cleft palate

cystic fibrosis, not dominant, as in Huntington's disease. She is a carrier, but not a victim.

Thus, Victoria had one "bad" X chromosome to pass on (I am referring only to the factor VIIIC gene area) and one good one. Albert had only one good X chromosome and his father's Y, which was irrelevant to the disease. The future Edward VII drew Albert's Y and Victoria's good X. Unfortunate Leopold drew Albert's Y and Victoria's bad X. Victoria's daughter Alice obtained a good X from Albert but Victoria's bad X. That same bad gene ended up in Czarevitch Alexis's pool two generations later. Another daughter, Beatrice, received the same chromosome X as Alice and was also a carrier. In contrast, Victoria's other daughters drew good X's from both of their parents. Their lines, like that of Edward VII, remained clear of the disease.

Of course, this analysis will apply equally well to other X-chromosome diseases. More conditions have been assigned to this chromosome than any other (see Table 1). In late 1989, the number was 160, ten percent or so of the total of all mapped genes. The roster of famous (or rather, infamous) X-linked conditions includes the most prevalent kind of muscular dystrophy (on the short arm) as well as fragile X syndrome, the most common inherited cause of mental deficiency. The Xq28 area, some 10 million base pairs long, is particularly rich in genes related to disease.

I selected one paper arbitrarily at the 1989 meeting of the American Society of Human Genetics and found that the location of the gene for the disease described in it, hydroencephalus, was just down the block from that of factor VIIIC.[10] The illness involves "stenosis of the aqueduct of Sylvius" (translation: abnormal increase of water in the head of young children due to the narrowing of an important passage).

A better-known gene neighbor of factor VIIIC is red-green color blindness. This condition was the first assigned to a specific human chromosome. The connection was made in 1911 by E. B. Wilson, a colleague of Thomas Hunt Morgan at Columbia University. (By contrast, no gene was assigned to a non-sex chromosome until the late 1960s.) In addition, the linkage between color blindness and hemophilia was the first to be demonstrated for humans. Typical data of the type that was considered are reported in the genetics text by R. F. Weaver and P. W. Hedrick[11]: A couple, both free of color blindness and hemophilia, have six sons. Two are normal, three have both color blindness and hemophilia, and one is color-blind, but not hemophiliac. What kind of pattern, if any, does this represent?

First, we should note that color blindness, like hemophilia, affects males almost entirely. In white populations, one in twelve males are color-blind, but only one in two hundred females. Both genes are sex-linked, and the mother in the above example must have carried both traits. (If the father had either, he would have shown it.) Further, the traits stayed together in three of her six sons while in two, the normal genes for color vision and factor VIIIC comigrated. Presumably, both genes were relatively close together on her X chromosome, as they were separated by crossing-over in only one of the six cases.

Using an analysis of this type with a much larger data base, J. B. S. Haldane and his co-worker, J. Bell demonstrated the linkage in 1937.[12] Now, a half-century later, we recognize that the color and blood genes are no more than 1.1 million base pairs apart in band Xq28. Actually, two color vision genes are located there.

A single gene that describes a red pigment resides alongside one to five copies (the number varies from individual to individual) of the gene for green. Spelling errors can cause either a red or a green type of color blindness. The gene for a third pigment, blue, lies far away on chromosome 7.

The point of these surveys is not merely to conduct genetic tourism. In the case of factor VIIIC, the knowledge has been put to work. In the 1980s, the human race, after millenia of suffering, was able to lay its hands on the gene that was the cause of hemophilia A and use that gene to produce factor VIIIC. The advance came in large part through the efforts of biotechnology companies that had not existed a decade earlier: Genentech and the Genetics Institute. The Genentech work will be described.[13]

Because of its low concentration in blood, factor VIIIC had been known more through the effects of its absence than by its own properties. Finally, a preparation of the version of that protein present in cattle was reported. The starting material used was twenty-five thousand liters of animal blood. This source was hardly suitable for those wanting the human factor, but partly purified preparations could be obtained using less abundant supplies. Just enough of the human factor was obtained to allow the sequencing of a few amino acids. That slim lead proved to be sufficient.

Using the protein code, Genentech workers could deduce a likely string of thirty-six DNA letters that would code for twelve amino acids in factor VIIIC. A probe was prepared that would be used for text-matching against the entire human genome. The Genentech team needed to improve the odds in their favor as much as they could, so they looked for some source that might be enriched in the part of the text they wanted. The team managed to obtain DNA from a male with an XXXXY constitution (he had four copies of the X chromosome, such anomalies can result from errors in the production of sperm and egg cells). The attempted match succeeded. Using this probe, the scientists systematically scanned shred after shred of DNA (each actually had been cloned within a virus), until they found one that contained a portion of the factor VIIIC gene. Having obtained a foothold, they could then expand their territory by chromosome walking, until eventually they had collected and put together DNA passages that covered the entire gene and found that their quarry was much larger than expected.

Some years before the discovery of the even bigger (250,000 base pairs) cystic fibrosis and gigantic (2 million base pairs) Duchenne muscular dystrophy gene areas, they had landed the first of the giant genes. It came in twenty-six separate protein-coding units, spread over 186,000 base pairs. Although it coded a protein of 2,332 amino acids, ninety-five percent of the gene was taken up by the twenty-five interruptions. The largest of these, 32,000 bases long, has recently been found to contain a small separate gene on its own. This gene within a gene has no interruptions, runs in the reverse direction (it is coded on the other strand), and produces a protein whose function is unknown. This represents no isolated instance. The gene for neurofibromatosis (elephant man's disease), discovered in 1990, has three smaller genes nestled within its huge confines. The smaller genes raised false leads in the search for the larger, disease-related one. How scrambled the old beads on a string idea has become!

The protein product of the factor VIIIC gene is itself of unusual size, but this does not represent its final form. Once made, it is carved up and altered by various enzymes. The active factor in blood contains less than half the amino acids, divided into three separate pieces. It would seem a monumental waste to have only five percent of a gene specify the protein product and then throw away half of that product before the active form is attained. As we have seen, however, DNA stretches can fill many other functions in addition to the direct protein-coding one (control, structure, or, as above, coding for unsuspected auxilliary proteins). Any alteration in the existing text must be considered the possible source of a disease, which in this case is hemophilia.

In fact, the text of a number of hemophiliacs has been examined, and diverse sources of error found: simple spelling changes (ninety percent or more of the cases), deletions, the duplication of a part, and, in two cases, the intrusion of a LINE-1 sequence into the gene. Hemophilia, unlike sickle-cell anemia, does not bring compensating advantages with it, but the large size of the gene makes it a vulnerable target, and new mutations are constantly being introduced in each new generation. Queen Victoria and Suzanne Massie, in their separate centuries, both suffered events of that type.

The variety of changes that can lead to hemophilia makes diagnosis no easy matter, despite the fact that the gene is in hand. We must patiently accumulate data until the entire list of harmful changes has been brought to completion. Until then, partial reliance must be made upon linkage studies (with their need for family data) or even upon the older method of fetal blood-clotting analysis. Possession of the gene, however, brings an even more desirable prospect closer: effective treatment and ultimately cure.

By the 1980s, hemophiliacs could receive treatment not just with blood, but with blood preparations greatly enriched in factor VIII. With such remedies available, their life span could be brought almost to normal. However, new hazards were carried along with the life-giving blood preparations: hepatitis and AIDS. In a 1990 editorial by Marsha F. Goldsmith in the *Journal of the American Medical Association*, it was estimated that sixty to eighty percent of the 20,000 hemophiliacs in the United States had been infected with the AIDS virus.[14] Additional precautions were taken in purifying blood concentrate to eliminate these hazards, but these steps served to drive an already hefty price up to new and dizzying levels, ten thousand dollars or more per patient per year. At this level, the factor VIII preparation could only be made available in emergencies and not as a preventive treatment.

Conventional science had come to an impasse, but a new technology had come of age to open other routes to a solution. Once a gene has been captured, it may then be put to work. To do this, the gene and its control areas must be spliced skillfully into a suitable location of the DNA text of microbes or mammalian cells. If that job has been done properly, then the altered cells will manufacture the protein described by that gene. A material that initially was available in trace quantities can now be produced in amounts large enough to satisfy the needs of the public.

In the case of factor VIIIC, its production in mammalian cells came immediately after the isolation of the gene. This clotting substance could be now prepared by a route that avoided any possible contamination by AIDS or hepatitis. *Nature* in two articles praised these events as "a technical triumph without parallel"[15] and felt that "one of the most ambitious and exciting goals of the biotechnology industry has been achieved."[16] The companies were praised for their "fleetness of foot."[17]

Unfortunately, the feet of the biotechnologists ceased to be fleet at that point. The first isolation and expression of the protein came in 1984, but additional technical obstacles that have come up since then have prevented the production of a moderately priced, genetically engineered antihemophilia product. Ugly patent battles between those who hold the rights for the blood purification method and those who would use gene splicing have also obscured the nobility and humanity of the larger goal.

In time, however, all of these dust clouds will settle. The challenge is visible; only further ingenuity is needed to set things right. In the future, efforts to produce an ample and secure factor VIIIC supply will compete with those attempting to insert a good gene for that substance into the patient through gene therapy. We stand much better off than those of Queen Victoria's day when no path to combat the disease was around at all.

We have gained the ability to manufacture, as desired, all of the proteins of the human body, even the fragile, sparse, and ephemeral ones. Possibilities have opened up for the improvement of the outlook of many affected humans; not just those suffering from the famous diseases, but from others not yet named or even recognized. A taste of things to come can be had if we briefly sample one additional protein: human growth hormone.

11

AN OPPORTUNITY FOR GROWTH

A CLASSROOM ENCOUNTER

"Why should this be done? What practical use can possibly be made of this substance?" the young woman asked. She was in the habit of provoking my coteacher and myself with acidic questions framed around her theme that modern science was harassing the human race. The locale was the science survey class in the graduate science journalism program run by my friend Bill Burrows at New York University. Most of the students in the program come with an open mind, intent to learn what science is doing before they decide upon its value. A few, like the young woman I have just mentioned, arrive with preestablished ethical agendas that they would like to prop up with a sprinkling (not too many, if you please) of scientific "facts."

In this case, the response was dramatic and came from an unexpected quarter. In addition to the rotating group of scientists who pass through the course, two at a time, a journalist presides, grading the students and occasionally asking questions on his own. In this particular semester, we had a seasoned free-lancer who has written for the *New Yorker* and other magazines. He leaped up from his seat and faced the questioner, standing at his full five feet, four inches of height. "How do you think I would have felt if I were a foot shorter," he asked? "I would have been that if I hadn't been given human growth hormone."

We had been discussing the production of this hormone in bacteria by Genentech scientists using gene-splicing methods. Normally, it is made in the pituitary, a small gland at the base of the skull. The gene in question resides on the long arm of chromosome 17 in a landscape that has grown more familiar to us: It clusters with four brethren (active only in the placenta) in a 66,500-base pair stretch of DNA that is also inhabited by a LINE-1 sequence and forty-eight Alus. The growth hormone is a protein of average dimensions, 191 amino acids in length. If a child fails to produce this substance at the appropriate time, and no external supply can be had, he or she grows up to be about four feet tall, a pituitary dwarf.

The remedy is not a simple one. In the past, the deficiency could not be diagnosed until the early teen years. At that time treatments with the hormone would be necessary twice a week for several years. Unfortunately, animal hormones cannot be substituted; the substance is species specific. Until recently, the only supply came from extracts of pituitaries dissected from human cadavers. Fifty thousand corpses would yield enough hormone to treat perhaps two thousand children for a year. Until about 1979, the NIH performed the service and administered the product without charge. Then two European firms began to supply the hormone, but the cost was $45 per milligram, or about $9000 each year for one child.

The cost was not the only problem; an unrecognized hazard existed. The human pituitary extract supplied in the 1970s was not fully purified. In 1985, four people who had received the extract died of a rare viral infection, Cruetzfeld–Jakob disease. The incidence was much higher, and the age of onset lower, than in the general population. Was it due to contamination of the extract by virus from an infected donor? That question could not be settled. Although purer preparations were now in use, the Food and Drug Administration chose to be cautious on the side of safety and ban the procedure.

This step might not have been taken if another source of the hormone were not waiting in the wings. In 1979, Genentech scientists and collaborators had isolated the messenger RNA for the hormone, made a DNA copy of the message, and inserted it into a bacterial plasmid. Suitable bacterial control messages were spliced alongside the protein code, and the bacteria were induced to produce the human hormone.[1]

One small problem existed in the product: The growth hormone had been made with an extra amino acid, methionine, at the start of its chain in accordance with bacterial custom. It was anticipated that the bacteria would snip off this extra bit as usual, but they declined to do so. Fortunately, the 192-amino acid product had the same effect on growth as the natural substance. A modest fraction of those who received it in tests showed an allergic response of their immune system, but this led to no ill effects. The route by which it was made allowed for no contamination by disease agents from human donors. Production of the hormone was scaled up to the industrial level, and it was christened Protropin. The product was undergoing the usual slow review by the Food and Drug Administration when the cadaver-isolated hormone was banned. Not surprisingly, Protropin received accelerated approval within months of the ban. The cost was still high, about $35 per milligram, but now ten thousand to fifteen thousand children per year could be treated.

For Genentech, this proved a bonanza, its first true commercial success. Revenues from the hormone were $41 million in 1986, $60 million in 1987. Such a lucrative opportunity was not likely to pass unnoticed. Eli Lilly prepared its own product, Humatrope, with the correct number of amino acids. Patent wars followed. Much more interesting in the long run were indications that these events represented just the first rumble of an avalanche, biologically, commercially, and even ethically.

A variety of moral questions have come up concerning the use of human growth hormone. Gina Kolata had written in *Science* in 1986: "In the near future, it is almost certain that many affluent parents of short children will have their children treated with human growth hormone as a matter of course"[2] (that is, to improve the height of someone in the normal range rather than cure dwarfism). She reported an anecdote concerning one father who brought his boy into University of California at Los Angeles for testing. When told that the predicted height of his son was five feet, seven inches and that no treatment was necessary, he roared, "That's absolutely unacceptable."

Dr. Thomas Murray, in *Genetic Engineering News*, has carried the speculation further. He considered the supposed social advantages of height and the cost of human growth hormone and asked

> Will the children of the rich, on top of their other advantages, also be made taller than the children of the poor, and will that increase their ability to compete with and dominate generations to come?[3]

A moment of consideration may convince us that no new ethical problems are involved here. The right of the wealthy to purchase superior education and even private tutors for their children has been accepted in the United States. Education would certainly increase competitive and social advantages even more than height. It also has been established that expensive cosmetic procedures may be sold: nose jobs, tummy tucks, breast expansion or reduction, and so on. Yet no one has made an issue of the firm breasts and chins of the rich versus the saggy ones of the poor, if such a correlation exists at all. These issues could be brought up and debated once again, but there is no need to place them in the special context of genetic engineering.

Let us jump forward to two stories by Natalie Angier in the *New York Times* in July 1990. She quotes the wife of one Frederick McCullough, age sixty-five, on the attributes of her husband: "He does appear to have abnormal energy for a man his age. He never complains of being sick. Every day he feels fine." He was happily mowing the lawn, climbing on the roof for repairs, and washing the siding. Now all of us know of someone that age that shows comparable vigor, but Mr. McCullough had displayed none of it until he started his experimental therapy: six months of treatment with human growth hormone. He was not alone; twenty other men aged sixty-one to eighty-one had also had this treatment, and all showed some reversal of aging effects. The participants reported greater strength, better ability to work, less wrinkled skin, and measurable improvements in the distribution of their muscle size and body fat. They had lost almost fifteen percent of their body fat and gained nearly nine percent in lean body mass. One of those who organized the test, Dr. Lester Cohn of North Chicago Hospital, said, "The results are quite amazing."[4]

The men were not chosen at random, however. They were picked from the one-third of the population who have "almost worthlessly low levels of the hormone." The remainder of us continue to produce it at some level throughout life. We might not become more spry if we were given an additional supply, but then again, we might.

Other obstacles must be cleared before this hormone can be released, even for the use of those who will benefit from it. The usual search for unpleasant side effects must be made. (When a drug is manufactured in the laboratory rather than isolated from nature, however, an almost endless series of variants can be made that might get around any problem of the above type.) Further, some scientific error may have been incorporated in the study, although the fact that it was published in the *New England Journal of Medicine* and not just released to the media gives us some confidence. If we accept the possibility that the results are typical and correct, then we can suddenly envision new possibilities for the human condition.

The sight of a listless, old man would not ordinarily produce the same reaction in us as that of a dwarf, hemophiliac, or sickle-cell anemic. The latter three are thought of as victims of disease, the former is something that we accept as normal. Yet we understand little about the true role of human growth hormone. It may well be that its primary purpose is to govern the balance of muscle and fat rather than control growth. If so, then the less than robust aged may be the victims of a correctible genetic malfunction, unrecognized until now. Articles in the media hint at another application for the hormone as an antiobesity substance with a potential market of billions per year.[5]

We will not have a proper idea of what optimal human health means until we have fully explored the properties of this substance. However, it is not unique. A series of less explored growth factors have come to light recently with effects on the healing of wounds, fractures, burns, and ulcers. Other substances of this type restore immune defenses in cancer patients and even kill cancer cells. Dr. Michael B. Sporn of the National Cancer Institute commented recently to a *New York Times* reporter, "I think that [the growth factor field] is going to be the new frontier of chronic degenerative diseases, but that's going to take several more years."[6] He was talking about heart disease, arthritis, and other chronic diseases.

Other classes of vital human factor also await exploration. These substances have always existed within us and have great potency: They can kill cancer cells, affect our emotions, dissolve hazardous blood clots, stimulate the heart, perhaps delay aging, and bring us other benefits that we have not yet recognized. Many are ephemeral and present in low amounts, however, and so have eluded discovery.

Nature has been kind enough, however, to prepare a complete list for us with the exact amino acid recipe for each item included. Where is this complete compendium kept? In the cells of each of us, in the DNA of the human genome.

THE BODY PARTS LOTTERY

The 3 billion A's, G's, T's, and C's of the human blueprint will appear at first to make up one of the most monotonous arrays imaginable. Scattered among them, however, will be a few hundred million letters that furnish recipes for the one hundred thousand proteins of the human body. Experience and some clever computer programs will be needed to tell the protein-coding parts from the interruptions and other types of text. Ultimately, we will have our list of new proteins, and the gigantic lottery will then begin, a lottery not for cash, but for the ingredients of the human body.

I have called it a lottery because some items may represent treasures, others, trivia. No instruction manual governing its use is supplied with each new item, unfortunately. An anticancer factor or a mood-altering substance may lie alongside a structural component of toenails. At the start, we will not know which is which.

The properties of each protein are governed by the three-dimensional shape into which it folds after our body has made it. Our science is not yet good enough, however, to predict shape directly from amino acid order, and the shape alone may not be enough for a prediction of its role in the human body. In some cases, a comparison with the recipe of an established substance may provide some clue as to the function of a new one. However, many substances will be unique or the first member of a class to be discovered.

The list of new human proteins may provide a strong temptation for medical researchers and industrial speculators. They may wish to play the lottery. To enter, they need only select some unexplored substance described in our DNA text and prepare it for the first time. They may produce it by isolating it from human cells, or they may decide to prepare it by using gene-splicing techniques as we described for factor VIIIC and human growth hormone. More money, ingenuity, and labor may be needed in some cases than others, but if they want the product, they should be able to get it, one way or another.

Once they had it in hand, they could then explore what use it might have. Of course, the new substance might have only routine properties and be of little commercial interest. With thousands of items from which to select, however, they could try and try again.

If they were lucky, their selection might aid learning, prevent baldness, slow aging, abolish obesity, or do something else that was both wonderful and marketable. Even in such a case, they could not immediately collect their prize. Current patent laws would hinder them, according to N.H. Carey and P.E. Crawley of Celltech, Ltd.[7]

Under the present regulations, any patent application would have to be filed before the DNA sequence that described the product reached the public data base. The ultimate use for the protein would also have to be specified at that time, a circumstance that these gentlemen thought was "most unlikely." However they also felt that the "good sense of those who interpret patent law" would find some way around this difficulty.

Some observers have been less than thrilled at this prospect. Erwin Chargaff, as we have seen, has found many developments of modern molecular biology not to his taste; it would be out of character for him to change his viewpoint here. He described the proposed development of human proteins as "a gigantic slaughter house, a molecular Auschwitz in which valuable enzymes, hormones and so on will be extracted instead of gold teeth."[8] The metaphor is so spectacular that I regret having to quarrel with it. We are not slaughtered, however, when we donate blood or provide cell samples for scientific study, and the hospitals and biotechnology corporations that extract and use the information in them do so not to take human life, but to rescue and improve it.

Professor Chargaff's prose will prepare us, however, for the greater debate that took place, and continues, on the Human Genome Project. It is not a subject that invites detached discussion.

12

A PUBLIC UNDERTAKING

Walter Gilbert once wrote,

> The total human sequence is the grail of human genetics. It will be an incomparable tool for the investigation of every aspect of human function.[1]

In a similar vein, James Watson commented,

> A more important set of instruction books will never be found by human beings. When finally interpreted, the genetic messages encoded within our DNA will provide the ultimate answers to the chemical underpinnings of human existence.[2]

In this age of advertising, we learn to discount ambitious claims automatically. The proper cosmetics will rejuvenate our skin and remove our wrinkles, and miracle diets will melt away our weight without effort. We might not be swept away immediately by the seemingly self-serving claims of Gilbert and Watson, then, particularly when we remember that Gilbert was one of the discoverers of rapid DNA sequencing and Watson was among the two who had shown that human heredity is encoded in DNA. They could merely be promoting their own business. Let us now add a third comment by Senator Albert Gore, Jr., of Tennessee however:

> Many have said that the tools that will emerge from mapping the human genome will be the most important and powerful that science has ever provided, resulting in changes even greater than those brought about by atomic power or the computer revolution. I don't think those are overstatements.[3]

The last person cited has been a key member of the U.S. Congress, a body not normally known for its unrestrained enthusiasm in throwing money at new and large science initiatives. Of course, even the U.S. Senate may at times let in an eccentric or two, but Gore's comments were representive of general feeling. Biochemist Charles Cantor, a significant figure in the Department of Energy's effort on the genome commented: "I think Congress is very enthusiastic about the project. I have testified before the Senate and I was amazed at how knowledgable they were."[4]

The Senate's enthusiasm was backed by the commmitment of funds, and the establishment of a new center. The idea had first come up in public discussions in 1986; by 1989, necessary administrative structures were being put into place; in fiscal 1991, there is already funding in the vicinity of $100 million. The hoped-for level of $200 million per year has not taken place, but this delay may be just one more side effect of the financial chaos that has dominated budget discussions of the early 1990s.

Let us compare this history with that of another large science project: the Hubble space telescope. The idea was proposed in a 1946 report by astronomer Lyman Spitzer. (At that time, Oswald Avery had just shown that genes were made of DNA, and the Watson–Crick structure was still seven years in the future.) The National Aeronautics and Space Agency adopted the space telescope as a goal after 1969, but funds allocated for it each year were always insufficient, according to a report in *Nature*.[5] Only in 1983 was it taken seriously and properly funded. The *Challenger* explosion in 1986 and technical difficulties delayed its launch until 1990, however. By comparison, the Human Genome Project, which was launched in the same year, had done much better. Why?

One strong reason, as we have seen, is the obvious medical connection. By a trial-and-error process of groping about in the genetic text, the genes for factor VIIIC, cystic fibrosis, Duchenne muscular dystrophy, neurofibromatosis, and a number of others had been located. These discoveries provided a route to diagnose these diseases directly without reference to a family history. In the case of a positive test, the parents could opt for either life-style counseling or abortion as their own ethical systems permitted.

Once in possession of the gene, geneticists were also in position to produce the protein coded by it. In some cases, this material could be used to treat the disease. Looming on the horizon in many cases was the prospect of gene therapy to repair the defect within the patient's body.

Hemophilia, cystic fibrosis, Duchenne muscular dystrophy, and neurofibromatosis are controlled by a single gene; a flaw in a single protein causes the condition. They follow the rules of Mendel and Morgan, which permits them to be located by a fairly straightforward process. Geneticist Eric Lander of MIT commented recently: "Solving the single-gene diseases is straightforward—hard work but straightforward. . . . The frontier of diseases is more complex inheritance."[6]

He was referring to a larger set of maladies that are much harder to analyze. They are worth the effort, however, as they victimize millions rather than thousands at a time. They include cancer, heart disease, high blood pressure, diabetes, schizophrenia, and multiple sclerosis. Because they run in families, they have a genetic component. Several genes are involved, and the simple rules of genetics are not followed. Further, environmental factors can play an important role in determining whether the condition shows up or not. The same disease may strike at both of a pair of identical twins. This is not inevitable, however; it occurs twenty to fifty percent of the time.

Of these illnesses, cancer is the one that has been penetrated most deeply at the genetic level. Much study has been devoted to a class of genes, the *proto-oncogenes*, that work naturally within us to regulate growth. If they are damaged by certain types of chemicals, they may be altered in their functions. They switch and become members of a class called *oncogenes*, which can cause cancer.

Fortunately, we can defend ourselves against these hazardous chemicals. Other genes that we possess produce enzymes that can alter the chemicals and render them harmless. However, oncogenes may also be introduced into our DNA by viruses. Even when they are active within us, their action may be overridden by the activity of other defenses, the tumor suppressor genes, which act to keep growth within limits. One investigator commented, "The true size of this class of genes is unknown and at present unknowable,"[7] as they are discovered only when one of them is damaged in a patient and its protective action is lost.

It is hard to follow the progress of such a multigene game, particularly when no rulebook exists and it is not clear how many players there are and whose side each one is on. This does not mean that progess is hopeless, but just that it is slow. In order to connect a particular gene with a disease, more markers and greater numbers of family groups are needed. False leads abound and mistakes can be made, as happened recently in the case of manic-depressive illness.

To show that a complex net of threads of this type can be untangled, Eric Lander experimented using an unlikely organism: the tomato. He has deduced that six locations correlate with fruit mass, five with acidity, and four with the content of soluble solids in a tomato. He has also made a start toward locating the genes that are responsible. Experiments were run in which plants with desired properties were crossed, however, so his methods could not be used on humans. They could be applied to mice, though, when they suffer the same disease as we do, and the conclusions used to understand the human effects.

One stumbling block that has delayed research both for the simpler and more complicated genetic diseases has been the need to stop, search through, and sequence each new chromosome area that becomes linked to a disease. Daniel Koshland has estimated that $120 million was spent by the Cystic Fibrosis Foundation in four years on that disease alone.[8] If the human genome was read out, once and for all, there would be no need to repeat this struggle for each individual case. Research on all hereditary diseases, simple or complex, would speed up greatly.

The obvious medical rewards to be gained from the project have helped to protect it from any public outcry of the type that greeted the new gene-splicing techniques in the 1970s and the proposal to release genetically altered bacteria in the 1980s. The aim of the Human Genome Project has simply been to gather new information rather than to produce something or release an altered organism.

We have mentioned some of the ethical questions that have come up and the difficult choices that individuals may face when they are diagnosed with a disease that cannot yet be cured. Occasionally, individuals have wanted research to be delayed until the ethical questions are settled. Nancy Wexler of the Hereditary Disease Foundation, who is intimately familiar with this situation, has offered her opinion.

> If someone were now to suggest that the search for the HD gene be halted because finding the gene will expose many thousands of individuals who cannot now utilize current tests to the dangers of testing, the suggestion would be greeted with horror. People are free to choose the test or not but they are not free to choose if they will die of the disease. These families would be aghast at the notion of slowing down research. They don't have to know. They do have to die.[9]

Occasional letters of criticism have been written, but no Coalition to Preserve Our Unread Text has urged a moratorium on DNA sequencing, no laboratories have been vandalized, sequence compilations burned, or gels smashed. No city councils have felt their citizens imperiled and attempted to ban this type of work within municipal limits. The general public has either remained oblivious of the whole business or gone along with Congress.

This does not mean that the Genome Project has gone unopposed. Opposition has been voiced and continues to the present day, but it has come from a very surprising direction: other scientists. For example, Nobel laureate David Baltimore, now president of Rockefeller University, commented at an early stage, "The idea is gaining momentum. I shiver at the thought."[10] Other fears have been raised that I have brought together in a scenario to create a single bad dream.

THE GENOME CRITIC'S NIGHTMARE. Dr. Busychair sat in his office in the Midwest Sequencing Institute, shaken by the phone interview that he had just concluded. As associate director for science, it was his job to ensure that directives from the head of the institute, Dr. Grant Scooper, were carried out. His orders were clear: sequencing of chromosome 12 must be finished on schedule. Scooper would try to keep the results of the recent fiasco from the press or minimize it, but no time loss could be allowed. The budget for fiscal year 2004 was now being drawn up, and failure to come in on time might be disastrous. Pacific Sequencing, according to rumor, was ahead of schedule for chromosome 10. (It was so hard to get accurate information. The Pacific data base had proved impossible to crack, despite his colleagues' many efforts. By the time the actual information was released to Congress it would be impossible for Midwest to avoid an unfortunate comparison.) If Pacific *had* finished their job on time and this outfit had not, than one of Midwest's two unfinished chromosomes or at least an arm would be taken away from them and given to the West Coast Center.

This was made more likely by the growing importance of the West in next year's presidential elections. Even Midwest's long-term congressional patron, Senator Subsidy, was unlikely to rescue them this time. Subsidy had been instrumental in obtaining one of the coveted $60 million-per-year institutes for this state, which was a godsend after their defeat on the newest superaccelerator and the Mars Mission base. However, a display of incompetence now would threaten Subsidy's own reelection; he would have to distance himself. A scandal would result if the annual showpiece conference on the Conquest of the Gene had to be delayed.

Loss of even a chromosome arm would mean a twenty-five percent cutback in the permanent scientific staff that had been recruited with such difficulty. The monotonous, grinding nature of the work made it difficult to find good people. Fortunately, the genome cost overruns had caused Congress to mandate a twenty-five percent rollback in funding for all other NIH institutes. That caused the closing of a number of laboratories at medical schools, research institutes, and even in other branches of the NIH itself. Some of those unemployed had reluctantly signed on at Midwest.

Although he noted their dissatisfaction, Busychair could not really understand their complaints. All they needed to do was ensure that the various instruments were repaired and that the Sequencing Corps interns spent the required number of hours at their benches manning them. (The NIH had decreed that all biomedical Ph.D.s and M.D.s whose training had been supported in any part by government grants had to put in one year of public service as interns in the Sequencing Corps.) The staff scientists did not even have to examine the sequences. That was the job of the Computer Bureau who kept them secure until report time, free from prying by Pacific or Northeast. Above all, the scientists did not have to face pressure of the type he had now. Unless he got the work back on schedule, even some administrative jobs would be lost. His own would certainly turn over.

Not that he should be blamed for the fiasco that slowed them down. His time was so consumed in defending his budget from the politcal office that he could not be expected to oversee every detail. It was not his fault that the Second Batallion had spent months sequencing band 21q12 when they should have been doing 12q21. The typo was honest, and mistakes can happen. The sequence of the former belonged to Northeast and had been recorded two years ago. Even worse, the two determinations of the same sequence had proved to be ten percent different.

They had come from different individuals, of course. Yet the variation was too much, even if an Eskimo and a Hottentot had been compared. Northeast and Midwest had managed to hush their differences up; they would deal with the discrepancies later. What did it matter? Most of it was repeated sequences, with little or no meaning, anyway. Yet, there was the problem of band 12q21, which still had to be done. Busychair did not despair entirely as there was one remaining hope.

The Nobel laureate at the local university, Professor Melanogaster, had lost all her funds for fruit fly sequencing work, as the government now felt that only human DNA should be done until the project was complete. Silly of her to work on that, anyway. What more was there to learn about fruit flies? Even if it were worthwhile, why waste time working out the meaning of the sequences? The computer guys would get to that in due time. Yet Melanogaster was clever and was even said to have worked out some more rapid sequencing method. (Midwest could not consider its use, officially; too much had been invested in their current machines. They had to discredit any new technique for now.)

Suppose, however, that Melanogaster was given 12q21 under contract with no questions asked about how she got it done? Just so long as she did it by the deadline. She would probably refuse at first, but what if we told her that we would look the other way if she squeezed in some fruit fly sequences too. Perhaps even a bit extra support could be furnished from some private sources that Subsidy controlled. Busychair allowed a small smile to flicker across his face as he reached for the phone. . . .

In order to create this bad dream, I crowded together a number of different arguments and fears about the project. Each deserves some separate attention, so I have sorted them out below.

Argument 1. DNA sequencing is repetitive and boring and therefore not really science. One anonymous graduate student was quoted by James Trefil in the *New York Times Magazine*: "I don't want my life's work to be sequencing pairs 100,000 to 200,000 on chromosome 17."[11] At the San Diego Genome I conference, one professor noted that he assigned every medical student who came to be trained in his lab a certain amount of sequence to do. Sydney Brenner has joked that some others (he excuses himself) believe "that mapping and sequencing are a thankless task, suitable only for a penal colony where transgressing molecular biologists might serve sentences of up to twenty megabases [million bases]."

Comment. The first step in the business of doing science before you can figure out the pretty explanations *is* collecting the needed data. Sometimes lots of it may be there, as in the human genome, and collecting it may be monotonous, but who says that science always has to be fun? Certainly not Gregor Mendel, who laboriously counted thousands of pea specimens over many years.

I have sometimes given out research problems of a type that I could not stand to do myself. You run the same procedure over and over again, just changing the temperature or the amounts of the chemicals that you put in. Such studies can reveal the rate at which DNA is damaged by environmental chemicals. No student was compelled to take such a problem from me and do it for a Ph.D., but there were several who volunteered to do so. They accepted it, because valid results were fairly certain in a study of this type, and only a moderate amount of laboratory ability was needed. Those who preferred the wild-and-woolly problems could spend years at work and end up with nothing to show for it.

The people who have designed the Human Genome Project concluded though that it was wasteful to put hordes of people, as opposed to one, on the identical task. So the first part of the effort will include the full automation of as many procedures as possible. The machines and computers will not get bored.

Argument 2. Only the gene (protein coding and control) areas, which are five percent of our genome at most, need to be done. The remainder represents "junk" (repetitive phrases and Alu family members, for example), the sequencing of which would be a waste of money. The writing has been very florid on this point. For example, I will quote Robert Weinberg of Massachusetts Institute of Technology:

> Most of the genome seems to represent evolutionary detritus—the discarded drafts of essays that lost any meaning 100 or 1000 million years ago.[12]
>
> Seen from my perspective, a gene appears as a small archipelago of information amidst a vast sea of drivel.[13]

Sydney Brenner has clarified the definitions:

> We have the surprising result that most of the human genome is junk; junk and not garbage because there is a difference that everybody knows; junk is kept while garbage is thrown away.[14]

The monk Gregor Mendel in the 1870s. His genetic studies with peas were a generation ahead of the science of his time. From Hugo Iltis, *Life of Mendel*, translated by Eden and Cedar Paul (W.W. Norton, New York, 1932). (Original German edition published by Julius Springer, Berlin, 1924.) *(Courtesy of W.W. Norton and Springer-Verlag, Heidelberg)*

Statue of Mendel erected in Brunn, Austria–Hungary, in 1910. From Hugo Iltis, *Life of Mendel*, translated by Eden and Cedar Paul (W.W. Norton, New York, 1932). (Original German edition published by Julius Springer, Berlin, 1924.) *(Courtesy of W.W. Norton and Springer-Verlag, Heidelberg)*

Thomas Hunt Morgan, the pioneer of genetic maps, in 1906. From *Thomas Hunt Morgan et la genetique*, by Hilare Cuny (Editions Seghers, Paris, 1969). *(Courtesy of the publishers)*

Friedrich Miescher, who discovered nucleic acids in 1869. *(Courtesy of the National Library of Medicine)*

A recent photograph of James Watson. He is now director of both the Cold Spring Harbor Laboratory and the National Center for Human Genome Research of the National Institutes of Health. *(Courtesy of Margot Bennett and the Cold Spring Harbor Laboratory)*

James Watson and Francis Crick in front of their double helix model of DNA in 1953. From *The Double Helix*, by James D. Watson (Atheneum, New York, 1968). *(Courtesy of the Cold Spring Harbor Laboratory)*

Frederick Sanger, inventor of a method for reading, or sequencing, DNA text. Reproduced with permission from the *Annual Review of Biochemistry*, vol. 57. Copyright © 1988 by Annual Reviews, Inc. *(Courtesy of Frederick Sanger)*

Walter Gilbert, inventor of an alternative method for reading DNA text. *(Courtesy of Walter Gilbert)*

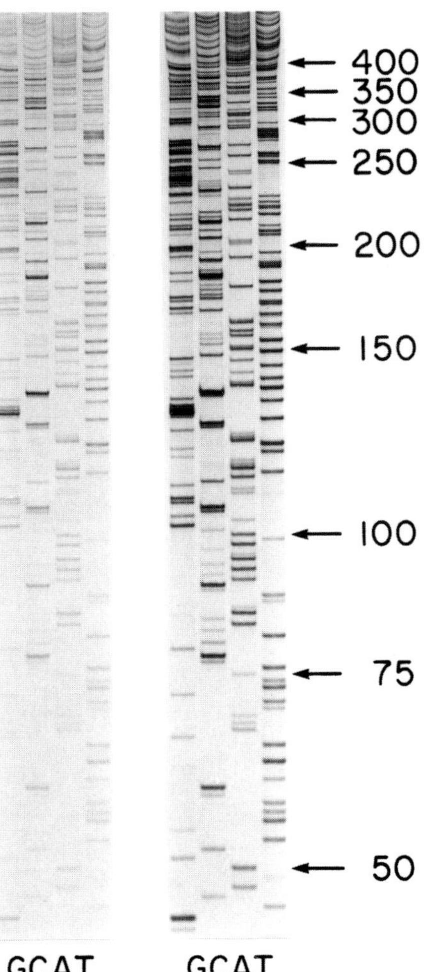

Two sequencing gels prepared by the Sanger method. The order of letters in DNA can be deduced by reading the radioactive bands in the ladders from bottom to top. From S.A. Williams, B.E. Slatko, L.S. Moran, and S.M. DeSimone, *Biotechniques*, 4, 138 (1986). *(Courtesy of the authors and the Eaton Publishing Co.)*

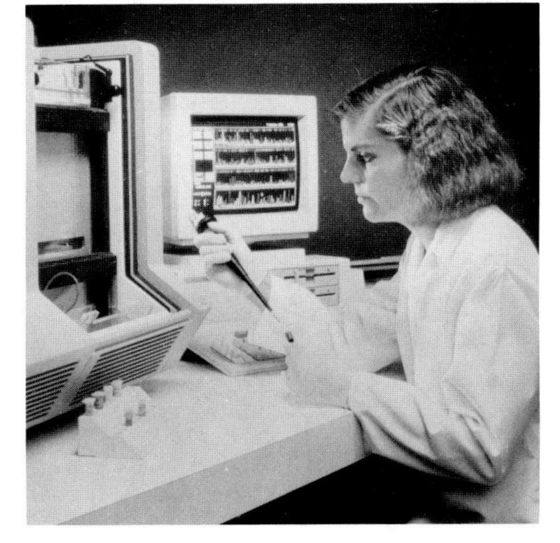

A technician with a first-generation DNA sequencing instrument. *(Courtesy of Applied Biosystems, Foster City, California)*

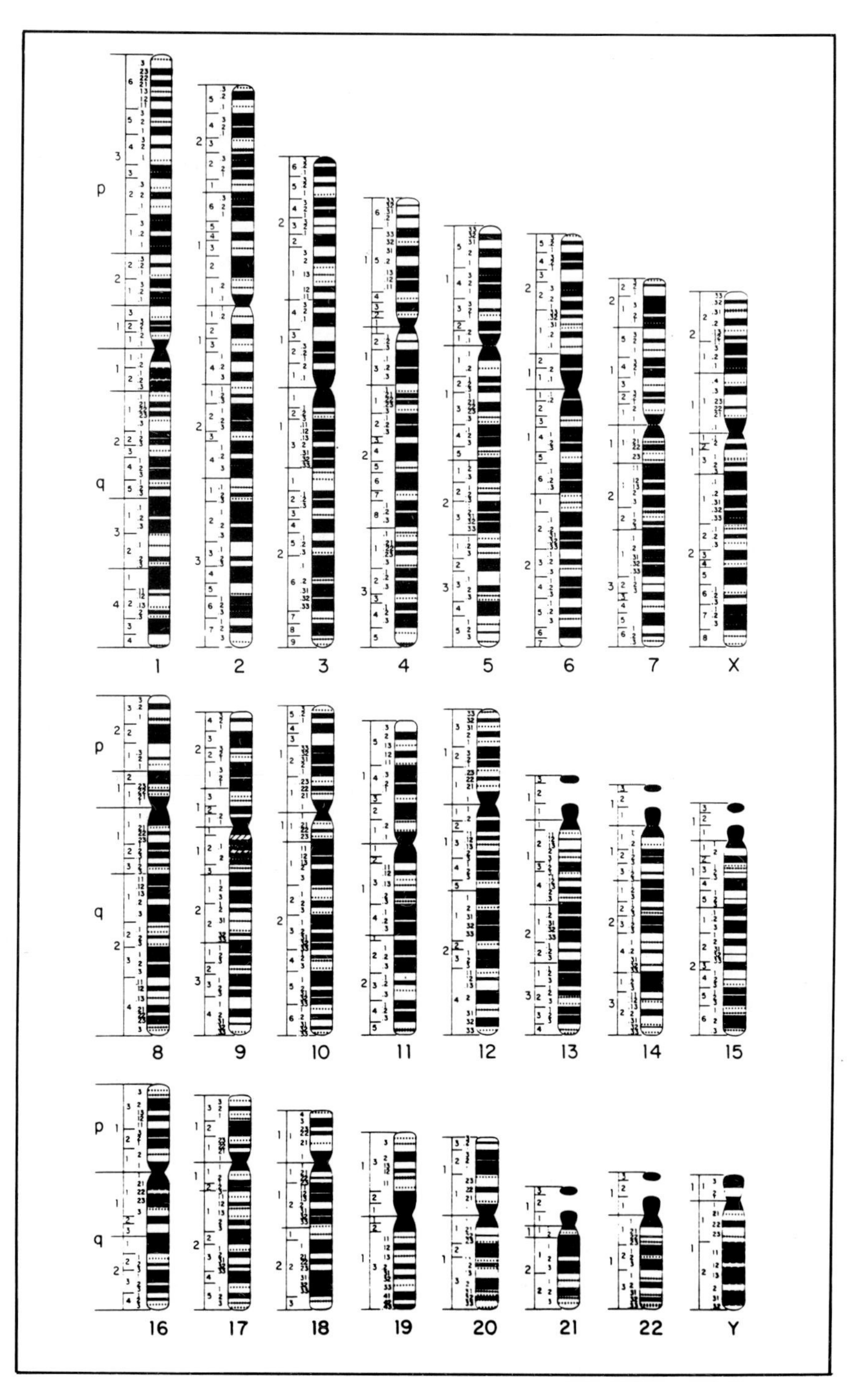

Schematic drawing of stained human chromosomes. The twenty-two nonsexual chromosomes as well as X and Y are represented. From Jorge J. Yunis, *Human Genetics*, 56, 296 (1980). *(Courtesy of the author and Springer-Verlag, Heidelberg, publishers)*

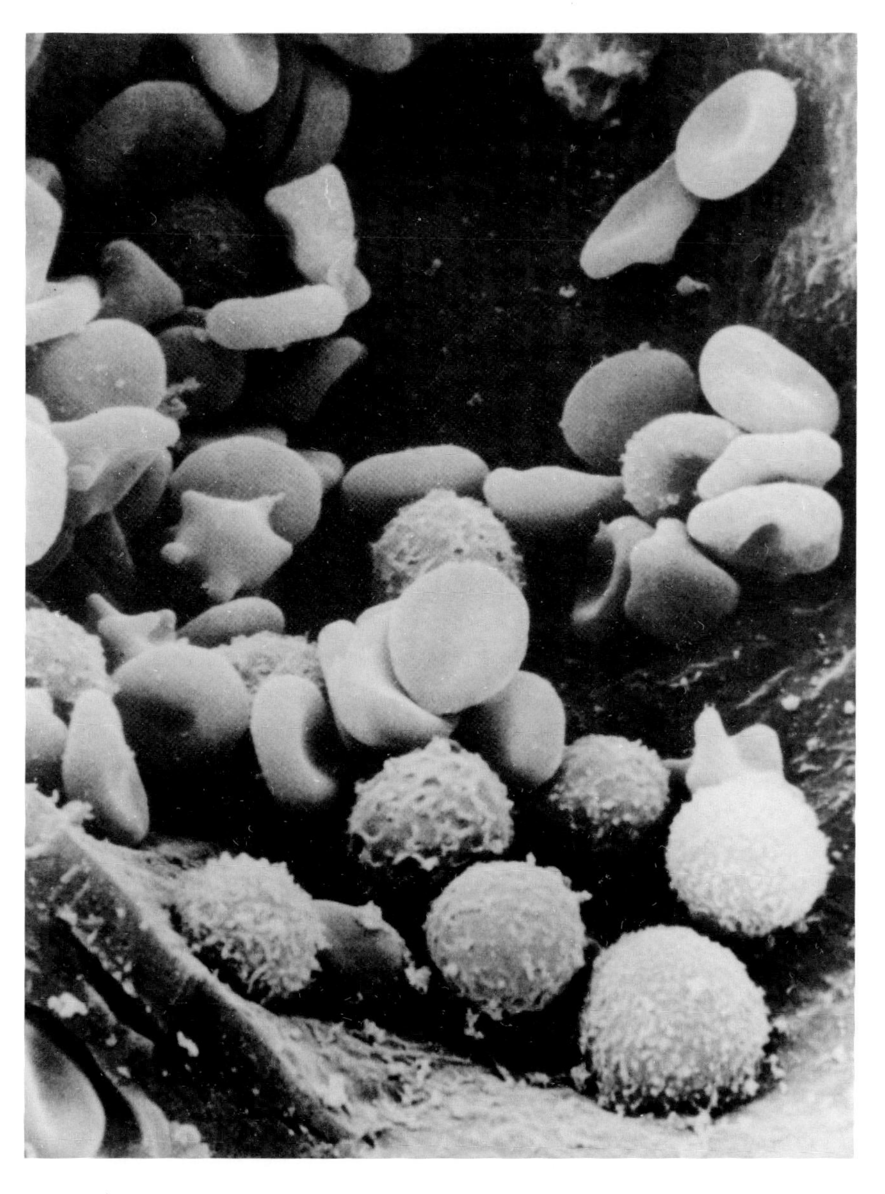

Electron microscope view of red (disc-shaped) and white blood cells in a small blood vessel. From Richard G. Kessel and Randy K. Hardon, *Tissues and Organs: A Text-Atlas of Scanning Electron Microscopy*. Copyright © 1979 by W.H. Freeman and Company. *(Reprinted with permission)*

Nancy Wexler, president of the Hereditary Disease Foundation, clinical psychologist on the faculty of Columbia University College of Physicians and Surgeons, and head of the working group on ethical, legal, and social issues of the U.S. Human Genome Project. *(Courtesy of Nick Kelsh, Kelsh Marr Studios, Philadelphia)*

Huntington's disease victim being transported across Lake Maracaibo in Venezuela. *(Courtesy of Nick Kelsh, Kelsh Marr Studios, Philadelphia)*

Charles Cantor of Lawrence Berkeley Laboratory. He is the principal scientific advisor for the U.S. Department of Energy Human Genome Project. *(Courtesy of Charles Cantor)*

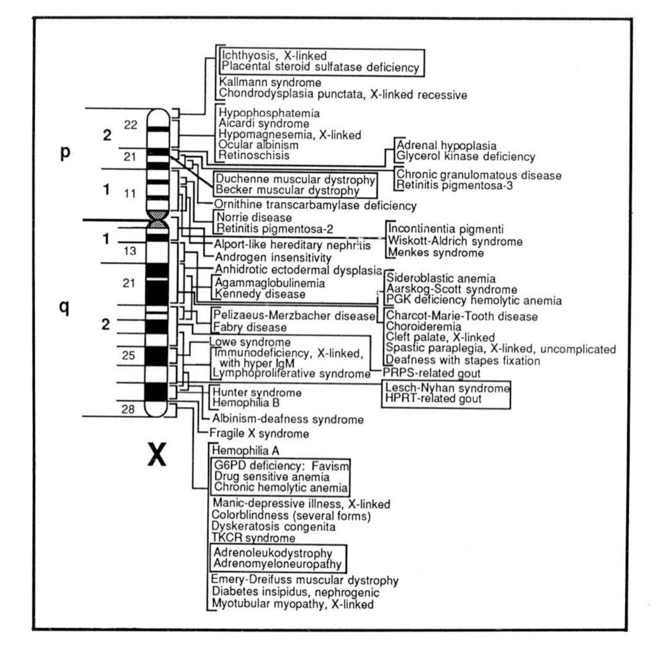

Diagram of stained X chromosome with the location of various disease-related genes indicated. *(Courtesy of Dr. Victor A. McKusick)*

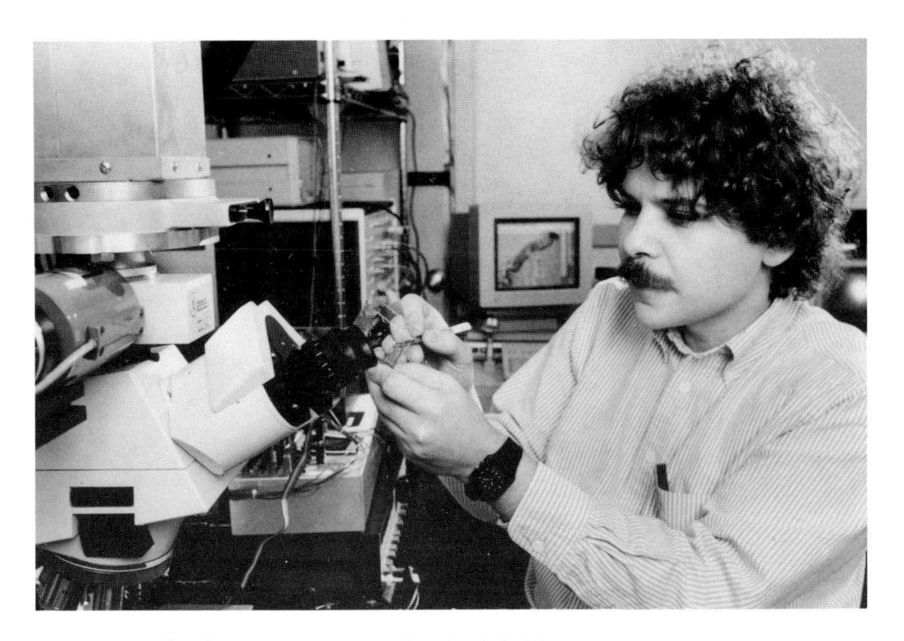

David Schwartz, inventor of pulsed-field gel electrophoresis, a method for the separation of large DNA molecules. *(Courtesy of NYU/Philip Gallo photo and David Schwartz)*

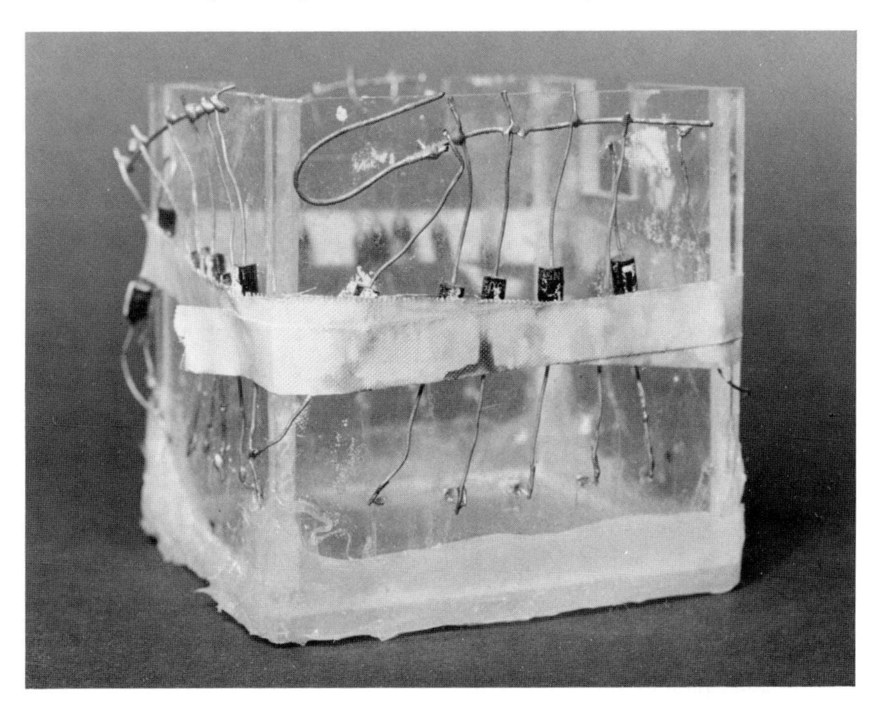

David Schwartz's first pulsed-field gel apparatus. He constructed it at Columbia University in 1981 with the aid of fellow graduate student John Welch, using a pencil box (four inches on each side), wire, and tape. *(Courtesy of David Schwartz)*

On the grounds of the Cold Spring Harbor Laboratory. *(Courtesy of Margot Bennett and the Cold Spring Harbor Laboratory)*

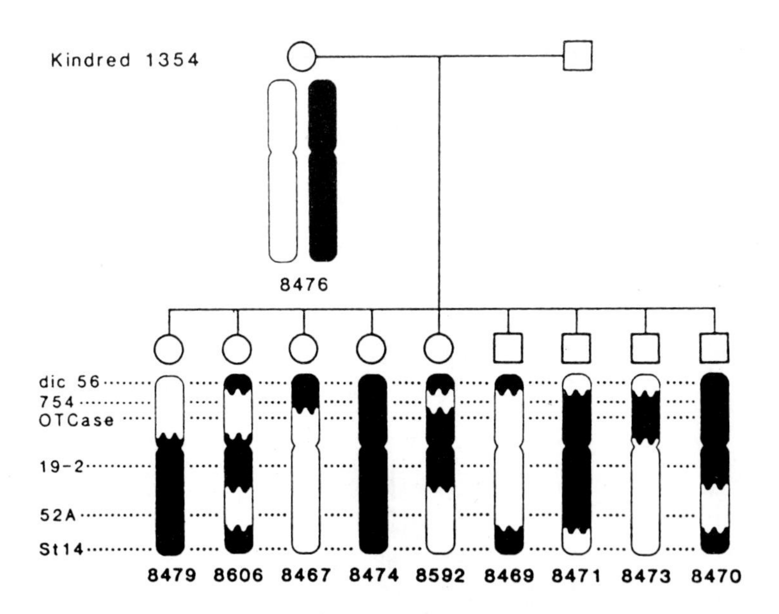

Diagram illustrating the inheritance by nine children of portions of their mother's two X chromosomes (indicated in black and white). From Dennis Drayna and Ray White, *Science*, 230, 753 (1985). Copyright © 1985 by the AAAS. *(Courtesy of the authors)*

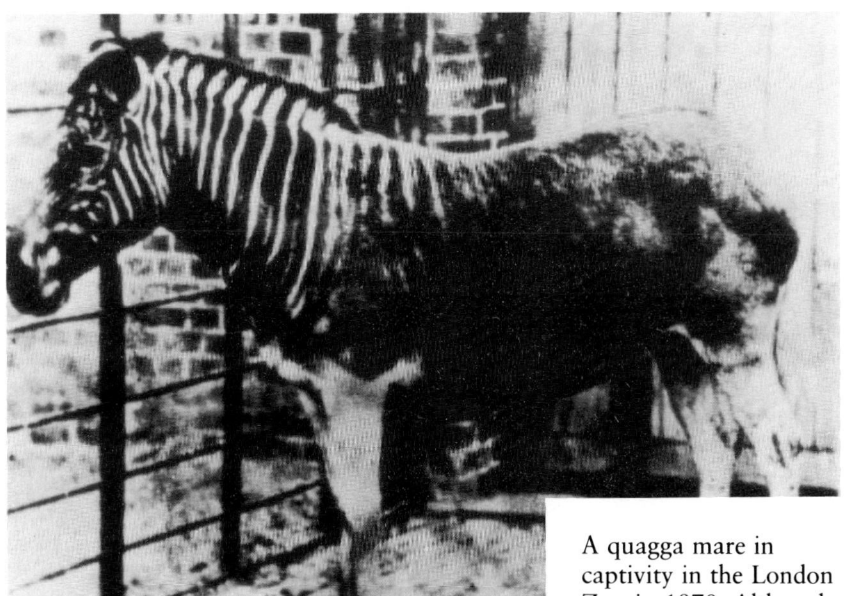

A quagga mare in captivity in the London Zoo in 1870. Although this species has been extinct for over a century, fragments of its DNA text have been rescued from a quagga hide in a museum. *(Courtesy of the Zoological Society of London)*

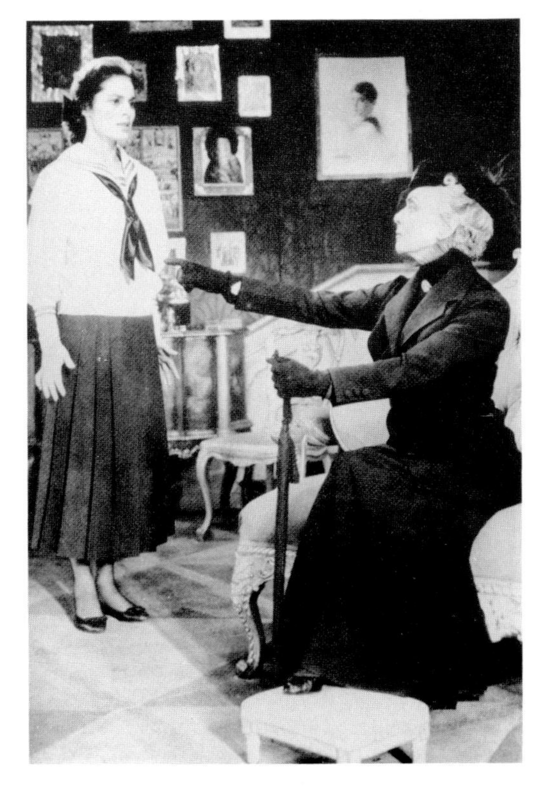

Anastasia (played by Viveca Lindfors) is confronted by the dowager empress (played by Eugenie Leontovich) in the 1954 Broadway production of *Anastasia* by Marcelle Maurette, English adaptation by Guy Bolton (Random House, New York, 1955). A dispute of this type concerning identity could be settled readily using today's genetic technology.

The British royal family, Queen Elizabeth, Prince Philip, and their children, Prince Charles, Princess Anne, Prince Andrew, and Prince Edward, at Balmoral in 1972. When a sufficient number of genetic markers have been mapped, the four children will be able to determine their exact degree of genetic relatedness to one another, if they choose to do so. This will also be true for other sibling groups. *(Courtesy of Camera Press, London)*

Queen Victoria, Prince Albert, and their first five children, as painted by Franz Winterhalter in 1846. The queen is holding the future King Edward VII, while Alice, ancestor of Czarevitch Alexis, Anastasia, and Prince Philip, is at the center of the five children. *(Courtesy of Her Majesty, the queen)*

(The Alus, LINE-1s, and other repeats certainly have been kept.)

Most outspoken of all were biologists James Bruce Walsh and Jon Marks in a 1986 letter to *Nature*. In their words, "Sequencing the genome would be about as useful as translating the complete works of William Shakespeare into cuneiform but not quite as feasible or easy to interpret."[15] A better reason for tackling this "alphabetic morass" would be to provide employment for molecular biologists. "The social consequences of keeping molecular biologists off the streets where they might develop into thugs, ruffians, land-fraud swindlers or worse, is obvious." The tone suggests that some other argument, the next one down on my list, may be the real one for them, but let's stay here for a bit. Should the junk be sequenced?

Comment. It depends what you are *interested* in, this being more a question of values than science. The pure biologists would feel it absurd to do humans at all at this time. Simple problems go before complicated ones.

For them, the crucial question is, How does life work? We can start with a simple creature like a bacterium. Its sequence of 4.5 million base pairs is partly done. Why not finish that job first? Then we will see if we can find out how it all works. Those principles could then be extended to a more complex one-celled organism like yeast. After sequencing yeast, we wold move on to a multicelled type, such as the flatworm, at "only" 100 million base pairs.

Beyond that, plants and finally mammals would be examined. It would be far better to sequence mouse than human. We can breed them, their life span is relatively short, large families can readily be found, and so on. The studies with humans, which involve ethical problems and other difficulties, would come last of all. (Francisco Ayala of the University of California at Davis has proposed essentially that order of priority.)

Our imaginary pure biologist would be terribly keen on problem solving, but not very involved with the progress of the human race. If she were forced to work on human sequences, however, she would undoubtedly select the gene areas of DNA for study. A different choice would be made by the pure physician. He is interested in humans, but primarily in terms of curing disease. Bacteria concern him only when they cause disease. He also would want only the genes and not the junk. But even a gene would lose interest if it only controlled, say, hair color and caused no illness.

Let me introduce a third investigator, whom I will describe as the humanist (although this term can be used in various ways). She enjoys problem solving for its own sake, as do the others I have mentioned.

However, she is also concerned about the quality of human life on ths planet, the fate of individuals, and the future directions of the human race. She wants the genes to be done, but she is also very interested in the junk. She has this interest not because some of it might not be junk after all (this actually is possible, these sequences may contribute to chromosome structure[16]), but for a very different reason. The junk contains the largest share of the spelling differences that distinguish one human from another and is, therefore, more important than the genes for studies on human identity, relationships, genealogy, history, and evolution.

The humanist is aware that the "alphabetic morass" is already being applied through DNA fingerprinting to the conviction of rapists and the reunion of parents with their biological children. For this person, nothing less than the whole genome will do for a start. When that is done (or even earlier), she will want to explore as fully as possible how individuals differ in their DNA. As biochemist Robert Sinsheimer wrote, "One man's garbage is another's treasure."[17]

In practice, the Genome Committee has tried to bring together all of the interests to some extent. All of the human genome will be done, but areas of medical interest will certainly move toward the front of the line. The bacteria and worm are likely to be done enroute for their own interest and because the practice will be useful in doing the human genome. Further, much of the mouse will be done at the same time, as this will aid in deciphering the function of human genes. These alterations of strategy (not of purpose) caused some critics to strike their banners. David Baltimore, by 1990, was ready to say, "I have no objection to the human genome project as it is currently organized."[18]

Argument 3. The money spent on the genome effort would be better spent on other research, for example increasing the funding of regular NIH grants. Nobel laureate geneticist Joshua Lederberg commented in 1987: "Is it worth the cost? Undoubtedly! Is it the wisest use of that level of expenditure? I have very grave doubts!"[19] Marvin Rechsteiner of the University of Utah Medical School has put it on a more personal basis, as reported to Natalie Angier of the *New York Times*: "The fat cats are getting the cream while I'm sitting here starving!" He launched a letter-writing campaign to oppose the project. In fact, most of the scientists that I have met who oppose the project, which is a substantial number, object primarily because they fear the effect on their own research grants. James Watson commented a few years ago, "I'm in favor of it and everyone else at Cold Spring Harbor is against it."[20]

The situation was particularly acute in 1990, when for various reasons the number of competing NIH grants to be funded had fallen to 4663 from a high of 6446 in 1987. Only twenty-four percent of those applying were funded, in contrast to thirty-eight percent in the earlier year. For a number of scientists in the biomedical field, this made it difficult for them to pursue their research careers. Those just starting were affected adversely. Dan M. Cooper of the University of California at Los Angeles Medical Center summed it up:

> As a researcher and physician, I daily witness the growing frustration of creative investigators for whom [finding] research funding has become the central theme of their work. This is ultimately the fault of the dwindling commitment of society to reseach.[21]

The Genome Project was not directly at fault, but it "appears excessive." It should not be given "unearned moral status." To many, it appeared that those in this project should feast while their equally worthy colleagues went without.

At the Human Genome II Conference in 1990, two of the dissenters, Bernard Davis of Harvard University and Donald Brown of the Carnegie Institute of Washington, were given time in the program and allowed to present their complaints. Brown in particular objected to the high visibility of the project and urged that it be "submerged" into the normal programs of the NIH.

Comment. Charles DeLisi, a computational biologist who helped turn the interests of the Department of Energy toward the Genome Project, has pointed out the dangers of regarding the science budget as a "zero-sum game."[22] That phrase is used by game theorists to describe a pursuit like baseball or chess where any gain for one side must automatically represent a loss for the other. In terms of past history, there is no reason to believe that if the Genome Project had not been conceived, those monies allocated to it would have appeared as general science support. Budget matters do not work with that rhyme or reason.

The American public seems very willing to spend large amounts on whatever catches its fancy and ignore what does not. A new gambling casino, the Taj Mahal, suddenly appeared at a cost of $1.2 billion, its funds supplied by bankers and Donald Trump. It was built in Atlantic City, a resort that seemed to have no shortage of such establishments. The huge amounts that are spent on items such as chewing gum and cosmetics are always cited by underfunded causes who wish a portion of those funds diverted into their own treasuries. The world, however, provides no mechanism for such shifts.

When public funds are allocated, there is constant pressure on the "ordinary" items, while the extraordinary ones continually step up and claim large amounts of funds. For example, a cleanup of nuclear wastes inadvertently spilled in weapons plants will cost $28.6 billion over five years. The sum of $5 billion will be spent by taxpayers on asbestos removal from schools, the necessity of which remains a subject of scientific debate. For example, a recent letter to *Science* on the topic of asbestos and cancer from two scientists of the National Heart & Lung Institute of London University ended with the following words:

> To those of us who have spent our lives in public health research it seems strange and sad that a country with one of the highest infant mortality rates in the Western world and no shortage of other health and behavioral problems should commit billions of dollars to the control of a miniscule or nonexistent health risk. Perhaps the real problems are too difficult.[23]

I could cite many other cases involving public decisions about science and technology where funds have been wasted because proper intelligence has not been applied. The nuclear power plant at Shoreham, New York, was built at a cost of over $1 billion and then shut down without operation. The United States has banned the sweetener cyclamate since 1970, while other countries such as Switzerland have found it safe for public use. Are the Swiss that callous about the health of their citizens, or is an erroneous and wasteful decision of the past being perpetuated through inertia? The job for science in the United States is to put its normal funding onto the same priority level as the above examples, perhaps by preventing such waste and redirecting some of the savings to itself. Scientists would be far more effective if they combined their efforts to increase the total level of support than if they accept the present amounts as if they were engraved in stone and spend their energy quarreling over the meager spoils.

Science and technology have every right to command a far greater investment by this nation if we consider that the continued prosperity of our population is dependent upon further technological advances. Imagine what the condition of the United States would be, for example, if all technology developed since 1776 were suddenly destroyed. Most of the wealth we value would go with it, and we would be able to support only a small fraction of our population in an agricultural existence far harsher than our present one. The remainder would perish.

Yet this country and others go on rapidly expanding their populations with the expectation that "progress" (that is, advances in science and technology) will provide the necessary means. In these circumstances, massive investment in research would seem a reasonable expectation. The marvel is not that the Genome Project has been funded, but that ordinary science does so poorly. It has not captured the imagination of the public or made the above circumstances clear to the nation.

In a sense, science resembles an army that has penetrated deeply into unknown and hostile territory, but forgotten to safeguard its supply lines. Too many wish to be in the vanguard and push ahead, and too few are willing to keep contact with the population that supports them. Scientists speak blissfully to one another about RFLPs, PCRs, VNTRs, and the like, while the uncomprehending public wastes fortunes on ill-advised technological decisions and amuses itself with fad diets, creation "science," and crystal power. Here lies the true competition for funding.

The remedy for the underfunded legions of science is not to "submerge" the Genome Project, but to imitate it and to reemerge to visibility on their own. If some of the "troops" could be redeployed from writing ever-increasing numbers of grant proposals to explaining to the public what it is that is exciting about science, then both public awareness and the funding of research might reach their deserved levels.

Argument 4. The genome project is one further example of Big Science, which is squeezing out Little Science. This argument overlaps the last one somewhat, but has a flavor of its own. Joseph Palca wrote in the May 18, 1990, issue of *Science* that scientists were "watching with horror the advent of scientific megaprojects with voracious financial appetites."[24] The Apollo moon-landing mission (cost: $30 billion) had been one prototype of such a project. Some current or projected examples include the superconducting supercollider (a high-energy particle accelerator; the projected cost, which has been rising continuously, is now $8.3 billion), the Hubble space telescope (cost thus far, $1.55 billion plus $200 million per year to operate; however, repairs are urgently needed, so this number will rise), and the space station ($23 billion).

Funds of such size attract political attention, and scientific considerations become secondary. David Baltimore, in his earlier phase, wrote: "A senator is unlikely to fight for a $100,000 grant to be awarded to his state but will certainly fight for a $50 million project."[25] Joshua Lederberg noted that the genome project "has attracted political constituencies who smell 'pork' and initiated turf battles among government agencies." "Are we likely to get good science out of such agencies?" he asked. A worst-case scenario was described in the earlier nightmare.[26]

Little Science is represented by the moderate-sized laboratory, examples of which were the groups of Thomas Morgan and Fred Sanger. At one stage in my own career, I headed a laboratory of about a dozen people, generally two or three postdoctoral fellows and the remainder graduate students. I had the freedom to select the research area, whatever caught my fancy, but also the responsibility of securing the grants that paid the salaries of the postdoctorals, some of the graduate students, and myself for the summer. My research ideas were judged by those who worked on the same type of problem in other places, but not by my department chairperson or deans in my university. This system is known as *peer review*. I met my peers at meetings, but not on a daily basis. I was the leader for my group, king in my own miniworld.

In an article in *American Scientist*, Charles DeLisi has pointed out that this type of organization "is as fundamental to the social structure of science as the family is to the structure of society." Furthermore, it worked. "It gave birth to the ideas that ushered in molecular biology and the atomic age. It spawned the semiconductor, the laser, and more recently, superconductors."[27]

Critics of the Genome Project have suggested that there was no need to organize the goal of reading the human sequence. It would inevitably be done, one part after another, by the cumulative efforts of individual investigators. Their collective judgment has worked beautifully in bringing molecular biology to its present point. Why not leave it to them to decide which parts of our DNA should be read and when?

Comment. The classification of the Human Genome Project as Big Science in the same category as the supercollider is misleading. Things become clearer if we set up two categories: Big Instrument and Big Concept. The supercollider and other projects mentioned fall into the first category. A Big Instrument allows new data to be collected; it does not in itself represent information. All of the project must be done, or there is no point to it. If half of the fifty-three-mile tunnel for the supercollider was dug, and the project was canceled, nothing would have been accomplished. No data could be collected until the job was done.

The Human Genome Project is very different. It centers on an idea.

The complete human sequence represents a goal of particular significance, a large concept. It has been named and special offices have been created for it to make sure that it takes place rapidly and efficiently, but it does not focus on one piece of equipment or an exploratory vehicle. Much of the work will still be conducted in Small Science environments. Individual investigators will still be able to make their own plans and apply for their own grants. The first center grants that were announced went to groups of accomplished scientists at the University of Michigan, MIT, the University of California at San Francisco, and Washington University, St. Louis. These are hardly prototypes for Midwest Sequencing.

Some additional instruments may be designed, such as dedicated computers and sequencers, and new facilities built, but these are not the reason for the existence of the project. Further, if the work was delayed or interrupted at any stage, the parts completed up to that point would still be valuable.

Insofar as significant funds are involved, the Human Genome Project will be vulnerable to political influence and corruption. However, this is no different than any other government undertaking, for example, the Tennessee Valley Authority, created during the Depresssion of the 1930s. Should dams not be constructed or social security organized because large sums are involved and misuse possible? An alternative is that the science community, in part, accept a role as a watchdog in the process and attempt to prevent such events.

One scenario that has come up in critic's discussions is unlikely. Individuals will not be selected for sequencing purposes to honor or flatter them. Tongue-in-cheek suggestions have been made about using philanthropists (for a cash donation, of course) or politicians or even that the body of Charles Darwin be disinterred for DNA sequencing, but nothing like this will happen.

In practice, many laboratories will be involved, and each will draw material from diverse sources. For example, David Ledbetter at Baylor College of Medicine, who is involved in mapping chromosomes X and 17 has commented, "We collect DNA from patients who come into our clinic for genetic testing, so each sample is from a different and unrelated person. Our cell culture collection contains a number of different human genomes."[28] The initial genome will represent a mosaic derived from individuals of many genetic backgrounds. As we shall see, the specific readout of some individuals, perhaps even Charles Darwin, will come later.

Why do it at all, though? The point of Big Instruments can readily be grasped, but why create Big Concepts? What is wrong with Small

Science with its thousands of small ideas? This can best be answered with a comparison: Small Science is similar to a librarian who has assembled a superb book collection and cataloged it well, but refuses to admit the public to the library. After all, they might disturb the arrangement of the books or even want to borrow some of them.

Small Science has worked magnificently well in advancing the limits of knowledge. It has not gotten its message across to the public. The average person has more than enough to cope with each day to consume her time and energy: She must commute to work, raise the children, maintain the home, pay the bills, and find a bit of time for relaxation to make it all worthwhile. She will not wade through abstract schemes full of complex terms just to keep abreast of scientific progress. To penetrate this, science reports must have an impact on her daily life or capture her imagination by addressing the largest questions that face the human species. Who are we? How do our bodies and minds work? How did the universe get here? What aspects of it will affect us most? What place does life have in this universe? Where are we headed?

Many questions that attract individual scientists today do not relate readily either to our daily experience or to large human questions. In the past, it was different. In Mendel's time, visible, familiar objects such as birds, rocks, peas, and the weather were part of the frontiers of science. The condition of science now can be represented by the career-long interest of a colleague in my department, cyclopropanes (a class of chemicals in which three carbon atoms connect together to form a ring). He could not have acquired this passion through collecting them as a child or watching them in his home with a telescope or microscope. He was inspired, most likely, by specialized lectures in college and graduate school, and he maintained his interest over the years by contacting others in his profession who shared it. He is entitled to follow this avocation in the same way that a devotee of chess, rock climbing, or string quartets would do, but the general public is unlikely to share that interest or support it with its own funds.

Despite my earlier imaginings, as far as I know, Gregor Mendel did not solicit grants for his pea-breeding experiments. Thomas Morgan's most productive work on the fruit fly was financed by his university and out of his own pocket. Watson and Crick made their discovery while ostensibly involved with other projects. They borrowed the time, materials, and even the data they required for their achievement. In these cases, no public explanations or solicitations were necessary.

Science today has entered another era where expenses have mounted, and public support is needed for some areas to move forward at all.

Unfortunately, some scientists feel that this should be provided as an entitlement with no explanation (except perhaps in jargon) provided.

I have mentioned the difficulties in 1990 when the NIH funded fewer than five thousand grants, twenty-four percent of those it considered. In 1980, however, about the same number of grants were funded, but this was forty-two percent of those who applied. What has changed? The number of grant proposals escalated from twelve thousand in the earlier year to nineteen thousand in the later one. Boyes Rensenberger, a science writer for the *Washington Post*, diagnosed the situation for the newsletter of a national biology society:

> Young people may not be choosing careers in science as much as we would like, but there are still many more new scientists competing for those dollars. Should every person who wants to be a biomedical scientist be guaranteed a grant or a piece of a grant? Just because you decide to be a scientist, does the government owe you a living? Should you be allowed to do any kind of research you want, no matter how removed it is from biomedical applications?[29]

The cutting edge of research is glamorous and involving, and the freedom of an academic scientist is enviable when compared to other occupations. However, communications to the public must be kept open to sustain this. Their imaginations must be captured, and we cannot do so just by packing more facts into school curricula. One way to accomplish this is by organizing larger endeavors, Big Concepts, that dramatize the long-range directions in which Small Science is headed. For this purpose, the Human Genome Project seems ideal. It can readily be explained and has applications that will enter the lives of all of us if we wish to let them in.

Norton Zinder of Rockefeller University summed it up aptly. The appeal of the Human Genome Project is not due to its potential contributions to medicine or evolution:

> No, the answer is both more simple and more profound. Determining the map and the sequence of the human genome is about us. And we are greatly interested in us. We can only consider these efforts as . . . the beginning of a

> great adventure in human biology. We have to do the sequence only once, and for all time we shall have for study the human genetic dictionary.[30]

The project will be a monument for our time. As writer Timothy Ferris pointed out in the *New York Times* of April 29, 1990: "Science is our equivalent of painting in Michaelangelo's day, of music in the time of Bach, of seafaring in the age of Prince Henry the Navigator."[31] The Human Genome Project will remain with such achievements of physical science as the Apollo and Viking missions to immortalize our era.

It will be easy to follow the progress of this effort, and it has a well-defined end point. I can even imagine a ceremony, the Sequencing of the Golden Base Pair, that might mimic the driving in of the Golden Spike, when the first transcontinental railway was completed in the United States in the last century.

A satisfying end point is not enough, however, as we learned from the Apollo missions. A Big Concept was involved there, but it got lost along the way. John Logsdon of George Washington University brought it back during a NASA twentieth anniversary symposium that celebrated the first moon landing:

> How does one put a value, after only two decades, on the initial steps toward man becoming a multi-planet species? . . . The Apollo astronauts were true pioneers in the outward movement of humanity.[32]

That spirit was echoed in the first communication from Tranquility Base on the moon. Bases are needed for the future if we are moving outward.

Somehow that message got lost in the subsequent outpouring of data interesting to Small Science: seismic levels, magnetic fields, chemical compositions of rocks, and the like. The communication lines to the public were clogged. With no larger concept to guide follow-up activities, Apollo became a "dead-end project." (I am using the words of Walter McDougall, Pulitzer Prize-winning historian of the University of Pennsylvania.) The last two flights were canceled to save money, and the component instruments were consigned to museums rather than used as a starting point for further explorations.

The Human Genome Project can escape this fate, as at least two follow-up challenges will emerge upon (and even before) its completion. One involves the full interpretation of the text: defining and reading the protein-coding, control, and structure areas and others of importance. Beyond that lies yet another endeavor: to determine how these parts of our hereditary message interact with each other and with environmental influences to shape the properties of the human body. (In parallel to this will be efforts made on other species that matter to us, such as crop plants, domestic animals, and hazardous microorganisms.) Philosopher René Descartes anticipated all this in the sixteenth century, although he overlooked the environmental contribution:

> If we really knew all the parts of the seed of any particular example, man for example, we could deduce from that alone, by certain and mathematical reasoning, the shape and conformation of each of his limbs.[33]

Another enterprise will branch off in a different direction. We will want to extend sequencing to ethnic groups and individuals, so that each of us can master, if we choose to do so, the information that lies beneath our biological identities. We will discuss this at length in Chapter 17, "Roots, Twigs, and Branches."

Taken together, these projects announce an even larger and more general purpose. As a species, some of us will choose to say, Thank you, Mother Evolution, for having brought us thus far. We now have the plans and will take responsibility on our own for the maintenance, improvement, and redesign (if we wish) of the bodies that you were so kind as to provide us with.

This statement may raise additional objections, not within science, but from those who object to the overall direction in which modern science is leading us. They will be considered separately in Chapter 19. At this point we can assume that the first phase, the reading of the sequence, is going to happen. Immediate questions then come up: Who will do it, and how will it be done?

13

WALLY, CHARLIE, JIM, AND HUGO

In 1986, after Renato Delbecco's letter was published in *Science* and a key conference on Molecular Biology of Homo Sapiens held in Cold Spring Harbor, it became clear that a human genome sequencing project was likely. Some questions that emerged were, Who will do it? What organization would have the primary role? What individuals would symbolize the effort and have the most to say about it? At various times, four names have come up, each of which represented a different group. We have met two of them already. Their colleagues call them by nicknames and so will I. So, let's meet or remeet Wally, Charlie, Jim, and Hugo.

WALLY

Wally, or Walter Gilbert, wanted to do it on his own. In 1987, the Department of Energy (DOE) was the group most interested in human DNA sequencing, but he felt they would never get around to it or at least not quickly. Yes, the government was moving, but in slow motion. "While everyone else is fussing," he told the press that February, "I might as well do it."[1]

How would he get it done? Some observers had felt that a large improvement in technology was needed before large-scale sequencing was started. Wally disagreed: "DOE wants to develop the technology until a single graduate student can do the whole genome as a thesis. I believe the technology is here. Just take it and apply it to produce the sequence." Whether right or wrong, he certainly was in a position to make that assessment. With Fred Sanger, he had earned the Nobel Prize for inventing modern DNA sequencing methods.

Wally envisioned a group of some three hundred individuals, perhaps housed in one location, each turning out 1 million letters of DNA sequence per year. In ten years, they would have completed the 3 billion bases of the human genome. He hoped to hold the cost down to perhaps $300 million overall. His proposal aroused a great deal of discussion and animosity—not because he wanted to do the job via a massive, dive-right-in approach, but because he wanted to do it for profit.

He proposed to found a new private business organization, the Genome Corporation. They would determine the sequence and put the data into a data bank "where it would be available to everyone—for a price."[2] The rights of such a company would be protected by a copyright. Customers could use the information as they chose, but could not reproduce it or sell it. Of course, others could obtain the same information on their own by doing their own sequencing, but Gilbert felt that it would be simpler and cheaper for them to buy it from him.

Lawyers disagreed on the legality of his scheme, but most of the heat that developed concerned the morality of it. Some people felt that "this information is so important that it cannot be proprietary"[3] and that the genome "doesn't lend itself gracefully to a venture capital thing, the avarice of Wall Street."[4] Tabitha Powledge, the editor of *The Scientist* wrote:

> This is a truly appalling idea. . . . If there ever was a development that violates the majestic cooperative human intellectual endeavor that is the moral core of science it is this: that information about the genes of the human species should be privately hoarded, doled out bit by miniscule bit, for gold.[5]

Gilbert, however, was not persuaded by these objections:

> Why do we have a society in which hospitals make a profit? . . . Is doing it for profit right? Our society has given us its answer, "sure it's right."[6]

He was no stranger to the business world, and the Genome Corporation was not just an academic daydream. At the end of the 1970s Gilbert phased out his Harvard laboratory; the *Midnight Hustler* ceased publication. He went on to play a key role in the founding of a new international biotechnology corporation, Biogen. By 1982, he had left Harvard to become Biogen's chief executive officer, working as a businessman rather than a scientist. Robert Kanigel wrote in the *New York Times Magazine*: "He plainly came to enjoy the rough and tumble world of business."[7]

Gilbert was thrilled to see what was once an idea transformed into new carpeting and people to answer phones. According to him (as quoted by Stephen Hall), the motivation for the change came from "a variety of reasons: wanting to do something socially useful, wanting to create an industrial structure, wanting to make money."[8] When Biogen opened a headquarters in Cambridge, Massachusetts, former Mayor Alfred Vellucci (he who earlier feared what might crawl out of such buildings) was now willing to cut the ribbon.

Gilbert saw himself as a link between business executives and biologists: "In a sense I became a spokesman for the scientists. I had also, in some ways, the greatest sympathy or affinity [for] the way in which the business was structured."[9]

Others did not share that opinion, according to writer Robert Kanigel, "Investment analysts complained that Biogen was being run more like a lab than a business, one lamenting that Gilbert apparently regarded the company's capital as a big research grant."[10] Robert A. Fildes, one executive, left after two years, complaining that Gilbert was more arrogant "than anyone I've encountered in 20 years of managing people." According to the article, one business analyst noted "Gilbert was such a strong person. It was difficult to run the company without him but difficult to run it with him, too."

As losses mounted and the price of Biogen stock fell from $23 in March 1983 to $5 a year and a half later, the company leadership chose the first option cited by the business analyst. He was asked to step down as chief executive officer. He retained a position on the advisory council and 5 million shares of stock. It had taken longer than expected, he felt, to bring products to market.[11] He might have added, as he did after his childhood chemistry set mishap, "I know what I did wrong!"

The Genome Corporation venture did not prove successful either. According to Wally: "I couldn't raise the venture capital to do it."[12] Ten million dollars had been mentioned as a starter sum. The stock market crash of 1987 was cited as one factor in his failure to raise that money, and business executives also wondered whether a company could make money just by sequencing.

As time passed, it also became increasingly apparent that the U.S. government was going to get the job done. In a sense, that development rendered the morality debate irrelevant. If the human sequence is a public resource and its determination is both feasible and useful, then it is the ethical duty of human government to carry it out. This in fact is what will happen. Had the government failed to do so, then it would be better to have the job done privately than not at all.

Upon departing from the business world in 1984, Walter Gilbert had reestablished his academic research career in the biology department at Harvard University. It was there that I went to interview him in July 1990. Although I had finished my own Ph.D. degree thirty years earlier in the chemical laboratories only one block away, I could not recall ever having entered the biology buildings. I stood before them now: a factorylike assemblage in red brick, which occupied three sides of a courtyard on Divinity Avenue.

Two large bronze rhinoceroses stood before the entrance to the building; fitting guardians for the approach to Wally. His spacious and comfortable office reflected his varied and unusual career. The shelves contained two poetry books by his wife Celia, a book on superstring theory, and *Invisible Frontiers* by Stephen Hall, in addition to the expected biochemistry material. A framed offering for 2.5 million shares of Biogen stock lay on the floor. The marble table at which we sat held a polished fossil, a treatise on art of the ancient Near East, and Randy Shilts's best-seller on AIDS, *And the Band Played On.*

His plans for now were academic, although ambitious. He had found business "fascinating for a while," but ultimately repetitive; "not as enjoyable as the constant change of scientific research." He felt that in the future, sequences would offer the best route for understanding biology.

A change in paradigm, in mindset, would take place. The sequences, stored in data banks, would now be the starting point for a new, theoretically based approach to that science. Biologists would look at the information in DNA, think about its meaning, and then plan an experiment.

Before that era could begin, the sequences would have to be recorded, of course, and Wally planned to take part in the action. For now, he would not be emphasizing the human genome. In his own laboratory he hoped to work out the text of the simplest free-living organism, mycoplasma. ("It's a perfectly good small bacteria.") He felt that his group could complete its sequence of 750,000 base pairs in a year and a half, using a large-scale strategy that he had devised. Then he could address the question of how a DNA text defines an organism. After that, he would attack some chromosome of a more complex type of cell, one more like our own (a *eukaryotic cell*).

Was there yet a role for business in the determination of sequences? Yes, but as a diagnostic method for individuals after some technological improvement had taken place.

CHARLIE

When Charles Cantor gives another of his continuing series of talks at conferences and dinners on behalf of the Department of Energy, one question invariably comes up: Why is Energy involved in the Human Genome Project?

In reality, that involvement traces back to August 1945, when atomic bombs were dropped on two Japanese cities, starting a lengthy experiment on the effect of radiation on human beings. The forerunner of the DOE, the U.S. Atomic Energy Commission, was founded the next year initially to encourage the peaceful uses of atomic energy. In 1954, it was also authorized "to conduct research on the biological effects of ionizing radiation."[13] That mission was still being carried out in December 1984, when its successor, the DOE, organized a conference at Alta, Utah. The development of methods for detecting mutations in humans was the central topic with the descendents of the Hiroshima and Nagasaki bombings obvious candidates for any techniques that were devised.

During the five days of the conference, the approximately twenty participants were "isolated by repeated blizzards," according to historian Robert Cook-Deegan. This circumstance allowed those present to concentrate on skiing and the intense exchange of exciting ideas.[14]

Cantor, then the chairperson of the Department of Genetics at the Columbia University College of Physicians and Surgeons, was among those present at the conference. He presented new results from his laboratory on the separation of large (over 1 million base pairs) pieces of DNA, using a new method of alternating electrical fields (*pulsed-field gel electrophoresis*, PFGE).

This technique had made a strong impression on him when it was suggested to him by a graduate student, David Schwartz, a few years earlier. "It was clear to me that it was going to have profound implications for a lot of genetics."[15] In fact, it pushed him toward a change in his research direction. Cantor, a DNA chemist by training, was then located in the Columbia University Chemistry Department. The development of the new method made it attractive for him to accept the genetics post when it was offered to him, because he would have many more interactions of the kind he was now interested in. At the Alta meeting, Cantor showed how a collaborator in his laboratory, Cassandra Smith, had been able to slice up the chromosome of the bacterium *Escherichia coli* with text cutters and separate the pieces. By breaking a large problem up into a number of well-defined smaller ones, they were preparing the way for an eventual sequencing effort.

This result and others presented at the conference made it clear that the best way to detect mutations was to read the DNA spelling changes produced by them. However, we would have to know the original text first so that we could tell what had been changed by the mutation.

Among those who read the report of the Alta conference and got the message was Charles DeLisi, head of the Office of Health and Environmental Research of the DOE. If the human genetic sequence was a necessary data base for the detection of mutations, why shouldn't the scientific community just go and record it? Why shouldn't DOE be the one to do the job? After all, this would be a large project, and DOE was quite comfortable with large projects. Several of the national laboratories run by DOE would be ideal sites for such activities, he felt. In addition, the agency had developed expertise in two relevant areas: computers and instrumentation. This had already earned it two assignments that were directly relevant to the human genome sequencing effort: Genbank, the computer data bank that compiled DNA sequences, was housed at Los Alamos National Laboratory in New Mexico. Chromosome sorters, advanced instruments that mechanically separated the individual human chromosomes and made them available for further analysis, were in operation at Los Alamos and at Lawrence Livermore National Laboratory, near San Francisco.

A meeting to consider the possibilities was held in 1986, and by 1987, a subcommittee of the department had given the proposal its strong endorsement: "DOE can and should organize and administer this initiative." It "should not delay implementation of its plan or defer to some other organization."[16] The suggestion was made that $1 billion be spent by 1995 to map and sequence the human genome. Among the keen supporters of the idea was Senator Pete Domenici of New Mexico, who could not have helped notice that one key location for this activity would fall within the area he represented.

The obvious candidate for the role of "some other organization" in genome sequencing was of course the NIH. As we have seen, the connections between the human DNA sequence and public health run far beyond the topic of mutations caused by radiation and chemicals. The NIH at that time was, however, "awed by its scale and nervous about diverting funds from other research," according to Leslie Roberts in *Science*. "In contrast to DOE, NIH seems very much the reluctant bride, unenthusiastic about an all-out effort but unwilling to turn the project over to DOE."[17] James Cassatt, an official of the National Institute of General Medical Sciences (an NIH component) wrote "at the time, few in NIGMS took the idea seriously. It seemed to me at least like a ludicrous idea; who would want to undertake such a project?"

The DOE had the right impulses and could appropriate a bit of money in that direction, but seemed, ironically, lacking in the energy needed to get things really moving. Walter Gibert started to think of a corporation to do the job. Charles DeLisi left DOE at about that time and wrote, eighteen months later, "I don't think they've done anything much beyond my memo."[18]

One cure for these doldrums was the recruitment of an energetic, nationally known scientist to represent the proposed project. An obvious and willing candidate was Charles Cantor. A new DOE Human Genome Center had been established at the existing Lawrence Berkeley National Laboratory on the Berkeley campus of the University of California. Cantor would head that unit and, in addition, assume a professorship in molecular biology at the university.

Cantor would have preferred to live in New York rather than return to his Ph.D. alma mater. However, the chance to direct a very large research operation proved a temptation. A group of fifteen research workers would move to Berkeley with him, where they would be joined by an equal number of new recruits. Ultimately, he would oversee a staff of one hundred, or even five hundred, in the final phases of sequencing. His operating budget for the current year would be about $3 million, well beyond the amounts available to conventional scientists running their own laboratories. Another aspect of his new role also proved inviting.

He had been a chairperson at Columbia University, and had he stayed on, he most likely would have become a university administrator. However, in his words, "I wouldn't have enjoyed that the way I'm going to enjoy this." Part of his role with the DOE was "to act as glue" between a lot of the projects and "to act as a major spokesman for the DOE and for the genome project for the US. At least right now, I'm the visible leadership for the DOE program."[19]

Unlike that of Jim Watson in the NIH, his role was unofficial, Cantor pointed out. By title, he was simply one of the three heads of National

Laboratory Units of DOE that are involved in the genome project. At that time, his peers were Anthony Carrano at Livermore and Robert Moyzis at Los Alamos. Above them was DeLisi's successor, Benjamin Barnhart, a kindly, somewhat elderly scientist-administrator. Recently, an effort by DOE to recruit a more lustrous superstar had fallen through. The vacancy was at Los Alamos in 1989. "We had a person of equal, or even higher status [than Cantor] lined up," George Bell, a DOE official, told *The Scientist*, "but he turned us down." An accompanying photo caption of the actual appointee ran alongside with the caption: "Moyzis: second choice."[20] Cantor was allowed to represent the DOE genome project to the public.

I went to interview Charlie at Columbia University in 1989 just prior to his move to Berkeley.[21] It was more the renewal of an old social acquaintanceship than a meeting, as I had known him since the 1960s when a friend of his had joined the New York University chemistry faculty. The use of the name *Charlie* instead of the current *Charles* labeled me as someone who knew him long ago, Cassandra Smith explained later.

Charles's office on a high floor of the Columbia Medical building in upper Manhattan commanded a sweeping view of Central Park with the mid-Manhattan skyline in the distance. Otherwise, it had the usual scientist's collection: books, journals, reprints, and computer output everywhere with an Escher-like print, a metal sculpture on his desk, and a tall plant. One departure was the complete shelf given over to travel books. Charles commented:

> The two of us [Watson and he] are constantly going off to strange places to push the genome program. It's interesting. It's very important for the two of us—and for other people—to go around and explain to people what this is about.[22]

His itinerary had included a conference in Sardinia a week and a half earlier and an IBM dinner in Puerto Rico the day before with visits to the Soviet Union and possibly Japan in the future.

He was fortunate in having a close collaborator, both on the professional and personal levels, to help him maintain this lifestyle. Cassandra Smith and he shared their large research group together and lived together. They had come from different research backgrounds, biophysical chemistry on the one hand and genetics on the other, but had now combined their interests.

The arrangement worked. "We complement each other very well," Charles said to me.

> She likes to watch the experimentalists very closely—deals with them all the time. I'm accustomed to much more distant relations with people doing science. Also, I'm away so much. I've had so many commitments on the national and international scene.[23]

The last time that I spent any period of time with Charles was at a conference at Banff, a resort in the Canadian Rockies, in the early 1970s. He was very tense and driven at that time to complete manuscripts in spare moments during the meeting and to ski every free afternoon. He was then also involved in an unhappy marriage. Now he seemed more relaxed and mellow, almost merry by comparison with his earlier self. With his dark hair, slim build, modest height, and moustache, he almost reminded me of another Charles: Chaplin.

The windowless office of Cassandra Smith, a few feet away, was very different. With its dark filing cabinets, Mexican wall hangings, cactus plants, Grecian mask, Rousseau-inspired painting, and stuffed cats (not all of them were stuffed; one ran off later), it seemed more organic and vital. She was short (that's why she needed a big name, Cassandra, not Cassie, she explained), dark haired, olive skinned, intense, sassy, and attractive. Her speech (in a husky voice, with sentences separated into fragments as her thoughts ran from one topic to another) reminded me of girls I had met while growing up in the Bronx.

She had partly been raised in Brooklyn, but with sojourns in West Virginia and Texas. Her background was diverse: a Russian Jewish mother and a father who was Irish on the maternal side and American ("Indian, black, whatever") on the paternal side. When I remarked that she would make an ideal candidate for genomic sequencing, she mentioned that when their current chromosome project was completed, she would like to compare the same chromosome in a variety of populations to see how different we all really are.[24]

One current project of Charles and Cassandra was to map (and perhaps sequence) the smallest chromosome: No. 21. They were not alone in that effort. Charles had commented earlier that there were too many groups there: "It's beyond the point of healthy competition." The small size of this chromosome, together with its association with important diseases such as Alzheimer's and Down's syndrome and its overall gene density, had proved attractive to many researchers. Cassandra was not deterred by the rivalry, stating at one point that they would blow the competition away. Her work style mirrored her intensity. She was routinely accustomed to working from 10 AM to 8 or 9 PM, seven days a week. On a visit to New York University once, she had commented to me that she rarely saw city streets in daylight.

Cassandra felt that they were not far from completing their map of chromosome 21. To borrow a somewhat hackneyed phrase, there was light at the end of the tunnel. After it was done, she and Charles would select another chromosome to map. (She was also to become a professor of molecular biology at Berkeley.) They only planned to work on two chromosomes. The choice was not settled, but No. 1 was one possibility. After the smallest, why not the largest?

Their plans did not work out, at least in terms of their chosen location. A little more than a year later, Charles was removed as the director of the Lawrence Berkeley Laboratory. An article by Paul Selvin, in *Science* (September 14, 1990) bore the title "Charlie Cantor Gets Kicked Upstairs. Amid charges of absenteeism and scientific inefficiency, the head of DOE's Lawrence Berkeley human genome lab departs." In the article, Berkeley professor John Hearst voiced the following criticism: "From Charlie's behavior, it appears he lost interest in the day-to-day running of a scientific lab and was more interested in flying around the country giving talks." Outside laboratories "were doing so much better that it looked like the Berkeley effort had stalled."[25] The other DOE laboratories had made progress in cloning chromosomes 16 and 19, but Lawrence Berkeley had barely moved on 21, another observer reported.

What had gone astray? Charles cited the absence of promised space and the difficulties of moving an active working group across the country, but acknowledged that he was "overwhelmed" with his multiple responsibilities. In his absence, much of the work in running the day-to-day lab operations fell to Cassandra Smith.

She was not well liked as a manager, *Science* reported. Scientists and technicians were ordered around and moved from project to project. According to one scientist, "Projects are likely to appear or disappear almost momentarily. She organizes and disorganizes groups on a virtually day-to-day basis."[26]

Cassandra denied that projects were changed that rapidly, although she felt that abandoning some efforts and concentrating on others was needed in successful science. Problems arose, she told me, because the authorities at Lawrence Berkeley had not provided the space, personnel, or support that had been promised. Bureaucratic regulations delayed them in setting up their specialized equipment. Also, the isolated location of the facility made it difficult for work to continue into the evening, as the bus transportation shut down at 5:45 PM. "You can't do science on eight hours a day," she added.[27]

Not all of the outcome was negative. As the headline in *Science* suggested, Charles was given new responsibilities. (Cassandra, however, maintained that all of that had been arranged *before* he was relieved of his position at Lawrence Berkeley.) He assumed a new position as principal scientist for the entire DOE genome effort and could now represent the entire agency officially on the national and international level. This honor and opportunity may have been worth the price of a chromosome or two.

JIM

The efforts of the Department of Energy to take the lead in the Human Genome Project were not greeted universally with delight, even by those who favored the project itself. James Watson later summarized some of those objections in the 1988 Annual Report of the Cold Spring Harbor Laboratory:

> The intellectual competence for managing the human genome project might never exist within a DOE whose leaders were inevitably physical scientists and where biology of necessity always occupied a low position on its totem pole of priorities. . . . There was also strong reservation about any project where the ultimate control of resources lay in the hands of administrators, like those who rule the DOE, as opposed to control by the scientific community itself. DOE's known propensity for overruling peer review panels had created much unease at the thought that they might direct the project. . . . I was afraid DOE would get all the money and totally misuse it.[28]

Plans were organized for an alternative, NIH-led program. Again, let us rely on Watson's words in *Science* on April 6, 1990:

> At the Reston [February, 1988] meeting, I strongly argued that the [head of the new NIH Office of Human Genome Research] position be filled by an active scientist, as op-

> posed to an administrator. I argued that . . . only a prominent scientist would simultaneously reassure Congress, the general public, and the scientific community that scientific reasoning, not the pork barrel, would be the general theme in allocating the soon-to-be-large genome monies. I did not realize that I could be perceived as arguing for my own appointment.[29]

He had functioned as an administrator for many years, directing Cold Spring Harbor Laboratory and raising its annual budget from $600,000 to $28 million per year. "Whether I was still a real scientist was not at all clear," he commented to *Science*.

When offered the NIH job, he did accept on a part-time basis, retaining his Cold Spring Harbor position. "I realized that only once would I have the chance to let my scientific life encompass the path from double helix to the 3 billion steps of the human genome." By the fall of 1989, the office was upgraded to a full center with Watson as director. It was now on a par with the other institutes of the NIH, with full funding authority over grants and training programs. (*Science* reported late in 1990 that Watson was "admittedly stretched thin" and suggested that he might hand over scientific responsibility for Cold Spring Harbor to a colleague while continuing in fund raising.)[30]

With the appointment of Watson, the NIH program made what was called "a quantum jump" in its credibility. A figure whose name was associated with the theory that had founded the field more than a generation ago would now head its most ambitious effort. It was as if Louis Pasteur had accepted the leading position in the war on AIDS.

An imbalance soon appeared in the roles of the two agencies. In fiscal 1989, the genome budget, in millions, for NIH was 27.5, for DOE, 17.5; in 1990, the score was NIH, 57, DOE, 27; and in the initial Bush budget for 1991, the request for NIH, 108, DOE, 46. (Congress would later reduce the NIH allocation to 88 million.) The president's proposed 1992 budget requested $169 million, total, for both agencies.[31] Senator Pete Domenici had commented that "NIH has won a turf battle," but told a DOE research director, "You shouldn't settle for anything else than leadership in this field."

A Joint Memorandum of Understanding had been signed between the two agencies earlier, defining their respective roles. NIH would run the actual sequencing effort in humans and direct all studies in model organisms. DOE would concentrate on those areas in which it had expertise: automation of existing technology, the development of new technology, placing DNA fragments in order on each chromosome, and assigning known genes to those fragments. In those areas where the interests of both agencies overlapped, suitable joint committees would be organized to coordinate efforts. Further, the two agencies would draw up and submit to Congress a plan for the next five years.

The person with the most to say about the project would thus be Jim Watson. It would be run by a skilled administrator who also had a famous scientific name. Furthermore, he was one of the most provocative and outspoken men in science. Two results became very likely: that the job would get done and that things would not get dull.

On November 22, 1989, I drove out to interview him in his Cold Spring Harbor Laboratory office. The trip from my home was an easy one—a drive of some forty minutes. As I drove, the scenery changed: A traffic-laden suburban business street lined with gas stations and stores gradually metamorphosed into an almost deserted country road. Finally, I turned down a hill and saw a bay ahead. On the left, a small sign marked the entrance of what appeared to be, at first glance, a country resort: a combination of old and new buildings that ranged from cottages to modern laboratory facilities set among the trees and paths. Immediately beside the highway was the Victorian house that was the residence of Charles P. Davenport when he ran his eugenics center on the premises. Watson's office was in an unpretentious bungalow. The space was light and airy, however, with a view of pine trees and Long Island Sound. The room held many memorabilia, including a photograph of Francis Crick. Watson at sixty-one had aged considerably from the boyish, long-haired youth who had posed for a photograph before a newly constructed double helix in 1953 and who had lectured to me at Harvard University six years after that.

He was now bald on top, with a fringe of grey and white scruffy hair surrounding his head, a weathered and worn look on his face, and lines under his eyes. Many of his earlier trademarks were still present, however. He mumbled much of the time, put on intense pained expressions, grimaced occasionally, sucked in his breath in a half-chuckle, and now and then flashed a wild look in his eyes. Above all, he spoke with candor.

We talked for over an hour and I found that the very outrageousness of some of his opinions made me like him. He could be quoted with relish. In a world where the official statements of high officials generally signify nothing, his were refreshing and highly provocative. I decided that they deserved some sort of collective enshrinement in print. I have assembled a sample on my own, based on my own interview, but enriched with some of the more noteworthy opinions that he has given elsewhere.

Inspired by my memories of a similar work of the 1960s called the *Quotations of Chairman Mao*, I would like to call this collection the *Discourses of Director Jim*:

Topic: Francis Crick.
Jim: Francis is so much better than I am as a scientist. It's a pleasure that a few people like that exist in the world. Most people are boring. Francis is not boring.[32]
Topic: The recombinant DNA debate.
Jim: The majority of people who opposed recombinant DNA did so for political, not scientific reasons. . . . They either didn't like science, couldn't stand Crick or me, didn't like DNA, or were well-known malcontents. It was never genuine conviction. These people already hated our guts. We were enemies to start with. They were very unsuccessful enemies and rather miserable people.[33]
Topic: The fifteen-year period set for the human genome project.
Jim: Congress really isn't interested in anything which is going to be done in 25 years. You have to ask their age; my age.[34] (Robert Wright elaborated on this theme in the *New Republic* in July 1970. He wrote, "If the project proceeds on its present timetable, the genome will be completely wrapped up and sequenced in a little more than fifteen years, when Watson is in his mid-70's. It would make a nice 77th birthday present: thousands and thousands of pages . . . filled with the definitive description of human life's immortal essence, an essence that Watson was the first human to grasp. Many great scientists have been honored with their own encyclopedia entries; Watson would be the first to get a whole encyclopedia.")[35]
Topic: The early targets of the project.
Jim: Congress would like to see the gene for Alzheimers. So would I. We'd better get it done in 10 years.[36]
Topic: Charles Cantor and the search for the Alzheimer's gene.
Jim: Charlie will get it done much faster if there are no more meetings; he can't turn down meetings.[37] Stay home and get the Alzheimer's gene, Charlie![38]
Topic: The difficulties of traveling continually to Genome Project meetings.
Jim: Its not easy. My wife doesn't want me to be away. . . . I only serve a good function if I actually try and say the truth, and that upsets people, and that's no fun.[39]

Topic: Another genome project spokesperson who does enjoy the travel.
Jim: Some people are sick—he's sick.[40]
Topic: The fraction of the NIH Genome Center budget that will be spent on ethical issues.
Jim: Three percent of our money should be put into ethics. If I said one percent, people would think that was tokenism. If I said twenty percent, people would think that was crazy.[41] (A year later, the figure was raised to 4 percent.)[42]
Topic: The wisdom of the plan to spend up to half of that budget on large centers, three in 1990, and perhaps fourteen eventually.
Jim: Here the cottage industry approach involving small groups of individuals, each working at a different site, seems unlikely to succeed.[43] If we go along the way the NIH usually does, it could easily take one hundred years to get the sequence.[44] If you have to form a big group, those people who aren't good enough to form a big group won't get funded. . . . The Human Genome Project cannot be a WPA for the human genetics [group].[45]
Topic: The virtues of science centers in general.
Jim: The double helix was not found in a random place. The professor was Sir Lawrence Bragg and he was interested. Big groups are an inevitable consequence of the kind of things we're trying to solve.[46] [But] some units in the war on cancer have poor reputations: We all know how fraudulent most centers are.[47] (Norton Zinder, head of the NIH Program Advisory Committee on the Genome Project, has also endorsed this point of view, commenting: "In the past, centers were like werewolves—you couldn't kill them—and a lot of them go bad.")[48]
Topic: The claim that the Human Genome Project has been draining funds that are needed by scientists working on a smaller scale on other problems.
Jim: To what extent should we do charity to fellow scientists in trying to give them small grants to keep them going versus actively trying to do the best science we can do? To me its not a real choice. The latter is the only way the country will survive.[49]
Topic: The supposed dullness of genomic sequencing in comparison with the rest of science.
Jim: It is more exciting and challenging than the average Ph.D. The average person who works on a Ph.D. is probably on a pretty pedestrian project. There's some sort of fantasy world that everybody's doing terribly important pieces of science. Most everybody is doing something pretty dull.[50] Of course, if recombinant DNA hadn't come along, there

would be no genome project and there would even be money for the second class scientists.[51]

Topic: The statement by one NIH workshop participant: "Jim Watson is trying to change the social fabric of science. It's World War II all over again."[52]

Jim: We did very well in the Second World War. We did get the bomb, not the Germans.[53]

Topic: The size of the overall NIH budget.

Jim: It must be approximately doubled in real terms by the year 2000. The monies can and should come from the bloated military–industrial complex, which increasingly is producing weapons that either do not work as promised or have costs totally incomparable with any addition they may give to true national security.[54]

Topic: His own outspokenness.

Jim: I get no pleasure in giving public talks unless I say what I think. . . . Occasionally I lose my temper and say something I shouldn't. Most of the time its usually thought through. . . . It works.[55]

Topic: His motivation in taking the job as director of the Human Genome Office.

Jim: I didn't take it for the glory. I just took it because I wanted to get the job done.[56]

Topic: His duties in that job.

Jim: My chief function is to be ceremonial at functions like this. I sometimes wonder what I do.[57]

Watson's career appears fated to be forever enveloped by provocative prose—his own and his opponents'. In the end, though, he will be judged by his works. After I left our interview, I had the chance to sample another of them. I drove further along the road that had brought me from my home to the laboratory, around a corner of the bay and through the business district of the town of Cold Spring Harbor, a few blocks that were filled with quaint tourist shops. Just beyond that lay a smallish, red brick building, once a school, that now harbored the world's only museum devoted to DNA: the DNA Learning Center of the Cold Spring Harbor Laboratory.

Within were items illustrating much of the ground we have covered: a slide of Mendel; the Nobel Prize certificates of Thomas Morgan, Alfred Hershey (for the experiment that showed that the viral genes were made of DNA), and others; a photo of Avery and another of Davenport's Station for Experimental Evolution. White fiber threads of DNA itself lay in a dish, while exhibits explained how the two chains wound and unwound, as well as the role of DNA in criminal investigations. My favorite display, however, was one of the models of the man-eating plant, Audrey II, from the play and film, *Little Shop of Horrors*, a fictional example of DNA running amok.

Life with DNA may be frightening, but life without it can be worse. The brochure of the DNA Learning Center reminds us, "While new information flows out of laboratories, the majority of Americans remain biotechnically illiterate."

The Learning Center conducts courses and labs for high school students to help change this condition. They also operate a van that takes their show on the road. In 1990, a group of Soviet biologists were to visit, as a prelude to the establishment of a similar center in Moscow. "There is not a lot of understanding of DNA in the USSR," one of the biologists said earlier.

As we discussed in Chapter 4, the ignorance in the Soviet Union resulted from the temporary triumph of a brutish ideology, represented by Trofim Lysenko, a henchman of Stalin, over the patient progress of the scientific method. From the 1930s until Lysenko's downfall in the 1960s, the mention of genes, chromosomes, and DNA was suppressed in the Soviet Union in favor of a peculiar pseudoscientific doctrine that meshed well with the political philosophy of communism.

Centers such as the one in Cold Spring Harbor serve to guard against the recurrence of such horrors. This particular project of Jim Watson's may prove as pregnant in its own realm as the earlier one he and Crick did in genetics.

HUGO

If any rival to Jim in directing human genome efforts were to appear in the coming years, it would most likely be Hugo. I should write HUGO, in fact, as it is a group, not a person: The *HU*man Genome Organization. Its role, hopefully, will be to coordinate human genome efforts from various countries, a United Nations for the Genome Project.

The male name fits, as a group photo of the Founding Council at their organizing meeting (in Montreux, Switzerland, on September 7, 1988) shows thirty-one males and four females.[58] Y chromosomes are out in force. All but one of the men appear to be dressed in white or light-colored dress shirts. The exception, in a dark casual shirt, is labeled "observer." Wally, Charlie, and Jim are in the photo, of course, the first two in the front row and the last barely visible in the rear.

The first elected president was Victor McKusick, and the current one is Walter Bodmer, head of the Imperial Cancer Research Fund Laboratory. Walter Gilbert was elected treasurer (he resigned the next February), and Charles Cantor has become vice president for North America.

If HUGO was born in Montreux, it was conceived and named in advance by Sydney Brenner at a meeting in Cold Spring Harbor in April 1988. Jim was at least a godfather. Only in 1990, however, has it begun to nurse: a small flow of money permitted the establishment of offices in Bethesda, Maryland, and London. The funding did not come from any government, but from two private organizations: the Howard Hughes Medical Institute, which has been outstanding among philanthropies in its support for this research area, and the Wellcome Trust of the United Kingdom. A third office in Osaka is contemplated, but support has not yet materialized.

The idea for the Human Genome Project developed and was nurtured in the United States, but some other countries have been responsive to it. It does make sense that as many countries as possible join in, as the first complete sequence to be done will be valid for the human race rather than a limited subset of it. Jim Watson ruffled some feathers in 1989 with an offhand suggestion that responsibility for different chromosomes be parceled out to various countries. His comments were reported in *Science* by David Dickson: "The French might take several chromosomes and the Italians might take others." The USSR? "They might take a big chromosome." "The Canadian government might be more likely to put in money if they thought there was a particular Canadian chromosome" (he suggested No. 7) and so on.[59]

Responses were not enthusiastic, but few were less so than that of Jack McConnell, director of advanced technology at Johnson and Johnson: "I find the whole proposal cavalier and disturbing. Watson might take an odyssey around Europe and drop off chromosomes with cronies."[60] This experience may have contributed to a distaste that Watson feels for international diplomacy. He commented to me:

> We're not anxious to have a series of bilateral agreements with other countries. Our office would go crazy if it had to have a bilateral agreement with Brazil. If we have one with the U.K., what do we do about a Brazilian one? We've got to smile, they would be devastated. There has to be some well paid secretariat that will put up with all this crap.[61]

Jim Watson, of course, has been an advocate of HUGO, which could fill that role.

As for the various countries, they preferred to find their own directions in this area. Their responses have varied. I have taken the liberty of composing a line or two, on my own, to describe my impressions of them as of 1990. The information has been extracted from journal articles and a talk given in 1989 by Peter Pierson, a science officer of the European Economic Community (EEC).

USA: (Initial reaction) We'll pay all of the costs for this, just like we did for the Apollo project. We've always taken the lead on behalf of humanity when adventuresome science projects came up. (Later on, after the costs became clearer) Perhaps this should be an international effort, after all, with many nations providing financial support. We can't support such science as we did in the past.

EEC: We are willing to spend a few million dollars on this work in our EEC laboratory facilities. Of course, everything that we do has to be approved by our member nations, which will take time. We have agreed on an initial drive on yeast, however. The experience of our component nations in beer and wine making has given us a lot of expertise in this area.

United Kingdom: Have you forgotten? We were there from the beginning and deserve the lion's share of credit for all of this. We don't feel courageous enough to bite the entire apple right now, though, so we will settle for the worm.

Italy: We would like to choose a suitable area for our nation to study; part of a sex chromosome would do nicely. We will take the larger part of the X: the end that is richest in genes.

France: "We will eat the chocolate without the bread," Philippe Kourilsky of the Pasteur Institute commented to Peter Coles of *Nature* in October 1990.[62] He meant by this that the French effort would concentrate on the protein-coding regions of the genome. There were likely to be several hands on the chocolate, however, as the U.S. DOE announced a similar undertaking at about the same time.

The French would contribute in other ways as well.[63] They were collaborating with the British in developing automated sequencing instruments and had for years maintained CEPH, an international data bank that preserves and makes available DNA samples taken from several dozen very large families. In addition, they would sequence small bacterial genomes.

Germany (West Germany at the time): We must think this over: Our past experience with genetics has not been so good.

I must elaborate on this one. Much hostility has been directed against gene splicing with the Genome Project just an indirect victim. In 1989, for example, a West German court blocked the operation of an almost-completed plant to manufacture human insulin. In 1988, *Nature*, in the editorial "Greens against Genes" wrote: "The particular issue of genetic engineering, linked as it is in West Germany with still recent horrors at places such as Auschwitz, is bound to be difficult."[64] My Jewish father, who died of complications related to diabetes, would not have been impressed by this particular way of showing remorse.

USSR: Yes, we would like very much to be full members of this international effort. With Lysenko, we have already had our stupidity for this century. For the same reason, though, we will not be able to do very much at first.

Denmark: This will make a fine subject for discussion at public forums.

Japan: We are a wealthy nation and intend to contribute to basic science. The way that we choose to do this is our own business. Many different agencies have responsibility for sequencing efforts. Memoranda must be sent and discussions held to arrange any new organization. In the interim, funds are being spent for research in this area by the old agencies. For the same reasons, we cannot give money for HUGO now. We must have the details of any proposed activities before we can request government funds.

The Japanese position, in particular, aroused Watson's ire. I quoted in Chapter 6 his remarks at the Genome I meeting about fighting war, if necessary. He has felt that "it is against the American national interest to work out the human genome and pass it out free to the rest of the world." Leslie Roberts reported on the controversy in *Science* on November 3, 1989. The particular target of Watson's wrath was the Japanese, who he wished to provoke into contributing to HUGO and ultimately paying for up to one-third of the genome costs. He wrote a letter to the relevant Japanese official in which he canceled a scheduled visit and asserted that they should pay up "if Japan wants to be considered a great nation." He added: "Japan should no longer expect to benefit from the generosity of other nations if it remains outside the HUGO sphere."[65] In particular, he proposed that DNA sequence information be withheld from the Japan unless they joined.

The Japanese were not pleased. Watson was accused of "blackmail" and "Japan-bashing," while others called it "shocking," "uncalled for," and "shooting from the hip." If data were withheld, the Japanese would be "extremely annoyed." Watson retorted: "You never get anywhere in the world by being a wimp." But his friend Norton Zinder reported that the letter sent to Japan was a final draft: "You should have seen the original letter. . . . Had he sent that Japan would have withdrawn its ambassador. The trouble with Jim is he is often right but not very polite."

I have not seen that letter, but for historical purposes, I can substitute the following selection, quoted in my interview with Jim:

> The Japs better get realizing that just because they smile, we're not going to say they are good people. People who smile often have daggers. They're basically minor people. . . . The Japanese are very smug now in believing they're going to dominate the world. But if the United States got going, we'd beat the shit out of them. I personally right now would take all our missiles and aim them at Tokyo. I would first destroy most of the missiles, but the couple I had, I would let them know that I remember Pearl Harbor.[66]

In practice, the withholding of data would be almost as difficult as the repositioning of missiles. Most scientists believe in the free exchange of information in basic research. A spokesperson for HUGO, quoted in *Science*, said that it was the unanimous view of all members [if we ignore that Watson is a member] that "all data would be freely accessible to scientists and not used to secure narrow national interests."[67]

Watson had to retreat to a different position: Data could be withheld for a time, but only so that the scientists who had gathered it would have the first chance to examine it.

> We'll get better sequencing if the people who are doing it are interested in the answer. . . . It wouldn't be fun to let raw data flow into Genbank. Science has to be fun.[68]

How long might the delay be? Six months would be too short, one hundred years too long (although Norton Zinder has endorsed the former figure). The impression is left, however, that transmission would be by fax and computer for those who pay their bills, Pony Express for the others.

Watson's action may be a bluff, but could yet hit the target. Japanese officials were concerned because he is influential, while Japanese scientists, interviewed by *Science* in San Diego, supported his efforts to secure more funding for basic science in their country.[69] By September 1990, increased funding for genome research in Japan had appeared, but none was yet in sight for HUGO. On the whole, by early 1991, the Japanese effort still lacked sufficient funding.[70] Time will judge the virtues of the following maxim that I have composed:

Be not a wimp; you'll win the day,
But first discover DNA.

We may be wrong, however, to worry about the disposition of sequences not yet determined. Let us update another old saying: Do not count your sequences before you have read the gels. What remains to be worked out is the human genome, or at least 99.9 percent of it. How will this be done?

14

THE TIME OF THE GENOME

Imagine that a historian was given full information at the start of World War II about the strategic plans and armaments of England, Germany, France, and the other combatants. He was allowed to question Hitler, Churchill, and anyone else about their intentions and permitted to examine any fortifications and weapons that he chose. At the end of his research, however, he was asked to provide a detailed account of how the war would proceed, battle by battle.

Your reaction would be the same as mine: that the historian could not do it. He could summarize the strategic questions, guess which areas and weapons might be important, and perhaps even speculate as to the final outcome, but he could not, except as avowed fiction, write the actual history.

Similar problems are faced when I try to sketch the course of the Human Genome Project during the last decade of this millennium, and the first one of the next. The methods to be used, the actual course of events, and the date of its conclusion cannot be fathomed. One central feature remains likely: The job will get done.

The debates and committee meetings in the late 1980s that led to adoption of the project by the U.S. Congress did, however, affect the general course of future events, just as the surrender of France and Hitler's decision to invade the USSR affected the vicissitudes of World War II. At least three important questions were considered and answered at an early stage:

1. Would the sequencing proceed on the basis of existing methods, or would new technology be devised?

2. Which overall approach would be emphasized: "bottom up" or "top down?"

3. Would human DNA alone be sequenced, or would other organisms be included?

When Walter Gilbert proposed to form a Genome Corporation in 1986, his announced strategy (at least at one conference that I attended) featured the first named option for each of the above questions. The choice of those who drew up the Genome Project was the second in each case. We cannot be sure in advance that the second group of choices (or a combination of first and second) will be more efficient, but they certainly raised more enthusiasm for the venture. Scientists such as Nobel laureate David Baltimore who shuddered initially at the thought of a genome sequencing project came around to favor it when they saw what the overall strategic plan would be. Let us probe a bit further to see why this was so.

THE STATE OF THE ART

At the end of the 1980s, the principle sequencing methods in use were still those devised fifteen years earlier by Sanger and Gilbert. The sequence of up to five hundred bases in a DNA chain was "read off" at the end of a laborious experimental procedure, "running a gel," which took many hours to carry out. I have never run or supervised such an experiment on my own, but I was able to watch one simply by going to the laboratory of a colleague, Charles Grubmeyer, in the New York University building where I work.

In that research group, the centerpiece of sequencing work was a hard, clear plastic slab, perhaps a foot and a half high, set in a plastic base. To set up a sequencing run, one of the several students who shared the apparatus first had to prepare a fresh gel. The ingredients were mixed and then poured between two sheets of glass to solidify. The glass plates, which contained the gel sandwiched between them, were then placed within the apparatus.

The sequencing reactions were run in a separate procedure that involved the transfer of tiny amounts of liquids (one of which held the

DNA) between small vessels using a manual device called a *micropipette*. The needed biochemicals were taken from commercially available vials with pretty color-coded caps. The dideoxy compounds that Fred Sanger and Alan Coulson had first prepared with their own hands fifteen years earlier were now prepackaged in the right amounts in pink-capped containers marked A, T, G, and C.

After the transfers were completed, the mixtures were heated in a metal block and then put one at a time into special "wells" near the top of the sequencing apparatus. The power was then turned on, and the separation run for up to ten hours. This continued automatically, but the instrument had to be monitored to prevent overheating. At the end of the run, the gel was removed, and extra chemical treatments were used to "fix" it. Finally, it emerged as a thin, firm plastic sheet attached to a paper backing. The sheet was then exposed overnight to X-ray film to make visible the sequencing "ladder" of bands, from which the order of letters in DNA could be read, as illustrated in Chapter 5.

Each of the students had a project that involved reading up to one thousand base pairs of the DNA sequence of a virus or the bacteria *Salmonella* (sometimes this had to be done repeatedly in different mutant strains). To do a run, he or she would start to prepare a gel in late afternoon and work through the next day. The following morning, the sequence of some 350 or 400 bases could be read, provided that everything had been carried out properly. The work was demanding and repetitive.

If only a few hundred letters of DNA text were wanted in the course of a research project, then such a sequencing procedure would represent a unique learning experience, and the worker could move on to a different adventure. However, most DNA targets offered by nature come in much bigger chunks. Consider the problem of reading the fifty thousand base pairs of lambda virus, for example.

In the very best of circumstances, the sequence of lambda virus could be learned by running and reading one hundred sequencing gels. It could be done that efficiently only if we had put in a lot of work ahead of time. We would want the DNA to be broken up in advance into tidy five-hundred-letter (or less) chunks, each in a separate bottle. Further, we would want the bottles to be labeled: No. 1, No. 2, No. 3, and so on in accord with the position of their message in the complete lambda text. The one hundred segments of DNA, once read separately, could then be lined up to give the full sequence of the virus.

Of course, this advance work would represent a major project in itself as vexing as the sequencing part. It is called *preparing a physical map*. In the above case, we could do it by slicing up the virus into a few large pieces with a well-selected text cutter. The relative order of those pieces would then be worked out. (A number of different procedures are available for this.) Each of the large fragments would then be subdivided further using a different text cutter, and the subfragments aligned in turn. This would be continued until a full array of ordered gel-sized chunks had been generated. Such a subdivide-and-conquer process would be termed a *top-down* approach.

While this process is logical and elegant, it is not really necessary for DNA texts of the size of lambda virus. Fred Sanger and Alan Coulson chose a more direct alternative in the early 1980s that relied more heavily on sequencing and avoided the need for a physical map.

Using enzymes that had no preferred cutting place or sonic waves, they broke the viral DNA directly into small fragments of suitable length for sequencing. Fragments were then selected at random, and their text was deciphered. This is termed a *shotgun sequencing strategy*. As text accumulated, portions of overlap would appear. This would not have happened if only a single virus were torn up, but in practice, huge numbers were used in each preparation.

As an analogy, imagine that we were trying to reconstruct the text of this book from a pile of shreds that had been prepared from multiple copies of it. If we selected one piece at random, we might pick up a section that represented lines 1 to 8 of this page. Many selections later, we might get lines 5 to 13. If we systematically compared everything that we picked up, we would notice that the two had some lines in common, so they presumably represented overlapping or contiguous passages from the book. We could combine them to form a larger piece that covered lines 1 to 13. Such a combination is termed a *contig*. The contig would be extended further when additional pieces of adjacent, overlapping text were found.

In such an approach, which has been dubbed *bottom up*, information piles up rapidly at the beginning. As you move along, however, you start to pile up duplicates of sequences that you already have, and the last few stretches of text become hard to find. I faced the same problem when I was a boy and attempted to get a complete set of 256 baseball cards (they came with bubble gum) by random purchase from stores. In the end, I had to fall back on a different method, trading cards with another collector, to finish my set. In sequencing, however, there is so much work to be done that usually only one research team will be "collecting" a particular set of DNA fragments at one time.

When Sanger, Coulson, and co-workers sequenced lambda by this strategy, they had to run many hundreds of gels, not just one hundred, and invent some additional strategems to get the job done. The monotony of running the same procedure repeatedly also took its toll. "After we finished sequencing lambda," Alan Coulson said to me later, "I swore I would never sequence anything again." He then qualified the remark: "The actual repetition isn't very exciting but you have to accept it. Its not terribly painful."[1]

Most recently, Bart Barrell of the original Sanger team has pushed the limits further and obtained the complete text of cytomegalovirus, 240,000 letters long. This effort consumed twelve person-years, or an average of twenty thousand base pairs per person per year. The rate of progress was not uniform though. Toward the end, the text was acquired at up to five times that speed. According to Barrell, "sequencing is still an art. There are many failures and a lot of down time."[2] As we suggested above, a lot of effort was needed at the very end to get the last one to two percent.

Ellison Chen at Genentech had a similar experience when he determined the human DNA area that codes for human growth hormone. This stretch, seventy thousand letters long, may represent the longest continuous known sequence of human DNA. In principle, he found, a skilled technician could do 100,000 base pairs per year, but in practice, twice that time was needed because "the work is so boring."

Whether we found it tedious or not, by repeated application of the method we could ultimately sequence texts as large as the human genome, just as we could eventually empty a large reservoir using a single teaspoon. By the middle of 1990, over 30 million letters of DNA sequence from all sources in biology had been compiled and stored in computer data banks; more than 6 million of them referred to human sources. By a naive calculation, the total output recorded would have to be increased some five-hundredfold to acquire the 3 billion letters of the human genome.

If we could assemble a large group, and each worker maintained the maximum rate given above, then some thirty thousand person-years of labor would be required to complete the project. If we wanted the job to be done in fifteen years, then two thousand trained technicians would be needed. This calculation is not meant to be precise. It ignores many technical problems, such as those of closing gaps. On the other hand, many improvements due to automation that are now ready for use would speed things up.

While such a project would be possible in theory, its mechanical nature, its lack of interaction with other areas of biology, and its failure to provide inspiration to the public make it extremely unattractive. As we shall see, the routes to be followed will be far more scenic and very different.

A CALL FOR INNOVATION

Imagine the following scenario for the Apollo project: The president and Congress authorize a manned moon landing, appropriate the initial funds, and set up the needed administrative offices with NASA, but the scientists involved declare that the $30 billion price tag is much too high and decide that the first years shall be devoted to finding a much less expensive way to travel there. This, of course, did not happen, nor has any sign of such thinking been observed in typical Big Science projects such as the supercollider and space telescope. The Human Genome Project, however, chose exactly this strategy.

The cost of sequencing projects varies from lab to lab, but can be expressed as several dollars per base pair at the present time. The goal stated by James Watson and others has been to reduce that to perhaps fifty cents per base pair, so that the total cost for the whole project is limited to about $3 billion.

This goal appears unique not only to science, but also to society. No committee exists for the purpose of reducing the cost of a New York subway ride from the present $1.15 to $.25, for example. The DNA sequencing field, however, was virtually created from nothing by the remarkable simultaneous innovations of Sanger and of Maxam and Gilbert in the mid-1970s, a time still fresh in the memory of most of us. Their achievements, while ingenious, hardly were meant to represent the last word in the field. Shortly after the methods were announced, in fact, a review in a human genetics journal stated:

> It also seems clear, judging by the experience of recent years, that the methods described here will become obsolete almost as quickly as they have appeared, to be replaced by even more rapid and elegant methods.[3]

THE MOLECULE OF THE YEAR

The vast areas that remain open for innovation can be appreciated if we consider the impact of another remarkable procedure discovered during the 1980s. This invention did not alter the basic strategy of the sequencing methods, but vastly extended their applications. At the Human Genome meeting in San Diego in October 1989, Charles Cantor singled the new method out for special attention: "You are going to hear over and over in these talks of the central importance of PCR." As I have mentioned already, the letters stand for polymerase chain reaction. I shall use the more descriptive phrase "DNA amplification" when I refer to it. The vast utility of this method gained it not only verbal praise, but also one unique honor. It was named "The Molecule of the Year."[4] This award was created by *Science* at the end of 1989 "to force us to choose one such discovery each year that is likely to have the greatest influence on history."[5] DNA amplification was selected "as the major scientific development of 1989."[6] The technique had first been reported several years earlier, but *Science* felt that it was appropriate to honor it when its importance had become fully apparent.

By 1989, in fact, the number of papers citing the method had become immense, more than one thousand. *The Scientist* had identified it as the second hottest area in science and the most frequently cited one in biology during that year.[7] A monograph and a laboratory manual had also been published to teach beginners how to use it.

What could a scientist accomplish with this new technique? She could take a very small amount of DNA and by means of a simple procedure amplify its quantity by a millionfold in a single afternoon. This ability was vital if the goal was to obtain DNA sequences from a single sperm, a hair, a trace of blood left at the scene of a murder, or the dried tissue of an ancient mummy. The technique also allowed tests for genetic diseases such as cystic fibrosis to be performed on a single cell taken from a not-yet-implanted embryo in a test tube, for example. In another application, the presence of very low concentrations of AIDS virus could be detected where other methods had failed.

To use the method, the scientist would have to know short sequences at both ends of the DNA that was to be amplified and have the ability to prepare them in the laboratory. These bits of DNA, called *primers*, are usually about twenty bases long, although twelve will often do. Further, the DNA stretch that she wished to amplify could not be more than about six thousand base pairs long. These restrictions did not hinder many clever applications of the type I described above.

For example, suppose she wished to test an individual for sickle-cell anemia. She could use a DNA sample taken from any of the patient's tissues. She would not need to fractionate the sample or probe it to run the test. By selecting the proper short end sequences, she could choose to amplify only the beta-globin gene area of the DNA sample.

A sequence of DNA that originally represented less than one-millionth of the total amount present would now be the majority component. It would be quite easy to purify it and determine whether the sickle cell spelling change was present, using a text cutter or other method.

How would this be carried out in practice? Consider the beta-globin sequence in Scheme 6, for example, and suppose that this was the exact length of DNA that the scientist wished to amplify.

Her first task would be to prepare a short chain of DNA that contained the first twenty bases shown in Scheme 6 on page 120, TAGACCTCACCCTGTGGAGC. The preparation of this primer would have been very troublesome in 1970, but has become routine since about 1980, as commercial instruments do the job in an automated procedure. Furthermore, once it had been made, the same preparation would provide enough of it to run tests on many individuals. A second primer would be needed to mark the other end of the region to be amplified. The Scheme 6 sequence ends with GTCTATGGGACCCTTGATGT. The mechanics of the method require, however, that the second primer represent the other chain, not shown in Scheme 6, but easily deduced by the rules of Watson and Crick. The other chain runs in a reverse direction, so the needed second primer would be ACATCAAGGGTCCCATAGAC.

An excess of these two primers would be put together in water with a small sample of the DNA of the individual she was testing, a liberal supply of DNA building blocks, and an enzyme that can copy DNA (a *DNA polymerase*). A number of such enzymes are produced by different organisms, but the best one for the purpose proved to be one that was relatively heat resistant. It is not damaged by the repeated heating cycles that are used in the procedure.

With these ingredients assembled, she would only need to repeatedly heat and cool to room temperature the mixture. With each heat-and-cool cycle, the section of DNA marked off by the primers would double in amount until, after thirty cycles, it was amplified one-millionfold.

The sequence between the primers need not be known in advance, as we do in the case of the sickle-cell gene. One geneticist, for example, noted that Alu families often occur in pairs in the human genome. He wished to collect all of the sequences that occurred between such pairs. He used the end of an Alu as the left primer and the start of another as the right one and harvested all of the sequences that occur between closely spaced Alus, anywhere on any of our chromosomes.[8] Using the same approach, gaps could be closed between contigs in an incomplete DNA sequence.

Many other ingenious applications of DNA amplification have been devised. The sensitivity of this method and the short time needed are important elements of it. Earlier we saw that DNA can be amplified by cloning it within bacteria or other cells. However, this method is much more laborious, requires weeks or months, and requires larger samples.

THE VANISHING INVENTOR

The name of Gregor Mendel is forever linked with the rules of heredity, Watson and Crick with DNA structure, and Sanger and Gilbert with the DNA sequencing breakthrough. Who then would be credited with DNA amplification? The *Science* award editorial made no reference to an inventor, stating that "most of the discoveries of science . . . result from the actions of many individuals, one of whom may contribute slightly more than the others."[9] Perhaps so, but not in the cases I have mentioned above. To which class did DNA amplification belong?

John Bell, reviewing the monograph "PCR Technology; Principles and Applications for DNA Amplification," edited by Henry Erlich, wrote in *Nature*:

> No group is more suited to review the field than Henry Erlich and his colleagues at Cetus. It was from their laboratory that the idea arose and was systematically and rapidly developed.[10]

The key papers on DNA amplification had indeed come from the Cetus Corporation, a biotechnology company located just east of San Francisco Bay, with R.K. Saiki as first author and H.A. Erlich further down the list of coauthors. When I started my search for the inventor early in 1989, I wrote to Saiki and was invited to come to Cetus. At the company, I was greeted by both Randy Saiki and Henry Erlich.

Erlich, a dark-haired, clean-shaven, and husky individual, made it clear that he was the person in charge of this area of research. He described some fascinating new applications of the reaction to forensic science, but when asked about the original discovery, attributed it to a former employee named Kary Mullis.[11] Cetus had developed the procedure, however, and held the relevant patents. Mullis now lived in San Diego, but the firm could not supply his address or phone number. Fortunately, I was able to obtain them from another scientist at a meeting a month later.

SAN DIEGO, CALIFORNIA; OCTOBER 1989. The Human Genome I meeting had just concluded. I had spoken with several key people and collected enough information to occupy many winter evenings, but one important interview had eluded me. I had hoped to meet Kary Mullis. He had not showed up for our initial appointment, and the signs that I posted on the bulletin board of the convention hall suggesting a later rendezvous were not acknowledged. Yet he was in town and had been sighted in the halls of the building where the conference was held. Unfortunately, I did not run into him and now the meeting was over.

On a sudden impulse, I returned to the almost-empty convention hall for a last look. One individual stood below my sign: Tall, lean, rugged, balding, and dressed in western attire, he seemed to have stepped directly out of a Sam Shepard play. (I later learned that he was raised in the Carolinas. What does a New Yorker know about these things?) This was certainly my quarry. He apologized for the missed connections and commented that he had to leave in a few minutes to catch a plane for San Francisco. Having said this, he then sat down with me at a small table adjacent to one of the swimming pools at the motel–convention center and talked for hours on that sunny afternoon. And so I got the story of the invention of DNA amplification (PCR). He later published some of the details in an account in the April 1990 issue of *Scientific American*.

KARY MULLIS'S ACCOUNT. A chemist by training, with a Berkeley Ph.D., his interests had ranged far and wide. While in graduate school, for example, he had taken time out to prepare a paper on the physics of time reversal that was published in *Nature*! He doubted that he could earn a living in this area and so resumed his chemistry.

He had been hired by Cetus in 1979 to prepare short stretches of DNA by chemical means in the laboratory. When automated machines became available to simplify his work, he found himself underemployed. "I had lots of time," he said. "I started playing around (with DNA). I didn't know much DNA chemistry—very little about genetics. DNA was just sort of a stringy mess—it was huge."[12] He became interested in a problem that Henry Erlich and others were working on down the hall—the diagnosis of sickle-cell anemia through DNA sequencing. So in his spare time, he tried to devise new methods for that purpose.

He recalls vividly one particular evening in the spring of 1983: "I used to drive to Mendicino County every weekend. I had a little place. I was building a cabin." Driving one particular Friday, he recalled, "It was dark at night. The buckeyes were in bloom. They were smelling really nice. Jennifer [his female friend, a chemist] was asleep beside me." As he drove, he designed a sickle-cell sequencing experiment, changing one detail after another as technical obstacles occurred to him. Unwittingly, he was assembling the exact set of items needed for DNA amplification. "The whole thing was a comedy of errors, in a sense."

Suddenly he realized that his new combination would not give him a sequence, but would double the DNA in his sample. What if the procedure was repeated again and again? "Biochemists were not in the habit of doing things over and over again," but he had a lot of experience with computer programs where this was common. He calculated that by repeating it again and again, he could get a huge amplification. "It was definitely lightning striking." In his excitement, he woke up Jennifer and told her that he had discovered "something fantastic," but she just went back to sleep. According to Mullis: "It was driving me mad. I stayed up all night writing this down . . . why won't this work." He decided that it would.

Back at work, he received no more encouragement.

> I had talked to a lot of people at Cetus—50 people before I did it—and nobody had any excitement at all about it. Even people who would have directly benefitted—they kept on business as usual. I think my excitement about it maybe put some people off, but also the fact that it would have solved a lot of problems that people had been working on for a long time. People don't like you to solve their problems. They like to solve their own problems.[13]

Finally, on December 16, 1983 (he remembers the date because it was his third wife's birthday), he got to test his idea. His assistant was Fred Faloona, a mathematician whom he had met through his daughter. He reported the circumstances later to *Scientific American*:

> When everything was ready, I ran my favorite kind of experiment: one involving a single test tube and producing

> a yes-or-no answer. Would the PCR amplify the DNA sequence I had selected. The answer was yes. . . . Fred and I celebrated on the night of its success with a few beers.[14]

Fred was lucky. "His first [biochemistry] experiment had worked, and we had just changed molecular biology forever."

Mullis told the Cetus patent attorney, Albert Hallunin, of the result that same evening and found that he understood the significance. His other colleagues were less receptive. "I got a run-around. Nobody was interested in the whole thing." Finally, he got to demonstrate its use for amplification of the sickle-cell gene. Henry Erlich and another senior scientist, Norman Arnheim, became interested. But now things became "really nasty" in another way. "Norman started smelling some kind of a big deal."

The initial 1985 paper in *Science* that described DNA amplification surrounded the novel new method with the details of the method for sickle-cell anemia analysis.[15] The names of Mullis and Faloona were embedded among those of a host of coauthors. They "did all that without talking to me about it," Mullis said. "When I saw what they had published in *Science*, I was really furious." As a result of the first publication, an effort by Mullis and Faloona to publish a separate account of their own contribution was turned down by *Science* and *Nature*. The journals felt that the details had already appeared, and another publication was unwarranted. Mullis and Faloona did manage to publish the details of their procedure in a compendium of biochemical methods two years later.

A group patent was also contemplated. It occurred to him that "these guys have been slowly trying to take my invention away from me." Only by threatening a lawsuit was he able to have the invention in a patent under his own name, although the property of Cetus, and the applications in a separate one under group authorship. The matter of invention seemed settled, but a new problem appeared.

" 'Develop' is the word they started using . . . which sounded like they had done it," Mullis said to me. "Henry and Randy [Saiki] were trying as best they could to obscure it. . . . He [Henry] could not develop Polaroid film if he tried. He's not a developer of anything." Mullis felt that he had made the key innovations.

More friction developed when Arnheim began to supervise Mullis's research more closely.

> I grew sick and tired of Norman. I was supposed to do what he thought. I think I'd now done something. The company ought to take notice of it. I had to write reports and crap like that. They ought to at least let me have the same freedom I had when I invented it. I had free time. Nobody was bugging me. I was enjoying myself. I was playing. It wasn't an accident that I invented it when I had free time. Give me back the situation I had then and I'll do some more stuff for you.[16]

One nice thing, however, happened: "Jim Watson heard about it." Watson had made his own inquiries, then invited Mullis to the important 1986 meeting "Molecular Biology of Homo Sapiens" at Cold Spring Harbor to present his work. "Hearing me talk, it was clear to everybody that I had done it." This acceptance also gave Mullis the encouragement to present his own version of events. One person who accepted his account was a young physical biochemist, David Schwartz. He told Mullis, "That's the way it is. That's the way it always is."

His supervisor at Cetus was less pleased, and let Mullis know it. Mullis's response was: "I am out of here." His affiliation with the company ended in 1986. He found employment with a plastics manufacturer on the West Coast for a time, but that also did not work out. Then fortune proverbially smiled on him. As the importance of DNA amplification was recognized, biotechnology firms began to consult him about its applications and pay well for his expertise. "I'm not rich but I live comfortably," he told the *Wall Street Journal*.[17] "I have this really nice girl friend who helps me remember to pay the rent."

Ironically, Cetus is among the companies that employ him as a consultant. DuPont has challenged Cetus's patents on DNA amplification, and his services are needed to defend the claim. As part of their understanding, Cetus now acknowledges his contribution. In a December 7, 1989, letter to *Nature* provoked by the patent controversy, J.S. Price of Cetus wrote: "Cetus is confident that Kary Mullis will continue to receive the recognition that he deserves."[18]

Mullis, meanwhile, has grown "tired of PCR." His interests, as reported to the *Wall Street Journal*, include artificial intelligence, tunneling microscopes, science fiction, and surfing lessons. The DNA amplification discovery will remain his special pride, however. As he told me, "This one was reserved for me. My fairy godmother said: 'If you'll be a good boy and raise your kids, then I'll give you this one.' "[19]

Kary Mullis is no longer working at the lab bench. How many others may there be nestled in some remote corner of the scientific establishment needing, like Mullis or Gregor Mendel, only a bit of work space, some free time, and a measure of security to make their contribution? Science historian Robert Root-Bernstein has examined scientific creativity at length and reached the following conclusion, reported in *Nature* by the reviewer Walter Gratzer:

> His analysis of conceptual upheavals operating in science over the last century or so also leads him to the conclusion that a large proportion have come from unknown researchers, operating in the geographical wilderness.[20]

QUESTIONS OF STRATEGY

A revolutionary new method in the direct area of sequencing could so much reduce the time and effort needed in the Human Genome Project that it would no longer be considered as Big Science (in cost, that is, not concept). Such an advance, however, although possible, cannot be relied upon. The strategic choice facing the genome planners in this situation reminded me of one of the favorite science fiction stories of my childhood.

FAR CENTAURUS. A group of colonists have set off from earth to establish a settlement on a planet of the "nearby" star system Alpha Centauri. The voyage lasts many centuries; they must pass the time in suspended animation. When they finally arrive and are awoken by the ship's robots, a party of inhabitants comes out to greet them. The welcomers are not extraterrestrials, but normal earthlings. During the time when the would-be colonists slept, a far more efficient interstellar drive was invented. With the aid of this faster device, a later wave of settlers has already colonized the planet. The original expedition represents a relic of the past: a museum in motion. Its personnel find no better alternative than to return home.

A similar fate threatens those, of course, who invest time and effort into improvements of the current sequencing method. However, this result is not guaranteed, and a strategic choice must be made. The repetitive and tedious nature of the Sanger and Maxam–Gilbert sequencing protocols begs for automation, particularly if a massive increase in their use is expected.

Instruments that carry out the band separation and gel-reading step have already been put on the market. These devices use fluorescence rather than hazardous radioactivity to detect the bands. They cost more than a luxury automobile, but are gaining acceptance. For example, Alan Coulson in Sanger's old establishment at Cambridge has bought one.

At the same time, masters of innovation like Leroy Hood at the California Institute of Technology are designing workstations that carry the automation further: They prepare the DNA and run the chemical reactions.

Other workers are looking for ways to extend the readout in a single run from five hundred to one thousand letters. George Church at Harvard University has developed mutiplex sequencing, which allows dozens of runs to be performed at the same time, using the same equipment. Wilhelm Ansorge at the European Molecular Biology Laboratories in Heidelberg, Germany, has developed very thin and rapid gels and foresees that his apparatus will determine twelve hundred letters of sequence every hour. With these improvements in place, those involved forecast that many thousands of bases will be read per day at each instrument at a cost of fifty cents or less per base. Computer handling of the data may become the rate-limiting step.

With the tedious steps performed by machines, humans would be free to devote themselves to planning, dealing with unexpected contingencies, and deciding what the results mean. In counter to Sydney Brenner's comparison of a sequencing laboratory to a penal colony, George Church has offered an analogy to a video game arcade, a place where a basically repetitive operation is so enjoyable that the participants work voluntarily for hours on end and even pay for the privilege.

Yet all of this enterprise is threatened by what I will call the Centaurus effect: a bypass by totally different and vastly superior methodology. The DOE, in particular, is pushing "far-out" technologies. Richard Keller of Los Alamos Laboratories, for example, is working on a procedure that would read hundreds of bases every second. Stocky, balding, and grey haired, he more resembles the image of a skilled technician than an academic genius. Yet, as we have seen, the notable advances do not come from predictable directions or standardized human types. In his method, a single DNA molecule would be secured to a solid surface and dismantled from one end by an enzyme. As the bases were released, one at a time, they would be swept past a detector and identified by a hypersensitive instrument. A great deal of tinkering will be needed to make this possible, for example, the bases will have to be rendered fluorescent by some prior treatment.

An alternative innovative procedure involves the direct examination of the base sequence by a new instrument called the *scanning tunneling microscope*. This device is capable of "seeing" the shape of individual molecules. (Feeling, as in reading Braille text, would be a more appropriate comparison. The tip of an ultrasharp needle passes slowly over a surface, sensing it by exchanging electrons with it. The results of a series of horizontal passes are combined to produce a "picture.") Strands of DNA have been spread on graphite and examined, revealing the basic helical shape deduced by Watson and Crick. Thus far, the resolution does not permit a base sequence readout.

Will one of these ideas or another novelty carry the day? Opinions differ, even among Wally, Charlie, and Jim. Walter Gilbert commented to me about the new single-molecule techniques: "None of that has yet proven itself in the slightest."[21] Wally plans to modify his own methods so that many separations can be run at the same time using the same gel. He hopes to scale up in the next four years to a speed in which 4 million bases per year can be read in his own laboratory. (Wilhelm Ansorge, who followed him at the Genome II meeting, started his lecture with the line "I have to draw you down from the dream world of Walter Gilbert and show you how we have to muddle through on earth.")

That rate of production would be good enough for Jim. In calling for the improvement of already-existing methods, Watson said: "A factor of 10 can be enough. I'm too old to believe in 1000-fold advances. I've seen it once with Gilbert and Sanger." On the other hand, Charles Cantor predicted at a Florida meeting, late in 1989: "When the human genome is sequenced, it will probably not be by any of the technologies we now use for gene sequencing."

The genome planners are hedging their bets. Both the evolutionary and revolutionary approaches to sequencing are being supported. Many workers feel that the older methods will rule the roost for the next five years anyway. The joint planners of the NIH and DOE do agree, however, that now is not the time to launch massive sequencing efforts, although work on particular objects of importance will continue. The next five years of the project will be a time for innovation in sequencing.

During this time, there are other jobs to be done. The human genome has to be subdivided into pieces that would be suitable for sequencing, in accord with the top-down approach. Geneticists had been doing this for years in preparing their linkage maps, but this effort has to be stepped up. Another kind of map would also be helpful as well, one that measured distances in actual base pairs rather than frequency of crossing-over. Both types would be sought.

THE MAP OF MAN

We report the construction of a linkage map of the human genome. (It is) linked to at least 95% of the DNA in the human genome. . . . Together with the data from other groups constructing more detailed linking maps of other chromosomal regions, a truly comprehensive genetic linkage map of humans will emerge.

—Helen Donis-Keller and thirty-two collaborators in *Cell*, 1987.[22]

What is now published is only to be considered as a general map *of MAN, marking out no more than the greater parts, their* extent, *their* limits *and their* connection, *and leaving the particular to be more fully delineated in the charts which are to follow.*

— Alexander Pope, letter to Henry St. John, Lord Bolingbroke, introduction to *An Essay on Man*, 1733

Donis-Keller and co-workers in the announcement accompanying their 1987 paper called it the "world's first genetic map" for humans. They did not acknowledge Alexander Pope. Of course, what Pope published was simply a philosophic discourse on human nature rather than a genetic map, a listing of the order of genes and markers on various chromosomes. The latter concept arose from the work of Thomas Hunt Morgan and his co-workers and was developed for the fruit fly by Alfred Sturtevant after 1910.

As humans breed less rapidly and in lesser numbers than fruit flies, and we cannot control their mating, the preparation of genetic maps for humans has taken more time. Only in the 1970s and 1980s did human gene mapping really take off with the development of new types of markers and radioactive and fluorescent probes that could be observed after they had bound to particular places on chromosomes. By 1987, a human map could be prepared with at least *some* points of reference noted on every human chromosome. (By 1990, *Science* was using such a map, now brightly colored, as a foldout centerfold. No apologies were made to *Playboy*.)

What was done by 1987 (or by 1990) was hardly the final word. Much more detail was needed. Genetic maps are calibrated in centimorgans, a measure of the extent of crossing-over between two adjacent markers on a map. If two neighbors are separated in one percent of the progeny in a single generation, then they are 1 centimorgan apart.

Using a genetic map, a scientist can locate the position of the gene for a hereditary trait by studying whether it stays with or separates from markers that are already placed. Once the new gene has been put in position, it becomes part of the map and can now be used as a reference point in mapping additional genes. Mapping, of course, is preliminary to the ultimate step: the actual sequencing of the gene, as was done in the case of cystic fibrosis.

The markers in the 1987 map had an average spacing of about 10 centimorgans. As 1 centimorgan has been found to be approximately 1 million base pairs (it cannot be exact; crossing-over does not take place at random, but concentrates in some places called *hot spots*), the markers were about 10 million base pairs apart. If we wanted to isolate and sequence a gene, this map would not be detailed enough to help us find it. A text area of that size is too large for any thorough search with the techniques we have now. A much more detailed map is needed. In fact, a minor controversy or "map flap" broke out in 1987 over the claim that a map had been completed.

Raymond White, at the University of Utah, commented, "What they have accomplished is important. . . . It is a very useful collection of markers. But it is not what we would properly call a map." Donis-Keller, then at Collaborative Research, replied: "A map is a map. It is not Ray White's ideal, but so what? This is the beginning for us. How can one person set the standard for the rest of the world on what constitutes a map?"[23]

Whether we consider the 1987 achievement a map or not, the preparation of a detailed map of the human genome has become one preliminary goal of the project. Calling for a 1-centimorgan map, Jim Watson announced early in 1989, "I have staked my reputation in getting it done in 5 years." This has proved difficult in practice, although Watson's reputation hangs securely on other matters. The routine placement of genetic markers can be as dull as random sequencing, unless you are doing it to locate an important gene. One prominent mapper, Maynard Olsen, initially called Watson's goal "achievable but fairly horrendous." By the end of 1989, however, he felt "there is zero probability that we will develop a 1-centimorgan map unless there is a major change in policy."[24]

Such a change was forthcoming. By spring 1990, the goal had been altered: It was now felt that a 2-centimorgan map would do, and "chromosome leaders" were assigned to make sure that the job got done for each chromosome. Among those assigned appreciable numbers of chromosomes each were Helen Donis-Keller, now at Washington University, and Ray White. After their rivalries over the cystic fibrosis gene and mapping priority, they were at last on the same team.

A MATTER OF BITS AND PIECES

When they have a detailed genetic map, scientists will find it much easier to put their hands on the important genes connected with human disease, and the preparation of such a map is quite properly one early goal of the Human Genome Project. However, the project itself has a much more ambitious goal, the total sequence readout of the whole kit and caboodle: genes, control areas, structural areas, unknown functions, and all. (One recent book confused the lesser mapping goal with the all-encompassing greater one.) In the longer run, the genetic map will give way to the more valuable physical one. In terms of our set-of-volumes analogy for the human genome, the genetic map represents a series of index tabs indicating particular topics that are scattered among the volumes, whereas the physical map consists of a systematic division into chapters with the page length of each section determined.

Once the human genome has been subdivided into an ordered set of subunits of manageable size, then the remainder of the job would be straightforward, even if it needed a lot of work. The subunits would be chosen by or assigned by contract to various laboratories, while a central coordinator kept track of who was doing what and checked to make sure that each piece got done.

Nature has given us a head start in this process by dividing our genetic text into separate packages—the chromosomes. There are twenty-two different types of paired chromosomes within us as well as the X and Y sex chromosomes, twenty-four jobs in all. Unfortunately, they have not been stored by nature in separate places, so that a research group cannot easily get one of them in pure form. They lie tangled together for much of the life of a cell, becoming visible only at special times, such as that of duplication.

The correct number of human chromosomes was not worked out until the mid-1950s, and only in the late 1960s and thereafter were the staining techniques developed that allow them to be identified and divided into parts for reference purposes. Finally, within the last few years, an instrument has been devised that allows their preparative separation.

This instrument, the chromosome sorter, identifies stained chromosomes by their fluorescence as they flow by in a stream. The flow is subdivided into droplets, each holding a single chromosome, and an operating program selects two types out for special treatment. For example, droplets containing chromosome 1 might get a positive charge, and those with No. 21, a negative charge. Those two chromosomes would then be sorted out from the general flow and from each other using an electrical separation. In one run, two of the twenty-four types of chromosomes would be obtained in relatively pure form, although in limited amounts. On the next run, the program could be changed and two others selected until an individual supply could be had for most of them (not all of them are separated cleanly).

DOE teams at Los Alamos and Livermore have prepared supplies of separate chromosomes in this way. After running the separation, they then sliced up their preparations with a text cutter and increased the amounts of the cut-up pieces by cloning them (multiplying them within bacteria; we discussed this technique earlier). They have made their collections of chromosome-specific DNA fragments available to other researchers.

A size limit exists, however, on the DNA pieces that can be handled in this way. One commonly used reproducing structure, called a *cosmid* (a combination of a plasmid and a virus), can accept about forty thousand base pairs of DNA, which is near the maximum for a carrier that grows within bacteria. The DNA amplification procedure of Mullis is much quicker than cloning, but can only be applied to DNA of six thousand base pairs or less. Further, the end sequences must be known.

More recently, Maynard Olsen and others have developed an alternative, called a *YAC* for yeast artificial chromosome, which contains functioning parts taken from actual yeast chromosomes as well as the DNA to be cloned, that can accept hundreds of thousands of base pairs of DNA. Although this microscopic genetic contrivance has no relationship to the Tibetan yak, their closeness in spelling has still brought some measure of relief to genome meetings. The incessant procession of slides of gels and linkage maps may now be interrupted by one of a herd of these shaggy, horned oxen when some cloner cannot resist making the pun.

To obtain reasonable amounts of DNA for sequencing, then, purified DNA from a chromosome must be broken up with text cutters, and the pieces amplified by cloning. To hold a medium-sized chromosome of 150 million base pairs, three thousand cosmid clones, or three hundred YACs, would be needed. A collection of this type is aptly called a *library*.

Until about 1990, such clones were the basic points of reference for groups working on the same chromosome. Unfortunately, various groups might select different text cutters, which made comparison difficult. An investigator who wished to work in a particular area would be safest if he could obtain a supply of the relevant clone from other investigators; this was not always forthcoming. A scientist would have to have a saintly nature to supply a would-be competitor with a needed asset that had taken him months or years of his own time to develop. The maintenance of a central government-operated supply center for clones was an alternative, but this was cumbersome and expensive.

A solution to these difficulties has recently emerged; it depends on the invention of Kary Mullis. The workers who report information concerning a particular gene or chromosome region will be expected to publish the sequence of a small region of it, enough for it to be amplified by Mullis' procedure.

Let us suppose for example that Jones at Far Out University discovers a gene that prevents baldness. As my father was rather bald and my mother's brothers were relatively hairy, I might be curious to learn whether I had a good working copy of this gene. I would not need a DNA sample from Jones, but only a part of the sequence, which he would be obliged to publish. I could make pieces of DNA corresponding to the ends of this sequence in the lab using commercially available instruments and use it to amplify my own DNA. If the relevant stretch of DNA was present in my genome, it would now be present in the product in large amounts. I could examine it directly or use it to probe on a cosmid or YAC library that I had prepared from my own DNA if I wanted to isolate larger pieces of that gene area.

A short published sequence as described above has been called, in jargon, an *STS* for sequence tagged site, but I will call it an *area of known sequence*. In the new genome strategy, such areas will become the reference points of choice, augmenting and replacing text spelling differences (RFLPs), cutting points, and even genes in that role.

For the full application of a top-down strategy for sequencing DNA, however, scientists will want to work with pieces larger than cosmids or even YACs. Ideally, one would want to divide a chromosome into, say, ten pieces and align them in order. Each piece would then itself be split in ten, and these also placed in order. The resulting one hundred pieces, each identified by an area of known text, would be available for sequence work for any who volunteered or contracted to do so.

Several technical problems must be solved before such a scheme can be carried out systematically. Known text cutters give pieces of less than 1 million base pairs in size. No well-established methods exist for breaking up chromosomes at a few exact places (although many clever solutions to this problem have been proposed and are under development). There would also have been little point to this until recently, as there would have been no way available to separate the pieces.

DNA fragments usually have been separated in electrical fields, most often when the DNA was embedded in a viscous material, or gel (*gel electrophoresis*). The technique only worked if the pieces were about fifty thousand letters or less. This limit has now been extended to about 10 million base pairs, with the advance again largely due to the ingenuity of one individual working with minimal support. By 1990, he had become a colleague in my department at New York University.

NEW YORK CITY, 1990

In background and appearance David Schwartz is very different from Kary Mullis. His short stature and rounded shoulders, together with a boyish face surrounded by disheveled curls of brown hair, give David a cherubic appearance. Only a smallish moustache serves to counter this impression. His childhood was spent in the northern Bronx, no more than two miles from the area where I grew up, half a generation earlier. His parents, Jewish refugees of the Holocaust, had survived Nazi prison camps and subsequently had to escape from their Soviet rescuers. They arrived in the United States with $7 in their pockets and earned a living by working in the garment industry. As they could not afford expensive toys for him or his older brother, David made his own out of boxes, strings, and tape—"things that people would throw out."[25] This hobby served him well later on.

Schwartz and Mullis do have features in common, however, that overshadow their differences. Both are fiercely independent ("I hate working for other people," says Schwartz), both pursue science for the joy of it ("As soon as this job stops being fun, I'm gonna try to go to Wall Street or something"), and both have invented techniques of major importance in modern molecular biology. Although both of their methods have been widely adopted, the contribution of the inventor has been underappreciated in each case.

Schwartz's inspiration arrived while he was spending an undergraduate senior year visiting at Harvard University in 1975. He took on a formidable research project. We now know that common yeast has sixteen types of chromosome that vary in size from 220,000 to over 2 million base pairs. His task was to try to separate them with a method proposed by his advisor, which used electrical fields in solution. David spent three weeks in the library and came to the conclusion that the method would not work. His efforts to communicate this to his advisor failed: "Hogwash; just do the experiments, Schwartz" was the response he remembered. "I was just some punk and this was a Harvard professor."

The method did fail, and that autumn he devised one of his own. The DNA would be suspended in a gel rather than in solution. On a molecular level, this environment offers a mixture of channels and obstacles, somewhat like the spaces and trees in a forest.

We may think of the DNA as an extended millipede moving in this forest with each segment, however, having a mind of its own. When an electrical field is applied, the DNA is motivated to move in a particular direction. All segments try to go that way at once, which results in the molecule getting tangled in the "trees." After a period of confusion, the segments hit on a straight-line arrangement, which allows them to thread efficiently between the "trees," moving in the proper direction. Occasionally the rear end closes up on the front, giving a wormlike series of contractions and extensions. Even very long molecules of DNA can travel well that way, however, which prevents their effective separation from the somewhat shorter ones.

The key innovation introduced by Schwartz was to have the field alternate rapidly between two different, although not opposite, directions. In shifting their direction again and again, the longer molecules got much more tangled up than the shorter ones, and a separation could be made. Once again, this separation by alternating electrical fields is called *PFGE* for pulsed-field gel electrophoresis.

David wrote the concept into his senior thesis, but received no encouragement from his advisor: "What do you think? Do you know more than Bruno Zimm?" he said, referring to the advisor's own Ph.D. advisor, the acknowledged leader in this area. Two years later, however, Schwartz visited his now exadvisor and found him trying the idea on his own without success: "They never saw a damn thing. I figured that they had the wrong pulse times, but I didn't tell him. I thought that it was unfair that he went off and did this."

That same year, David started his own doctoral work in Bruno Zimm's laboratory in San Diego. Although he admired his new mentor, considering him to be the most brilliant man that he had ever worked for, he again received little encouragement: "He didn't think much of the idea. He had done some calculations . . . students in his lab would get into all kinds of exotic projects." Just the same, David built a simple apparatus and started some experiments on his own. His work was interrupted, however, when his brother fell victim to a cancer that killed him, and David had to care for him in the last stages. David returned to find that a co-worker in the lab had decided, on his own initiative, to work on David's concept. He was upset that this was going on "behind my back," but fortunately, these efforts also failed.

In 1980, when Schwartz was in the middle of his Ph.D., his mother also fell terminally ill, and he had to transfer his studies back to New York. He had to start from scratch at Columbia University. "I had never heard of Charlie Cantor before," said David, but Cantor was working in the appropriate area. "So I told Charlie my ideas. He told me they were kind of crazy, but I could work on them." According to Schwartz, Cantor discussed the project with Zimm. They agreed that the project could be moved from Zimm's to Cantor's lab, but also that it would not work. Even if it did, it would have no applications.

Schwartz went back to work on his own and built the prototype of the modern apparatus: The current "commercial version isn't better," he says. Encouraged by his initial results, he told his mentor, "Charlie, I am going to try to separate yeast." When he had some success a bit later, "Charlie couldn't believe what was going on." Later on, when Cantor's "medical school friends told him this was important," he took a strong interest in the area. During a period that extended from 1981 into 1984, however, only Schwartz was working on the method in Cantor's lab. During that period, by David's own account, he developed all of the important procedures and techniques, obtained patents with Cantor, explored the physics of the method, carried out the yeast chromosome work, and initiated new biological applications to globin genes, Huntington's disease, and the fruit fly. He described these results in his Ph.D. thesis.

In 1984, Cassandra Smith, assisted by a number of technicians, began to work in this field. "Her research in another area didn't work out," according to David. He found it difficult to get time on the apparatus he had built and took to working at night. He completed his studies and received his Ph.D. the next year. He then took on a position at a Baltimore branch of the Carnegie Institution of Washington, D.C., and in 1990, moved to New York University. Among his achievements at Carnegie was a study, using a microscope, in which he could actually follow the torturous path by which DNA molecules threaded through the "trees" during an electrical separation.

To reveal another, perhaps Rashomon-like perspective on these events, I will quote from my interview with Charles Cantor:

> A student walked in with the basic idea. He comes across as pretty wild. A pretty wild character. He talked to a

> number of the major biophysical chemists in the country, trying to get people interested. They all felt this was just not going to work. This was hopeless. The ravings of a maniac or close to it. The moment David described the essence of it to me, it was obvious to me that it had to work. It was important. . . . David was not a high throughput experimentalist. He didn't set up scores of experiments. Cassandra got interested and she's the opposite of David. "Let's do one hundred experiments and see which ones work the best."[26]

Subsequently, a flow of publications on this technique by Cantor and Smith served to associate their names with the method, enough so that Harold Schmeck could write in the *New York Times* in 1987, for example: "Dr. Charles Cantor devised the pulsed field concept about 1983."[27] Lois Wingerson in *Mapping Our Genes* reported that the method "was invented by Charles Cantor and a co-worker in 1984."[28] Cantor himself, as we have seen, does not claim the honor.

In 1990, however, David Schwartz is not spending his time in efforts to reclaim primacy in the field he established. He has larger game in sight. He wants to develop new ways to map and sequence DNA. "It's stupid to use current methodology to try to do this. . . . It's like Stone Age man trying to go to the moon. . . . We'll make mapping fun." About the competition, he quips: "We're gonna scoop these guys. They'll be amazed."

His New York University laboratories had several of the square plastic electrical separation boxes adorned with hoses and wiring connections of the type that I had seen everywhere in the Cantor–Smith laboratories. However, they were outnumbered by the representatives of a smaller type, perhaps 9 inches square. Schwartz had invented a new, improved version that he called *POE* for pulse-oriented electrophoresis. He had used this to separate the smallest fruit fly chromosome, which has 4.5 million base pairs of DNA, from the other much larger ones. This separation had failed with his earlier method for some reason. The remainder of the room and an adjacent one were filled with other improvisations meant to turn DNA technology on its head and by a mammoth tangled wiring board.

Large concepts in science can provoke inspired and enthusiastic responses. From the perspective at the start of the Genome Project, it is not clear whether the best answers will come from visionaries like

Schwartz and Richard Keller, others even less known, or the established leaders of the field. I will again venture the following predictions, however:

1. The work will be difficult.
2. The job will get done.
3. It won't be boring.

THE TURN OF THE WORM AND SOME OTHERS

The conclusion of the human sequencing effort will set the stage for an even more ambitious one: the effort to understand and interpret the resulting data. Common sense dictates, however, that a study of such magnitude be preceded by, or at least run in tandem with, one that uses less complicated models. The policymakers for the Human Genome Project decided (quite wisely, I believe) that certain other organisms be included in the effort. In doing so, they solidified their support in other parts of the biological community and ensured that simpler models will be available for prior study in areas where the human version might prove difficult to unravel, such as chromosome function, the mechanism of development, and the organization of the nervous system. Some of the organisms to be included follow.

Bacteria. *Escherichia coli*, a simple inhabitant of our gut, has long been a favorite for biological investigation. Its circular chromosome contains about 4.7 million base pairs. A complete physical map has already been prepared by Charles Cantor, Japanese workers, and others. With one text cutter, for example, it can be sliced into twenty-two pieces whose positions with respect to one another has been worked out. About twenty-seven percent of its sequence was known by March 1990. Within a few years, it will surely be completed, although some other bacterium may overtake it and claim the honor of first organism to be sequenced. To commemorate that event in advance, I will quote from Jim Watson: "A total understanding of *E. coli*, of course, will not fall out immediately from the possession of its instruction book, and hundreds of years are likely to pass before *E. coli* poses no further scientific challenges. But the mere statement that we will one day know completely how *E. coli* functions is an extraordinary scientific assertion."[29]

Yeast. The best known variety of yeast, *Saccharomyces cerevisiae*, has been of course a favorite of the beer, wine, and bread-making industries. Yeast cells have nuclei and many other structures absent in bacteria that are present in our own cells, so they provide a simple model for our type of cell. Further, its chromosomes are much smaller than ours, so the way in which a chromosome is put together can be studied with less effort in this system. Yeast chromosome 3 with 360,000 base pairs is likely to be the first one from a higher organism to be sequenced (by the European Economic Community laboratories), according to one report. In the United States, David Botstein and Ron Davis at Stanford hope to establish a center to do the entire yeast genome.

The worm.

The worm is not to be trusted save in the company of wise people. . . . I wish you all the joy of the worm.

—Shakespeare, *Antony and Cleopatra,* act 5, scene 2

The creature that will be sequenced is a tiny (1 millimeter in length), soil-dwelling roundworm or nematode, *Caenorhabditas elegans* by name. This simple organism came up in conversations between Sydney Brenner and Francis Crick in Cambridge thirty years ago, as a worthy model for study. Subsequently, Brenner wrote to Max Perutz: "I would like to tame a small metazoan," that is, an organism containing more than one cell.[30]

This worm has 959 cells to be exact, but these suffice to build muscles, an intestine, other organs, and a 302-cell nervous system. Brenner felt that an intensive study of such a miniorganism would be an important prelude to the understanding of complex ones. As described in a *Science* summary: "It has nerves, muscles, intestines; it reproduces. And if you hit it, it reacts." The initial scientific reaction to Brenner's suggestion, summarized by his colleague, John Sulston, was hardly encouraging: "It was seen as just Sydney's madness, and not to be taken seriously."[31]

The madness spread, however, and now encompasses nearly one hundred laboratories around the world. The cells have been counted, and their development from a single one followed in detail. A complete wiring diagram of the connections in the nervous system has been prepared. The interested community, sharing a common intellectual origin, is also well wired: They share data freely by electronic mail and chronicle their progress in the *Worm Breeders Gazette.*

The study of the genetic organization of the roundworm was an obvious extension of this effort. It began before the Human Genome Project, but has now been fully incorporated in the overall scheme with Sulston and Alan Coulson at Cambridge University and Alan Waterston at Washington University, St. Louis, as coleaders. The mapping of the six chromosomes is nearly complete, and the leaders have more or less decided to go for it to attempt to gather the 100 million letters of genetic data by the end of this decade.

The fruit fly. This creature has been a historical and sentimental favorite since the time of Morgan. An immense amount of genetic data have been compiled on it, which would provide a head start in making sense of the sequence. Its 170 million base pairs, spread over four chromosomes, will afford a challenge intermediate between that of a worm and a mammal.

The mouse. Many human genes will turn up whose function is unclear. If the sequence of an experimental animal were worked out at the same time, then the comparable gene could be examined in it. The gene could be altered or removed, which would not be ethical in human studies. Eric Lander at Massachusetts Institute of Technology will head a center to map the mouse genome.

A plant. We will want to see how the other kingdom lives, although the lead here may be taken by the Department of Agriculture and the National Science Foundation. *Aribidopsis thaliana*, a member of the mustard family, has the inside track. This small weed is of no use whatever, but has a large seed production, a six-week life span, and a genome of "only" 100 million letters. Others will follow. At the 1989 Human Genome I meeting, Eric Lander pointed out that preliminary work had been done with corn, papaya beans, lettuce, and tomatoes and quipped: "Salad is virtually completely mapped."

Many of the above groups will be supported as part of the Human Genome Project. The selection will depend upon the overall level of financial support, technical advances, and the success of each community in organizing itself to formulate a coherent plan. As a human chauvinist, however, I want to turn my attention to our own species and, passing over the interim tactical struggles, contemplate the prospects when the human prototype sequence has been assembled.

PART III
TOMORROW

At its start, a period of fifteen years was suggested for the Human Genome Project. It would wind up in the year 2005. No compelling reasons were given for that date, but some folks suspect that Jim Watson, born in 1928, would like to be around when the job gets done. The original plans called for expenditures of $200 million per year by the NIH and DOE. Congressional budget cuts (which are gradually acquiring a reliability that would put them up there with death and taxes) are likely to reduce that sum. On the other hand, other countries will probably support the project. Above all, ingenious improvements in technology could really speed things up. For now, I will assume that the target date is reasonable.

In the next chapters, I want to look ahead at the human picture after the first genome has been safely stored in existing computer banks at Los Alamos, Heidelberg, and elsewhere and made available to any investigator who can hook in on-line. How will things change for the rest of us? I have attached the arbitrary date 2020 on the advances for a few reasons:

1. Another fifteen years will have gone by, a time equal to the planned length of the project. This would not only allow for some delay, but also give time for implementation of some of the possibilities after the initial data are fully in hand.

2. I have a delectable quote from Walter Gilbert that opens Chapter 15 that uses that date.

3. Through most of my life, my eyesight has been rated at 20/20, a descriptor that indicates excellent vision. Excellent vision of a different type is essential to any effort to peer into the future, so I have selected the year 2020 to invoke a symbolic blessing.

In looking ahead, I run the hazard of coming out wrong thirty years from now. Other perils are more immediate. Many scientists find any effort to speculate beyond the length of their current grant proposal distasteful. (Even in their grant proposals, they often list results that are already in hand as possible future advances.) Someone who authentically looks forward may be viewed as a crank.

Many of the discoveries that will be made in reading the human text will truly be surprises, items that cannot yet be foreseen. I have no ambition to anticipate the unforseeable or to portray, in words, the indescribable. Some developments that will occur, however, are just extensions of items that already exist in preliminary form. I do not think it foolish to presume that such extensions will take place and anticipate what use may be made of them. The increased expenditure of public money upon sequencing has already begun. It seems only fair to give the public, even now, some idea of what they may get for their money.

15

THE PERSONAL GENOME

In 2020, you will be able to go into the drugstore, have your sequence read in an hour or so, and given back to you on a compact disc so you can analyze it.

—Walter Gilbert[1]

The place does not really matter, of course. The actual service may be performed in a clinic, or the sample may be taken by your physician and delivered to a service center for analysis. It matters little also whether the service takes two weeks instead of twenty minutes. Further, as is the case with X rays today, you may only get an analysis with the actual data sent to you if you ask for it. The central message is this: For each of us, our genetic readout, or at least important parts of it, will be routinely available and affordable. The Genome Project will become personal.

Once the text has been read, two distinct paths of investigation, now linked by common need, will separate and follow their own ends. The first is theoretical, bio*logical* if you like. How does the human body, as governed by its genetic plan, actually work? By studying the text, we will discover thousands of hitherto unknown genes (that analysis itself will keep a generation of computer experts happily employed). Each of them will contain the plan for a protein or one of the RNA molecules that does useful work in the body. Biochemists will want to learn how each protein folds up to occupy three-dimensional space and what task it performs in the body. A single protein offers enough puzzles to keep a laboratory occupied for decades, so biochemists will for a time inhabit a type of paradise that most other scientists only dream about. Each will have the chance to select and explore virginal terrain of fundamental importance almost unimpeded by competition.

We will want to know more than the structure and function of proteins. Control and structural areas will have to be located and their functions mastered. The "junk" areas will be scrutinized carefully in search of some hidden purpose. Finally, we will have to figure out how these DNA regions interact in the complex checks and balances that control our existence. The final goal is, in my opinion, admirable: to figure out how and why we develop, feel, remember, sense, grow old, and die and to gain some control over the experiences we do not like. Some results will come in early, but the overall effort will take centuries. Over this period, human biology will represent a vital frontier of science, perhaps the most exciting one.

The Genome Project, on completion, will be in a better place than the Apollo project. It was important, symbolically, to set foot on the moon. Once there, however, it was not clear what we should do next or why. The Genome readout, on the other hand, will set the stage for even more challenging undertakings. At the end, we will have gained a far deeper understanding of our workings. How deeply we will penetrate cannot be foreseen, but geneticist Francisco Ayala has caught the spirit of the quest:

> The meanings of a novel like Tolstoy's *War and Peace* . . . go much beyond the literal meaning of the words and sentences. My own prejudice is that we will never be able to understand complex organisms, certainly not human beings, simply by understanding their genome's sequence, but we have to try.[2]

The other path will ask about human differences. As each new area of text becomes available, we will want to learn how *yours* differs from *mine* and what difference that makes in terms of our lives.

Curiosity about human differences goes back, of course, as far as the race itself. With the followers of Hitler and Stalin perhaps excepted, the remainder of us today recognize the importance of two separate factors, heredity and environment, or Nature and Nurture. The particular phrase can be traced back to Shakespeare's *The Tempest*, where Prospero, a nobleman, has the following to say about his slave Caliban:

> Abhorred slave, which any point of goodness will not take, being capable of all ill! I pitied thee, took pains to make thee speak, taught thee each hour one thing or other. . . . But thy vile race, though thou didst learn, had that in't which good natures could not abide to be with.

Prospero later repeats the theme, describing Caliban as "a devil, a born devil, on whose nature Nurture can never stick! on whom my pain, humanely taken, all, all lost, quite lost!"

Caliban, of course, came from an unusually bad background: "Thou poisonous slave, got by the devil himself upon thy wicked dam" (she was a witch). Given that background, we might even say that he was doing well. For the remainder of us, the issue has been far less clear. We have seen how, through human history, hollow theories have been put forward and political atrocities excused under the cover of self-proclaimed insights into the mechanism of heredity. Finally in our generation, the hard scientists have arrived at the catacombs where the innermost secrets are stored, and they have the needed tools in hand to gain access to these secrets. When the text of nature has been read and digested for many humans, the differences that cannot as yet be explained as due to Nature may safely be put in the camp of Nurture.

The task of reading the 6 billion characters (both sets of instructions will be wanted) in the genome of each of the billions of humans on earth does seem formidable. It dwarfs immensely the initial task of the Genome Project: determining the basic set of 3 billion letters for one prototype. Some scientists have even assumed that the costs and work involved would be proportionate. The noted population geneticist L. Luca Cavalli-Sforza wrote recently:

> Unfortunately, the effort of producing the whole sequence just once is so great that, at the moment, the idea of sequencing even two individuals instead of one cannot be seriously entertained.[3]

Fortunately, the analogy does not hold. With the prototype in place, one would need to learn how each individual differs from it, which would amount to perhaps one percent of the total text. It is far easier to determine how a DNA passage differs from a closely related one than to work it out for the first time. Some clever methods that determine individual differences readily in the beta-globin gene and other known areas have already been published, and further improvements may be expected.

In addition, not all the genome need be taken. With experience, we may find that perhaps ninety percent of the important information resides in a few percent of the genome. We will want to focus, initially, on two key aspects of individuals:

1. Text areas that cover important functions and vary among individuals. These will afford information on diseases and disease susceptibilities, as well as the genes that affect bodily, emotional, and mental characteristics of humans.

2. Selected regions spaced fairly uniformly along the various chromosomes, which show the greatest variation among humans. These areas may be "junk," that is, they may not affect how our bodies function. Yet, as we shall see, they will contain a rich store of information, important to some of us, about our biological relationships and geneology, both recent and further back, to the start of the modern divergence of the human species.

We have already started the task of gathering individual sequences, of course, with the analysis of the genes for sickle-cell anemia, cystic fibrosis, and other diseases. Population biologists have started to sample isolated ethnic groups. Cavalli-Sforza has already suggested that one ten-thousandth of the genome for one hundred individuals be collected now and preserved for future sequencing use. Such efforts will undoubtedly increase as costs diminish and more of the prototype has been determined. Allan Wilson, a Berkely geneticist whom we shall encounter again, wrote recently:

> The sequencing of multiple genomes is neither likely to occur nor necessary in the present decade. But biologists should recognize that the sequencing of multiple human genomes in the succeeding decade could produce conceptual and technical revolutions in many fields of biology. By anticipating the synthesis that could occur in that decade, all the way from molecular to population biology, we hasten the onset of the revolution.[4]

FORWARD TO THE FUTURE

Let us move forward to the year 2020 and the instance of a young man whom I will call Gregor Morgan in honor of the pioneers in heredity. Born in 1999, Morgan has turned twenty-one and can now acquire as much of his own text as he fancies. What will he learn?

To start, the hereditary conditions from which he suffers will be listed. The most important ones may already have been read prenatally, so that appropriate health measures could be taken. He will escape the plight of an adopted child described by Deborah Franklin in the *New York Times Magazine*: "Laura had so much pain and went undiagnosed for so long. She didn't need just family therapy. She needed lithium."[5] Laura's diagnosis was corrected when her biological parents were located, and schizophrenia and manic-depression disorder were found in her family background. Gregor's diagnosis requires no relatives. The information will be found in almost any cell from his body.

The list of conditions will include more than the famous diseases of today. Added to it may be others, less serious but unpleasant nonetheless, for example, baldness and a tendency to obesity. Yet others may be unrecognized at present due to the absence of visible, localized symptoms and yet take a chronic toll in aches and pains, unpleasant moods, and loss of vigor. Not every condition could be treated, of course. Some might be improved through changes in life-style, just as a susceptible person today might avoid a substance to which she is allergic or sunlight if her skin reacts badly.

Gregor, at age twenty-one or earlier, would also have a chance to learn of his disease *susceptibilities*: text changes that make him more likely to be stricken by certain diseases than other individuals in the population. We have learned about tumor suppressor genes recently, where both copies must be damaged for a certain type of cancer to develop. They could be disabled by environmental factors, but an individual born with one defective copy would be at greater risk. Alcoholism, diabetes, and a number of other diseases are believed today to have a genetic component. If the vulnerable individual were alerted, he would have the chance to alter his habits to minimize the risk.

Suppose Gregor learns, for example, that his body is very adept at "activating" certain cancer-causing chemicals that occur in cigarette smoke and some polluted environments. The chemicals are harmless in their native form, but are converted by our own enzyme systems into a more hazardous version that could start the process leading to cancer. Gregor's enzymes are more efficient in doing this than most other humans, so he stands a greater risk of cancer if exposed to the chemicals. Gregor smokes occasionally, but could give this up. On the other hand, he is about to graduate with a degree in chemical engineering with a special interest in petroleum. His intended work would give him some exposure to the chemicals. A difficult life-style decision will have to be made. Perhaps he will take such a job, but have his lungs examined periodically.

Mark Bitensky, a physican working at Los Alamos, provided a doctor's view of the opportunities to come in *Biotechnology and the Human Genome, Innovations and Impact*:

> There are profound differences in the way in which we respond to drugs, and profound differences in our susceptibility to disease and physical agents. . . . Having this remarkable tool of the human genomic sequence will put us in a position to customize medical care, to become familiar with and to address the individual features of our patients. We would be able to advise them about their own risks rather than the risks of the population at large. We would be able to diagnose and treat their illnesses and prevent them from exposures which are inappropriate in a much more sophisticated, focussed and humane way. . . . There is the opportunity to deliver an extraordinary level of medical care, a level which addresses individual differences and one which encompasses the complexity of humans. . . .[6]

Some of the unfortunate genes that Gregor discovers could not, strictly speaking, be blamed on his ancestors, but would represent biological intrusions in the form of hidden viruses. He might find, for example, that a virus responsible for a sexually transmitted disease was slumbering quietly in the midst of his chromosomal DNA. Nobel laureate Joshua Lederberg has commented:

> From the perspective of the virus, the ideal would be a nearly symptomless infection in which the host is oblivious of providing shelter and nourishment for the indefinite propagation of the virus' genes. Our own genome carries hundreds or thousands of such stowaways. The boundary between them and the "normal genome" is quite

> blurred. . . . Intrinsic to our own ancestry and nature are not only Adam and Eve, but any number of invisible germs that have crept into our chromosome.[7]

With the advent of DNA amplification, even such symptomless viruses can be detected once the existence of each species has been recognized and a sequence obtained. A partial list exists today, and when a number of human genomes have been sequenced, a more complete list will be compiled. The job for twenty-first-century medicine would be to ensure that Gregor's virus remained in its shut down form and did not suddenly spring to activity.

Other information would not affect his health directly, but would come up when he considered parenthood. We have considered the present difficulties with genetic counseling in the case of diseases such as cystic fibrosis, where a wide variety of genetic changes can cause the disease. By 2020, the likely strategy will be to sequence the entire gene in such cases and compare differences observed from the norm with an extensive list of known harmful and benign changes.

Let us suppose, though, that a new alteration in the cystic fibrosis gene area is observed in Gregor's case. The protein chemists would then get the job of predicting, on the basis of the shape and function of the enzyme and the nature of the change, whether it would be harmful. In some cases they might choose to prepare or isolate the protein and study it directly. Gregor might learn, at the end, that he was a carrier of cystic fibrosis and a dozen other recessive diseases (Dr. Aubrey Milunsky has suggested that we may each carry about twenty).[8] Genetic reading, or at least screening for diseases, would be much more common then than now, and the recognition that this number was normal and expected might protect Gregor from the stigma that such disclosures can bring today.

Not all of the information that Gregor obtained from his genetic readout would be morbid. Some would provide information useful in creating a life-style. I myself, for example, enjoy drinking wines and prefer some types, while disliking others that have been widely praised. I often attend tastings in order to sample a variety and find those that I like. The preferences of others is not always a useful guide.

While at tastings, I am always amazed at the disputes that erupt among the patrons, arguing in absolute terms about the quality and taste of a wine. The tasters of 1990 seem unaware of the genetic–biochemical differences that make our experiences differ from one individual to another. The ability to smell a characteristic chemical odor in urine after eating asparagus is controlled by a dominant gene, for example. I am not sure which variant should be considered the desirable one.

In any event, I can anticipate a time when food and wine critics might direct their recommendations toward different genetic subtypes. Gregor's readout would tell him that his responsiveness to certain tastes and smells was above normal, whereas others were below par. The relevant books might then direct him to cuisines that he had never tried, from caviar and cognac to chutney and curry.

The genetic items that I have mentioned so far seem likely; I will now move on to some speculations. Data collected by eugenicists and others suggest that musical and mathematical abilities run in families. Although the political philosophies that grew from the movements were dreadful, some of the data may have been valid. Although environmental and cultural influences can contribute, there may be a genetic component underlying such abilities.

Genes, of course, do not code directly for higher mental functions, but for enzymes and biochemical factors. It will be decades or even centuries before our understanding of the interactions of biochemical reaction cycles and nerve function have advanced enough so that we can directly infer potential mathematical and musical talent from gene sequences. In the interim, however, a different strategy might yield some provocative clues. Let us suppose that significant parts of the genome of a number of gifted mathematicians had been determined and the differences from the average sequence compiled. Computer analysis might (or might not) detect a common deviation from the genetic norm in a number of the cases. The genetic areas located would then be flagged as particularly worthy of further study, while Gregor might glean some vocational clues from the data.

Gregor, for example, might expect to find his vocational selection of engineering supported by the presence of genetic sequences associated with enhanced mathematical ability. If instead he found the DNA variations present in those with musical talent, this alone might not cause him to switch careers, but he might be tempted to explore music as a recreation.

I will take this line no further, but another matter begs for our attention. We have discussed Gregor's text thus far as if it were of concern to him alone. This will not be the case in the real world. At the present time we can get very interested in the love lives, finances, and personal habits of those we interact with directly or observe as public figures. This curiosity will very likely extend to their genetic makeup.

THE QUESTION OF GENETIC DISCLOSURE

Let us extend the plot by considering Gregor's female friend, Millennia. Born in 2000, she has not yet reached the age at which she is legally entitled to her full genetic readout. Furthermore, she does not care. More emotional than Gregor, she wants to be accepted or not for herself, not her DNA. She wishes to make her important life decisions on her feelings, not her sequences. She and Gregor are compatible; she shares his new interest in classical music, and she would like a deeper commitment from him.

Gregor, on the other hand, is undecided about their relationship. He is impressed with the way in which his DNA predicted a musical interest and curious whether she shares this genetic feature. He was also a bit unsettled by the list of diseases of which he was a carrier, despite the reassurances that it is normal to be a carrier. It would be a nuisance if she carried some of the same recessive genes, and they would need medical help to ensure that they did not have a diseased child. To be brief, Gregor would like a look at Millennia's DNA before they got more deeply involved. She would be unlikely to agree, however, even if it were legal (she is under 21). Would there be some other way to get the data?

The same question might come up in research studies on the DNA of highly gifted mathematicians or musicians. Suppose that some extremely noteworthy individual, an Einstein or Mozart of her time, refused to participate. Could an unethical person read her genes without her cooperation?

Imagine also a nominee for the Supreme Court. An uncle and a cousin have suffered manic-depressive disorder, and his brother has had a nervous breakdown for unrevealed causes. No health problems have surfaced in the nominee's previous service as a judge, but congressional spokespeople from the opposite party are concerned. They suggest that he may nonetheless be predisposed to emotional instability that would make him unsuitable for the high court. They ask for full genetic disclosure to reassure them that he does not carry the relevant genes, but the candidate refuses. Can he maintain the privacy of his DNA if his political opponents are determined to have it one way or another?

I remember one movie columnist who always reported whether the actors and actresses that he interviewed slept in pyjamas or in the nude. I have never had to protect my own privacy in this matter, because, with the exception of the few who knew the answer anyway, the rest of the world did not care. Most of us will face the same situation with regard to our DNA sequences, but I have tried to conjure a few exceptions above. In such cases, will an individual be able to protect the privacy of his or her DNA?

I believe that the answer will be no. Not if someone else really wants it and we live in a fairly open society, as we do now. If I wanted to obtain DNA sequence from a person, what I would need would be (1) a sample of her DNA and (2) access to someone who has the expertise to do the sequencing job and the necessary chemicals and apparatus.

For the first part, I would only need some small sample of the person's bodily material. Most analyses to date have worked with blood, but this might be tricky if the person is uncooperative or is to be kept from the knowledge that she is being sampled. We carry our full DNA text in almost every cell of our bodies, however, so the possibilities are vast. (Sperm, egg, and mature red blood cells are exceptions.) Saliva, hair (provided the root is attached), or a few skin cells will do. The DNA amplification procedure has already brought the sensitivity down to the point where analysis of a single cell can be done. At Cetus, Henry Erlich told me that DNA sequences have been obtained from the flap of an envelope that someone had licked before sealing!

A recent report in *Nature*[9] from a laboratory in Hertfordshire, England, gives some idea of the ease of obtaining someone's DNA. In this case, the input was a nuisance, as samples of DNA from the technicians were interfering with the analysis of viruses in infected monkeys using DNA amplification. The contamination took place even though the laboratory personnel wore lab coats and gloves with the opening between the gloves and cuffs sealed. Only when the technicians also wore disposable caps, face masks, and goggles could the monkey samples be kept clean. Without this protection, material, presumably dead skin cells, was shed from the exposed face and head of the human operators. This was confirmed when four open tubes "were deliberately contaminated by shaking operator C's exposed head and hair over them." All four tubes picked up some nonmonkey DNA.

With this precedent, endless opportunities open up for Gregor to swipe some of Millennia's DNA. He need only take a few hairs from her hairbrush. Alternatively, he might simply wipe his mouth with tissues after a tender kiss and preserve the tissues for the analysis. (It would not be hard to sort out her DNA from his, if he provided a separate sample of his own.) Perhaps he could "borrow" a used handerchief or some used inner clothing, stockings or underwear. (I am reminded of the primitive hexing ritual in which some hair, other body sample, or close personal item of the victim is required.) Some garments today are described as "revealing," but they will never be able to uncover as much as the underthings of tomorrow.

Gregor, of course, might find the tables turned on him the next week. Let us suppose that, despite his misgivings, he went on an employment interview with a petroleum company. If the company had a good supply of well-qualified applicants, they might opt for a deeper analysis before making a choice. They might wish to screen out those with a high susceptibility to cancer due to exposure to petroleum products, for example, to keep down their employee health costs. If they were legally prohibited from asking Gregor for his DNA information, they could easily work around that obstacle. For example, he might be given a warm handshake by his interviewer or asked to open a door whose knob was covered with a slightly sticky substance. The interviewer's hand or the doorknob might then be swabbed to get a DNA analysis. After the results were in and his cancer-prone gene detected, some other pretext could be found for denying Gregor the position.

A different situation would come up if the target were aware of the interest in her DNA and determined to protect it, as in the case of a temperamental genius composer or privacy-minded Supreme Court nominee. She would face the need for formidable security measures. Access to her residence would have to be restricted to close associates, and she could travel in public or visit other homes or offices only if she used the protection adopted by the Hertfordshire lab technicians. She would need to form close alliances with her hairstylist, launderer, doctor, and dentist. Unless she had the foresight to avoid licking envelopes, she would have to track down her past correspondence.

All of these precautions (which might require the resources of a dictator or at least a multimillionaire) would be to no avail unless she also had the cooperation of her children. Each of them carries one set of chromosomes derived from her with information on half her DNA. As we shall see, chromosomes from two children would reveal on the average seventy-five percent of her DNA. At the present time, we would also need DNA samples from the children's father or at least a close member of his family to sort out the source of each individual chromosome in the children. By 2020, we will have learned enough about the meaning of gene sequences to make this unnecessary.

Even if the children were cooperative, it would not suffice for them to refuse to provide DNA samples. They would have to adopt the same full set of precautions that their mother had. Again, massive resources would be necessary.

Of course, someone must agree to analyze the samples. DNA sequencing today is performed in a wide variety of laboratories. It requires some skill and training and the investment of some thousands of dollars in equipment. The capability needed is much less than that needed to build an atomic bomb, but more than that needed to distill your own alcoholic beverages. It may compare to the ability needed to prepare certain illegal synthetic drugs.

With further automation, the skill and training needed will drop, but the machinery will most likely still cost more than your microwave or television. The instruments would be found in commercial and university research labs, clinics of many types, and perhaps even medical offices and pharmacies. Even if they were controlled, however, some instruments would find their way into private hands and would be set up surreptitiously in basements. A black market could easily spring up for bootlegged DNA sequencing jobs. Concerns might also spring up in some foreign countries that offered anonymous DNA sequencing in accordance with local law. In addition, the sequencing of biological material in conventional facilities is likely to become so widespread that the insertion of a few extra jobs under assumed names would be a simple matter, easily arranged as a favor or by petty bribery.

Most ethicists who have considered the matter agree today that a person's DNA sequence should be his or her own property, and a societywide consensus on this in the United States and many other countries is likely. A bill to secure such rights was introduced into the U.S. Congress in 1990 by Senator John Conyers and endorsed by such unlikely allies as gene therapy pioneer W. French Anderson and antibiotechnology activist Jeremy Rifkin. However great the agreement in principle, it is likely to be unenforceable in practice. Bootleg DNA sequences may be inadmissable in certain legal proceedings, but the information will leak to the public anyway for individuals to use as they wished.

Perhaps it would be better if we moved the fortifications that protect our self-esteem to more defensible terrain. Even though our sequences were private in principle, we could learn to look at them as information that normally was not concealed, just as we do not usually conceal our faces or prohibit photographs. We attempt to prevent ethnic discrimination by passing laws, but not by keeping our faces and skin covered at all times.

BEYOND GALTON

Our genes, of course, can tell much more about us than a photograph does. To be able to share this information freely about ourselves, we must move beyond the hateful messages spread by early eugenicists and their political hatchetmen, such as the Nazis, that certain ethnic groups are inferior. Apart from the moral repulsiveness of that message, we can now see that the underlying science cannot support such concepts. They are rooted in the idea that genes are blood, not text.

Blood, as a liquid, can be contaminated in all its parts by the addition of a small amount of a poisonous or damaging component. If that component were normally present, then different samples could be graded as inferior or superior according to the relative lack of the substance, as gasolines are graded on their tendency to cause knocking.

Text, as we have seen, can harbor defective passages, with the single change of a letter enough to kill, as in cystic fibrosis or sickle-cell anemia. However, these changes are localized and can occur anywhere in our genome. A particular type of error may occur largely in one ethnic group, but none is free of them. Further, they do not define the group. Healthy members can lack the defect and have all their other ethnic features intact. With further medical advances, perhaps most or all such defects can be compensated by treatment or relegated to the rubbish bin by genetic manipulation.

The early eugenicists, of course, were not terribly interested in the well-defined genetic diseases. The royal families of England, Russia, and other European nations were not classified as inferior because they carried hemophilia. The genetic bigots selected much more sweeping behavioral categories for their rhetoric: shiftlessness, industry, promiscuity, various types of societal achievement, and the like. As I have mentioned before, no passages of text with those labels will be encountered, but only a list of activities whose meaning is defined on the biochemical level. Their interactions produce the complex effects we observe in our everyday world.

This does not mean that behavior, personality, and abilities of various types are determined by Nurture only. As Deborah Franklin wrote in the *New York Times* in 1989: "The question today is not whether genetics influences personality, but rather how much, and in what ways."[10] Pennsylvania State geneticist Robert Plomin, in a 1990 *Science* review on the role of inheritance and behavior (cognitive abilities and disabilities, personality factors such as emotional reactivity and sociability or shyness, schizophrenia, manic depression, and other emotional disorders), concluded that heredity contributed perhaps twenty to fifty percent with the remainder due to environment. "The role of inheritance in behavior has become widely accepted, even for sensitive domains such as IQ," he wrote.[11] A more recent study of IQ (intelligence quotient) in twins by a group at the University of Minnesota estimated that seventy percent of individual variance was due to heredity. However, multiple genes with small effects seem to be involved rather than one or two major genes[12] (and it will be a difficult, but not impossible task to identify them all and unravel their interactions).

The last item may provide a formidable obstacle for the would-be bigot. Our experience with DNA text thus far indicates that genes affecting the same function may often scatter on separate chromosomes. We may recall, for example, that hemoglobin functions through the close embrace of alpha- and beta-globin subunits. Yet the alphas reside on chromosome 16 and the betas on chromosome 11. The red and green color vision genes nestle together on the X chromosome, but the blue one can be found on No. 7.

The genes for the biochemical functions that underlie most of the visible racial and ethnic features, skin color, hair type, facial structure, and so on, have not yet been located, but they are also likely to be widely scattered. Efforts to demonstrate sweeping correlations between large sets of genes that determine physical appearance and other sets that specify behavior are likely to sink in a sea of complex data.

The widespread availability of data on personal genomes may have an opposite effect. Many profound human differences developed long before the geographical separation of races and are represented in all ethnic groups: A, B, and O blood groups, for example. When new, personal information on thousands of genes becomes available to each of us, we will discover profound, now invisible differences and similarities that run across existing ethnic lines. I have occasionally found that I had little in common in personality with someone from the same background as I, but a lot of empathy for the feelings of a stranger from a different land. Circumstances obviously play a role here, but the DNA texts may also show that more is involved in some cases. Our underlying similarities may prove more important than visible differences. Extensive genetic sequencing is likely to blur ethnic differences and make stereotyping difficult.

Humans have always sought reasons to feel more important than their neighbors, of course, and the availability of genetic text may lend new energy and direction to such efforts. An individual or group will always be free to declare that a particular local text variation (their own, of course) is the preferred or chosen one and organize socially on that basis. The huge stretches of genetic text whose function is not yet known will provide further fodder for the imagination. Alert mystics may attempt to dance to the tune of the new millennium by shifting their attention from palms and birth dates to uninterpreted genetic text, which has a more obvious relation to the fate of an individual.

A recent cartoon in *Science* showed Ms. Tena pushing palm and tarot card readings in a storefront while her avant-garde neighbor, Madame Rosa, was painting a sign that declared herself to be a geneticist! Long before scientists attach meaning to the almost endless strings of A, T, G, and C, the imaginative will have the chance to read into them what they will, as they do today for palms and crystal balls.

Some of us may feel, on the other hand, that the existence of a comprehensive genetic text diminishes us, reduces our individuality or dignity. We must keep in mind that each of us is much more than the set of letters he or she was born with. Ethicist Thomas Murray has considered the problem for all humanity in terms that apply to individuals as well. His thoughts were summarized in *Biotechnology and the Human Genome*: "Will mapping, or sequencing, in particular, somehow diminish our dignity? Will it somehow interfere with the moral standing of humankind?" Are there some things, perhaps, that we should never know? After considering this idea, he puts it aside:

> If we could absorb Copernicus, if we could absorb Darwin, if we could absorb Freud, I have a feeling we will be able to absorb human genetics. . . . I think it is only a shallow and fearful concept of human importance that is threatened by this kind of knowledge.

He goes on to compare human existence to the performance of a symphony. I feel that his comparison applies to each of us individually as well.[13]

> Knowing that you can write down the notes of a symphony on a sheet of music, even knowing the specific pattern and organization of the individual notes, diminishes not a whit our admiration for the work as it is performed. It may even enhance our respect for it. The significance of the symphony, its beauty, its elegance, its capacity to move us, simply takes place at a different level than the sequence and patterns of discrete notes and music.

With this metaphor to guide us, we can turn our attention to a different aspect of human society that will be profoundly altered by the possession of our genetic text: the determination of identity.

16

A QUESTION OF IDENTITY

Every human being carries with him from his cradle to his grave certain physical marks which do not change their character, and by which he can always be identified—and that without shade of doubt or question. These marks are his signature, his physiological autograph, so to speak, and this autograph cannot be counterfeited, nor can he disguise it or hide it away, nor can it become illegible by the wear and the mutations of time . . . this signature is each man's very own—there is no duplicate of it among the swarming populations of the globe.

—David Wilson

These words are a product of nineteenth-century literature, but describe most aptly a coming fundamental change in human circumstances: Each of us will gain ready access, through the routine reading of our DNA text, to our social and biological identities. We will learn both who we are should we forget through disease or injury, and who our parents were, in cases where this is in doubt.

This was not a problem when human communities were limited to small tribes or villages where all members were known to each other. Even today, most of us are not troubled by it. The exceptions have been numerous, however, and celebrated in human tradition and literature. I have culled a few famous examples.

CONFUSIONS OF THE PAST

THEBES, GREECE, FIFTH CENTURY B.C. King Oedipus of Thebes could not have been pleased to discover, very late in the game, that he was not the natural son of the royal family of Corinth. His adoptive parents had neglected to inform him of this detail at a time when it would have been helpful. As a consequence, when he learned of a prophecy that he was destined to kill his father and sleep with his mother, he fled from Corinth to escape that fate.

He had qualified for his present position in Thebes by rescuing the citizens of his new hometown from the unwanted attentions of a monster, the Sphinx (a human head–lion body combination, as depicted in the famous statue in Egypt). As part of his reward, he married Iocasta, the current widowed queen of Thebes. Neither the legend nor Sophocles's play, *Oedipus Rex*, indicates that he was concerned by any family resemblance, and needless to say, no technical aids were obviously at hand to inform him of this connection. Far in the future lay the simple DNA tests that would have warned him: Hands off! This woman has a fifty percent genetic overlap with you.

To make things worse, he was the direct cause of Iocasta's widowed condition. He had killed her husband, Laius, in a quarrel that erupted when the royal party had attempted to push him aside on the road. Iocasta was unaware of this, and Oedipus also did not know that the person he had killed was not only the previous king, but also his father. The prophecy had been fulfilled.

Laius had heard of this prophecy when Oedipus was born and had him exposed on a mountainside, with feet tied together, soon after his birth. Some helpful shepherds saved him by conveying him safely to the king of Corinth. The same witnesses finally ruined things, however, by leaking enough information to Oedipus and Iocasta to allow them to puzzle out the entire chain of events. The consequences were dreadful in Sophocles's version, with the one blinding himself and the other killing herself out of guilt and shame. This was perhaps an overreaction, as their intentions were good, but their information was inadequate until it was too late. The real culprit was, perhaps, the oracle that made the self-fulfilling prophecy.

BERLIN, 1926. The deposed dowager empress of Russia, mother of the executed Emperor Nicholas, confronted a young woman who claimed that she was her granddaughter, Anastasia. She had survived the Ekaterinburg massacre by luck and wandered in a state of partial amnesia since then. Although suspicious that the woman was a fraud, the empress was ultimately convinced by Anastasia's recall of the most intimate details of palace life. Having proved her point, Anastasia, for personal reasons, disappeared once again.

These events took place in a play, *Anastasia*, that had its New York premier in 1954 with Viveca Lindfors in the title role. A film version with Ingrid Bergman was subsequently made. Although fictional, they had a basis in reality. A number of would-be Anastasias appeared after 1918, with one, Mrs. Anna Anderson, claiming much attention in a lifelong effort to prove herself the real item. Mrs. Anderson never met the dowager empress, but Emperor Nicholas's sister, Olga, interviewed her for four days and pronounced her false. Olga commented: "My telling the truth does not help in the least because the public simply wants to believe the mystery."[1] Later in this chapter, we shall see how this dispute could have been resolved within a week had the genetic technology of today been available earlier in this century.

A GALAXY FAR AWAY, A LONG TIME AGO. Luke Skywalker was luckier than Oedipus. He received warnings in the nick of time that saved him from the possibilities of patricide and incest. He was engaged in combat with the evil lord, Darth Vader, when Vader informed him, "I am your father." The circumstance was hardly one to bring about mutual trust, but Vader continued, "Search your feelings—you know it to be true." Although this method can hardly be recommended for general use, it worked for Luke this time, and he declined to kill his father when he had the opportunity at a later point.

Luke had also been involved in a romantic competition for the hand of Princess Leia, but discontinued these efforts when he guessed correctly, during a conversation with the ghost of his former mentor, that she was his twin sister. He thus avoided the other sin that Oedipus had committed, but not through any purposeful action that he took. The civilization depicted in George Lucas's *Star Wars* trilogy of films was very advanced in the technology of transportation (particularly interstellar), robotics, and weaponry, but apparently molecular biology was not their strong point.

ABOARD THE H.M.S. PINAFORE, LATE NINETEENTH CENTURY. Events had taken an unfortunate turn for seaman Ralph Rackstraw. The admiral of the fleet had ordered him put in chains for the crime of proposing marriage to the daughter of Captain Corcoran. The society portrayed in the 1878 Gilbert and Sullivan operetta did not take kindly to such violations of class structure. Fortunately, a reprieve came from an unexpected direction. His former baby-sitter, Little Buttercup, revealed that she had switched, in confusion, two babies that had been left in her care at the same time many years ago. The infants, in fact, came from widely different social classes: "The well-born babe was Ralph, your Captain was the other." The admiral, rather trusting of this testimony, immediately released Ralph and proclaimed him captain, while Corcoran was demoted to the rank of seaman. Ralph was now free to make his marriage if he still chose to do so. Once again, fortuitous circumstances rescued a situation that technology was not yet ready to handle.

In other cases, identity confusion has been caused not by ignorance or error, but deliberately. One person pretends to be another to obtain an advantage or pursue a purpose.

ARTIGAT, FRANCE, ABOUT 1550. Martin Guerre, a farmer of this village, returned after a nine-year absence. He had left without warning to join the army, but on return makes amends with the wife, child, and neighbors that he had abandoned. After a time, however, rumors came up concerning the identity of the person who had returned. The returnee managed to keep doubts in check until he was undone by the return of the authentic Martin Guerre, who proclaimed his forerunner a fraud. The imposter was hanged even though he had acted out the identity of Guerre in a far more sympathetic way than the original Guerre. These events, recounted in the French film *The Return of Martin Guerre*, were said to have a basis in reality.

PARIS, 1794. In the climactic moment of Charles Dickens's *Tale of Two Cities*, the English lawyer, Sydney Carton, advanced to the guillotine, proclaiming "It is a far, far better thing that I do than I have ever done." His captors, however, were under the illusion that they were executing Charles Darnay, the descendent of a family of aristocrats. Carton, who resembled Darnay, had exchanged roles with him for altruistic purposes, permitting Darnay and his family to escape. We can understand, given the circumstances, how the possibility of an exchange had not occurred to the Revolutionary authorities. The swap succeeded.

THE PALACE OF ALIFARIA, SPAIN, FIFTEENTH CENTURY. The Count di Luna had many reasons to relish the moment. He had just had his rival, Manrico, executed. Manrico had not only served as a leader of the troops of an enemy country, Biscay, but had also captured the affections of the beautiful Leonora, a woman whom the count also desired. Further, Manrico was the son of the gypsy Azucena, who allegedly had killed the count's own brother, Garzia, in his infancy. This had been done in revenge for the execution of Azucena's mother, who had been killed by the count.

One treat remained for the count. Azucena was his captive, and he could now inform her personally of her son's demise. She spoiled the party, however, by responding that Manrico was not in fact her son, but actually the count's brother. She had changed her mind after abducting the infant Garzia and raised him as her son instead of killing him. Thus, the gypsy's revenge had finally been gained, she claimed.

The above tale was presented in Giuseppe Verdi's 1853 opera *Il Trovatore*. We can note today, of course, that the count had only Azucena's word as proof of Manrico's identity. Through most of human history, technology has been inadequate for the resolution of such issues of identity. The genetic advances that are about to take place will alter this feature of our existence. The films, novels, plays, and operettas of tomorrow are likely to turn on other plots. They could be set in the past, of course, but their authors, who will have been raised under a new paradigm about identity, may prefer to turn to other aspects of human existence to create ambiguity and mystery.

THE FINGER OF JUSTICE

The first profound advance in this area was anticipated in advance by an astute novelist. The David Wilson that I cited at the opening of this chapter was the central character in Mark Twain's 1894 novel *Pudd'nhead Wilson*. Twain had followed the efforts of a number of nineteenth-century investigators to develop an identity system based on the distinctive ridges, loops, arches, and whorls that all of us carry on our fingers. Among those involved was the founder of eugenics, Sir Francis Galton, who wrote, "Fingerprints are incomparably the most sure and unchanging of all forms of signature."[2]

Twain centered his novel around this method. Wilson had kept an extensive fingerprint collection on the residents of the town for his own pleasure. In the climactic scene, he uses them to demonstrate that two infants had been switched in their cradles many years ago. One was free and one, a slave, though only one part in thirty-two of African ancestry (this was in the South before the Civil War). The responsible nursemaid promptly confesses, and as in Pinafore, the two men are switched in their social positions.

The authorities were not as quick to appreciate the utility of fingerprints, but in 1901, Sir Edward Henry established the first fingerprint bureau in Scotland Yard. The United States was somewhat slower in adopting the technique, but it eventually gained general acceptance. Its principal application has been in criminal cases (we will get to this presently), but it also proved useful in some cases of identity as well. A number of sufferers of amnesia have been identified, in cases where their fingerprints were on file due to prior government service. For our purposes, however, we will want to look at certain cases where fingerprinting failed. In doing so, we will enter the area where identity questions most often arise: the identification of the dead.

SAN FRANCISCO, APRIL 1968. A trunk was received by postal express, shipped at a cost of $57.50 by a Ms. Fedala in Buffalo, to be picked up by Janet Farise (no address). Days passed, but Janet Farise did not appear, and the trunk took on a foul odor. When opened, it revealed the corpse of a 190-pound man, dressed in red trousers and a while silk jacket. He had been shot through the heart.

The employees of the Buffalo express office had no address for Ms. Fedala and recalled only a tiny woman with greying hair. The fingerprints of the man were taken, but matched nothing on file. The police speculated that this was a gangland killing, but the victim had never been arrested or had his prints taken for any other reason. With no known match, the prints were worthless, and when the case was described in Eugene Block's book *Fingerprinting* a year later, it was still unsolved.

NEW YORK CITY, MAY 1987. The badly decomposed body of an unknown man was found floating in the East River. Neither fingerprints nor a facial likeness could be obtained. He represented just one of the fifteen hundred unidentified bodies that turn up each year in New York and the police try to match with the list of fourteen thousand missing persons. Such identifications are difficult without a lead, but in this case a break occurred.

The case was reviewed by a police lieutenant a year later, and he noticed that a trademark and number were listed on the waistband of the underwear that the corpse had worn. By coincidence, the manufacturer was a special supplier of the Brooklyn Hasidic Jewish community and knew of a missing man in that community. "What a stroke of luck!" the lieutenant said, "It could have been J.C. Penney."

The body was exhumed and identified as the missing man with the help of dental X rays. He had been an engraver of gold jewelry with a wife and ten children. Thousands attended the funeral, according to the *New York Times* (May 14, 1988).[3] The identification was helpful in several ways. For one, his widow could not remarry under Hasidic Jewish law until the body was found.

ARLINGTON, VIRGINIA, TODAY. The Tomb of the Unknown Soldier stands as a memorial to America's war dead. The large block of marble that marks the site contains the remains of an unidentified soldier who died in World War I. Other unknown soldiers from World War II and the Korean War are buried in an adjacent crypt to give further honor to those who gave not only their lives, but also their identities to their country. In the future, scientists using new methods of genetic tracing may link the "Unknown Soldier" with existing family heredity lines and may, in fact, be able to identify him.

THE RECORD WITHIN US

I could cite such cases almost indefinitely but fortunately we have come to a stage in human history where such ordeals can end, for we have learned that every bit of human flesh carries within it the record of its owner, and we have gained the technology to read that record. It is written, of course, in the language of DNA.

Within each of us lies 6 billion letters or so of DNA script, about half from each parent. The complete DNA message will be read first for a prototype in the Human Genome Project and subsequently for individuals. What will we learn?

Suppose, for example, that both sets of my own chromosomes were sequenced and compared to the prototype. Perhaps ninety-nine and a half percent of each set might be the same. But the half percent that was different, if written out, would still represent 15 million characters for each set, 30 million in all. (Additional information would be needed to specify the location of each difference in the overall text, but we will ignore that for now.) Fifteen books the size of the one you are now reading or three volumes of the *Encyclopedia Britannica* would be needed just to record my differences.

The differences between the DNA texts of relatives would be less.

Identical twins represent the extreme case and would presumably have the same text. However, the differences between parent and child or two siblings would still amount to a *Britannica* volume and a half.

For simple purposes of identification, the information available in DNA is so overwhelming that only a small portion will be needed in ordinary cases. Even today, with the rather limited knowledge that we have of human DNA sequences, effective systems for the purpose of DNA identification have been devised. They are often called *DNA fingerprints*, but I will use the alternative *DNA profile*, as the first name may misdirect our attention to our fingers, which have no special significance here.

This information, the ultimate biological identifier, lies not just at the ends of our fingertips, but in every part of our bodies, in 1,000 billion or more copies. Skin from our fingertips would do as a source of DNA, of course, but so would other shed skin cells, blood, saliva, sperm, or hair (if the root is attached). Further, with the power of DNA amplification, tiny amounts of material will do.

The task for the future will concern questions of cost and efficiency. Which portions of the immense DNA text are best examined first to pinpoint individual variations? As the entire genome is sequenced, many areas will turn up that are rich in spelling differences and with experience, we will find the beset for each purpose. I can make one prediction already: The areas to be sampled will come from the junk, as these are richest in variability. One person's junkpile is another person's detective kit.

We can imagine a future procedure in which a small amount of blood or a saliva sample is taken, and DNA amplification is used to greatly enrich its content of certain key portions of text. The areas sampled would include the genes involved in important diseases, others that are predictors of physical appearance (we shall see the reason for this), and a selection of prime junk. The junk areas, which would be used for identification and kinship determination, would be spaced strategically across all of the chromosomes of the individual. In addition, some portion of the sample would be frozen and saved for future use should more extended analysis of sequences be needed at a later date.

A valuable technique used with white blood cells at the present time is to "immortalize" them. The cells are treated with a virus, for example, the one that causes mononucleosis, Barr–Epstein virus, which infects them and changes their properties. They can be grown indefinitely without becoming senescent, which ensures that their DNA will remain available for study in the future in whatever quantity is needed. When a study is done, a portion is removed, and the remaining cells are simply refrozen until needed at some future date. When supplies run short, the cells are allowed to multiply. Cost and effort are the limiting factors in applying this technique to human samples today. Hopefully these will diminish when this technique is applied on a broader scale.

Until the prototype human genome has been done, we will not have the information that we need to do an ultimate DNA profile of the type I have described and make full use of it. So important is identity in human affairs, however, that some more limited versions have been set up based on what we know today. The first, and most prominent one, was devised by Alec Jeffreys, a biochemist at Leicester University in England. His original method did not sequence a particular area, but rather sampled the entire genome using a text-matching procedure with a probe to explore the presence of a certain repeated phrase.

I can make this clear using an analogy in English. Let us suppose the phrase YOURETHEONLYONE was to be used as a probe. It would be strung together several dozen times by laboratory manipulations and used to explore a shredded sample of someone's text. Generally, a harvest of some forty or so different fragments might be obtained, each containing a repeated phrase containing YOURETHEONLYONE many times. In my genome, for example, we might find a fragment with YOURETHEONLYONEWHOKNOWS repeated thirty-two times in a row and another with ITHINKYOURETHEONLYONE-WHOCANHELPME repeated fifty times. Many more variations would also be found. In your genome, on the other hand, YOURETHEON-LYONEWHOKNOWS would be repeated twenty-eight times in a row and ITHINKYOURETHEONLYONEWHOCANHELPME sixty-one times. The same differences in repeat number would occur for many of the other phrases.

In Jeffreys' first method, the mixture of repeats was not analyzed in detail, but simply sorted out by length in an electrical field. In the analysis of my genome, the thirty-two-time repeat would give a band at a particular place and the fifty-time repeat would provide a separate band. Their locations would differ from those of your twenty-eight-time repeat and sixty-one-time repeat. For both your genome and mine, we would see a pattern of dozens of bands. It would not matter which was which, but it would be very unlikely that they would coincide exactly for two unrelated individuals. The pattern itself would provide the basis of identity.

The phrase YOURETHEONLYONE has no special claim to distinction in English, nor did the actual DNA repeat phrase that Jeffries first chose have any unique importance in DNA language. Many other repeats worked, including such simple DNA combinations as GATA repeated five times (GATAGATAGATAGATAGATA). Each of them gave a unique set of bands that varied from person to person and could be used for identification.

A different approach that is coming into vogue involves a closer look at a single repeating area. An elegant one was reported by Jeffreys in 1990 that again is best explained by an English analogy.[4] He chose a unique repeating phrase that was located on the short arm of chromosome 1 and devised a way to count the exact number of repeats in any individual. We will use the phrase THERAININSPAINSTAYSMAINLYINTHEPLAIN as an English model for the one that Jeffreys selected.

Now I might have 36 repeats of this on one of my two copies of chromosome 1 and 54 on the other, while you had 43 and 105 (the actual range he found for his DNA phrase was 32 to 161, in the samples he examined). You and I are clearly distinct, but we could have come out the same by coincidence. So far, this approach supplies much less information than the earlier one (which could display many more bands, even though I listed only two in my example), but we are not done yet. The repeats are roughly the same in their text, but may vary a bit in spelling from one occurrence to the next in the same individual. Jeffreys studied a single spelling change that we indicate by the underlined letter in the message THERAININSPAINSTAYSMAINLYONTHEPLAIN.

Consider the chromosome in which I have thirty-six repeats in a row. The first might end with INTHEPLAIN, the second with ONTHEPLAIN, and so on with either of the spellings possible at each repeat. Jeffreys found a way, using a text cutter, that he could determine them all in a single experiment. My run of thirty-six repeats could then be represented with a string of I's and O's, ignoring the rest of the phrase. For example, it might be IOOIIIOIOIOOOIOIIOIIOIOOOOIOIOIIOOOI. Your own run of forty-three would not only be longer, but would have its own pattern of I's and O's throughout.

The number of I and O variations that are possible is enormous, dwarfing the total number not only of human beings, but of all living things on earth (or even the number of cells in them). When Jeffreys sampled the chromosomes of eighty individuals from a variety of ethnic groups, however, he found that two chromosomes were the same in length and in pattern. One individual hailed from England, the other, from Venezuela; to their knowledge, they were not related. Patterns do not occur at random in humans; we use similar mechanisms to generate them and some favorites may occur more often than expected.

This scheme has not yet been used for DNA profiling, but could easily be expanded to yield more information, if we wanted to apply it that way. We could check the spelling variations at another letter of text or combine this with the analysis of another area. Remember, we are just sampling a few sentences from the huge DNA library present in each of us.

Let us pause now to sample a few of the applications that have already been made to human identity by the use of existing DNA profiles.

MIAMI, FLORIDA, 1988. Regina and Ernest Twigg learned the truth in an incredibly sad way. Their nine-year-old daughter, Arlene, suffered from a heart condition that required surgery. She died subsequently, but the blood tests that were taken revealed that Arlene could not have been the natural child of the Twiggs. Some switch had occurred in the nursery of the hospital where she was born.[5]

Baby switches are not just a device of fiction, but occur even though precautions are observed. In the case of my own son, Michael, I remember a band with the words *boy Shapiro* clasped firmly about his infant wrist. The Twigg child had also been labeled appropriately, and a footprint (unfortunately, smudged) had been recorded. Yet some mixup had occurred. The parents were now concerned with locating their biological child, and a possible candidate who had shared the same nursery was located. What could now be done to make the identification?

Fingerprints or footprints do not, taken by themselves, identify one person as the child of another (in the Pudd'nhead Wilson case, sets of prints made before the switch were compared with ones taken after it). HLA types and ABO groups, both common classifications that depend on the identity of certain sugars on the surfaces of blood cells, have been used for this purpose. These tests can exclude a parentage, as they did for Regina Twigg, but seldom provide the positive proof that one person has sired another. DNA tests can do that easily and go far beyond it, as we shall see.

In the first Jeffreys system, for example, suppose parent A showed bands of the following sizes (the numbers are arbitrary), 31, 38, 50, 52, 65, 71, and parent B had 23, 30, 45, 60, 81, 99. The bands of their child should represent a selection from the above lists, for example, 23, 38, 45, 50, 65, and 81. The presence of different bands would eliminate someone as a possible child, whereas a match only to parental bands would indicate that he or she was quite likely to be their offspring.

A year passed in legal negotiations, but at the end of that time, genetic tests showed that the girl in question, Kimberly Michelle Mays, was the biological child of the Twiggs. The legal father retained custody, but visiting arrangements were made for the Twiggs.[6]

LONDON, 1985. A Ghanian woman lived in England with two sons and two daughters. One son traveled to Africa to visit his father, then returned. The immigration authorities had reason to believe that the woman's nephew had returned in her son's place. How could this be determined? The father of her son was unavailable, nor was it clear that the son had the same father as his siblings. What could be done?

DNA profiles were taken on the woman and her three accepted children. Each of the children had some of the woman's bands, plus additional ones. The extras were combined to construct a set assigned to their father. The DNA profile of the boy in dispute consisted entirely of bands present in the woman or in the reconstructed profile of the father. It was extremely likely then that he was the child of the Ghanian woman and the same male who had fathered her other three children. The boy was readmitted to the country by the immigration authorities.

This case, reported in *Nature*,[7] showed the power of the new DNA methods. The biochemical tests that were routinely used in such situations could not distinguish a nephew from a son.

The application of DNA typing to simple determinations of paternity is straightforward, and it has been applied extensively. In India, for example, "paternity cases make up a significant percentage of the court cases," according to *Nature*. A center in that country has devised its own low-cost DNA profile method, and their procedure has been accepted by the Indian courts.

SAN FRANCISCO, JANUARY 1989. Mary-Claire King looked gorgeous in her formal, flowing, rust-colored gown. She was not attending a grand ball, however, but appearing as a featured speaker before a vast audience at the annual meeting of the American Association for the Advancement of Science (AAAS). In similar circumstances, many women scientists would choose to dress as drably as possible, but Mary-Claire had chosen that gown for a special reason. Her grandmother had worn it in the 1920s on the occasion when she was elected to an office in her church that no woman had held before. She had passed it on to Mary-Claire with the advice: wear it when you want people to notice you. This was the first time that her advice had been taken.

Mary-Clarie King is certainly worthy of notice, whatever she may be wearing. As a professor of genetics and epidemiology at the University of California at Berkeley, she has published widely on topics that range from evolution to breast cancer. Her doctoral training was also done at Berkeley, in the lab of Allan Wilson, a pioneer in the use of DNA sequences to study evolution. She came out with a strong desire to put her special scientific training to use not only in solving problems of research, but also those that arose in real life. "I'm very much a product of the Berkeley sixties," she said to me.[8]

The story that she presented at the AAAS meeting illustrated how much this type of idealism is needed in a world riddled with depravity. Her account was reinforced by presentations that had been made earlier in the day by anthropologist Clyde Snow and biochemist Cristian Orrego.

BUENOS AIRES, ARGENTINA. A group of Argentinian military officers had seized control of their country in 1976 and began to arrest those it considered subversives in "the defense of national security and honor." Their aims were the following, according to a military governor of Buenos Aires, cited by Clyde Snow: "First we will kill all the subversives, then kill their collaborators and sympathizers, then those that remain indifferent, then the timid."[9] The subversive list included the far Right and Left, homosexuals, Marxists, Zionists, Freemasons, and even the Rotary Club. They did not reach their goal, but in the few years they did murder perhaps nine thousand men and women, mostly in their twenties and thirties. Many bodies were disposed of at sea, while others were buried in rural cemeteries with the marking "no name."

Those arrested included pregnant women and mothers with very young children. The pregnant ones were kept alive long enough to deliver their offspring. For example, Liliana Pereyra, age twenty-one, was held in a military academy until her delivery in February 1978, then killed by a shotgun blast to the head.[10] About two hundred children were born in such circumstances or abducted while they were very young. Many of them were given to childless military couples, who passed them off as their own.

When the military dictatorship was deposed in 1983, efforts could begin to investigate the atrocities that had been committed and repair what was possible. Concrete evidence was needed to prosecute those who had committed the murders. "No name" graves were exhumed, and masses of bones and bullets were uncovered. Some, such as the murdered Liliana Pereya, could be identified by matching their skeletons to dental and medical records. Many other bones remained anonymous. Efforts to identify them continue to this day, even though the climate for prosecution is less favorable under the current administration. An identification would still allow a proper burial and bring some sense of closure to the relatives of the *desaparecido* (the disappeared).

One group of those relatives, *Las Abuelas* (the grandmothers), had more ambitious objectives. This group formed for the specific purpose of finding and reclaiming their missing grandchildren. Their children had been murdered, and they had no record of their vanished grandchildren, but they suspected that many of them were being raised under false identities. To prove their claims, however, a scientific test for grandpaternity had to be established.

Mary-Claire King entered at this point with the training, energy, and motivation to take on such a job. These qualities were needed as circumstances were far from ideal. Reagents and facilities were in short supply (she had to bring needed chemicals from the United States on her trips to Argentina), threats were received by laboratory workers, judges were dubious about the scientific evidence, and the military parents were far from cooperative. Using HLA blood groups, Mary was able to make some identifications, provided that most of the grandparents were alive. Court orders were usually needed to get the blood sample from the child.

The results were sometimes quite dramatic. Paula Eva Logares had disappeared at age two when her parents were murdered. She was adopted by a Sergeant Llabalen and his wife, but registered as their natural daughter. The birth certificate looked irregular to school authorities, however. Blood samples were obtained from the girl and her grandparents, and the HLA tests established, with 99.9 percent certainty, her real identity. The finale was reported by Jared Diamond in *Nature*: "By court order she moved to her grandparents house, where she went directly to her former room and asked for a doll that she had last seen at the age of two."[11]

Other cases proved more difficult to resolve in this way, especially those where only one grandparent was still alive. New devices were needed, and Mary-Claire King was inspired by her continuing contacts with the laboratory of Allan Wilson.

Her former mentor, who also spoke at the same San Francisco meeting, had found that human evolution could be traced in depth by studying spelling differences in a particular type of DNA. The name of this DNA is *mitochondrial DNA*, after a structure, the *mitochondrion*, which generates energy in the cells of all higher organisms. I will call this structure an *energy cell* and refer to its DNA as *energy-cell DNA*.

Energy-cell DNA is distinct in a number of ways from the DNA in chromosomes that we have talked about thus far. It is an extra bit of our heredity that occurs outside the cell nucleus with several copies present in each of the hundreds of energy cells that exists within a human cell. Although each copy contains only 16,569 base pairs (only one two-hundred-thousandth of the DNA information of the chromosomes) its presence in many copies gives it a bulk that amounts to about half a percent of human DNA. In terms of our metaphor for the human genome in which the chromosomes are represented by forty-six huge unabridged dictionary-sized volumes in a room, energy-cell DNA would be represented by a few thousand copies of a slim pamphlet, roughly equivalent to four pages torn out of this book, scattered about the same room.

Most biologists accept an evolutionary explanation for the presence of this unusual DNA outside of the chromosomes. At one time more than a billion years ago, predecessors of the energy cell existed as free-living bacteria with a full complement of DNA. In some way an energy-cell ancestor was taken into and merged with a larger bacterial cell. It continued to produce energy, but much of its DNA was no longer needed and could be discarded. Other portions of the DNA were moved to the nucleus of the enlarged cell. A few genes that were important to energy production were kept within the energy cell, however, and there they remain.

The origin and function of energy-cell DNA, however fascinating, is not what concerns us here. Several other features attracted the attention of Allan Wilson and Mary-Claire King: Human energy-cell DNA is readily separated from that of the chromosomes. Its sequence has been fully determined by Fred Sanger and his co-workers. Finally, human energy-cell DNA, unlike chromosomal DNA, is inherited only maternally. A sperm may have a few copies, but they do not enter the fertilized egg.

The fate of energy-cell DNA in inheritance can readily be followed. The copy that I have came from my mother and her mother before that. My son does not have my maternal DNA, he has my wife's. I have no daughter, but my energy-cell DNA will survive. My mother had several sisters, and they collectively have several daughters, who also have some daughters. I share the same energy-cell DNA with them.

Let us return to Argentina and the search for the kidnapped grandchildren. In the case where only one grandparent survives, that is most likely to be the mother of the mother (women marry at an earlier age and live longer than men, on the average). To establish whether a woman is the maternal grandmother of a particular child, then, only requires a comparison of their energy-cell DNA sequences.

The determination will be less exact than those that utilize appropriate areas of chromosomal DNA. The genes in energy-cell DNA are packed together tightly with very little of the junk present that is invaluable in comparisons of identity. One particular area of convenient size (a few hundred base pairs) in a region called the *D loop* does show some variability. By use of this area, Mary-Claire King and co-workers were able to identify the maternal grandmothers for several of the kidnapped Argentinian children with the chance of error by coincidence estimated at one in one thousand.

More than fifty of two hundred lost grandchildren have been located so far, and the work continues. The grandparents recognize that they will not survive indefinitely, however. Many of them have left samples of their white blood cells frozen in liquid nitrogen for future use. For now, they are looking for their grandchildren. In the future, they feel, their grandchildren will come looking for them.

The idea of a DNA bank where biological samples are preserved for the day when they may be needed is not limited to Argentina, of course. In the United States, such services are already offered by the Center for Human Genetics at the Boston University School of Medicine and the DNA Banking Service of the San Francisco Children's Hospital. Quantum Chemical has founded a subsidiary, Lifebank, Inc., to provide DNA storage on a commercial basis. The practice of DNA storage will grow.

The use of energy-cell DNA to identify a grandparent or a more distant ancestor also has wider implications. During the interview,[12] Mary-Claire King speculated about other possible uses of the method. The most dramatic example that occurred to her was that of Anastasia. "It would have been possible, knowing what we know now, to determine whether she was or wasn't—very straight forward." On considering this later, I recognized that Anastasia's great-grandmother in her maternal line was Queen Victoria. Any other descendent of Victoria in an all-female line would share the same energy-cell DNA sequence with the occasional difference of a letter change due to mutation. Anastasia's mother's sister, for example, survived until 1950, while others who bear that DNA are alive today. Prince Philip, for example, and Anastasia both descend in a maternal line from Victoria's daughter Alice.

Much more could be done to explore relationships using DNA once the prototype sequence has been determined. Before we examine that topic, we will look at another application of DNA profiles that has commanded much more media attention: its use in criminal cases.

THE DNA DETECTIVE

A key point of contention in many criminal trials has been the presence of the accused at the scene of the crime. The use of fingerprints for this purpose is so familiar today that no child's detective kit would omit the necessary magnifying glass and powder. In 1910, however, the validity of this type of evidence came into question. The objections disappeared as the public and the courts became more familiar with the technique, and legal challenges are no longer heard. A different problem exists in many cases, though.

Fingerprints can be invaluable when they are present, as in the Clarence B. Hiller murder case, but often they are not there. A criminal can simply wear gloves or take the trouble to clean off any surfaces that he has touched. He will have much more trouble cleaning up his DNA after him. With the use of DNA profiles, hair, saliva, or even skin cells shed at the scene of a crime might be used to establish the presence of a suspect at the site. The detectives of the future will include a vacuum cleaner in their gear along with the magnifying glass, whereas the criminal who is determined to thwart them might have to equip himself with a face mask and body suit, apparatus that in itself would be much more incriminating than a pair of gloves.

This battle of wits has not yet surfaced, as the technique is still new. Two more obvious bodily products, semen and blood, have taken center stage in the trials to date that have relied on DNA profiles. A few examples will show the power of the method in such cases.

NARBOROUGH, ENGLAND, 1987. The police had come to a point of high frustration. In the course of three years, two schoolgirls, each aged fifteen, had been raped and strangled as they walked along quiet footpaths near this small town. Every clue and eyewitness report of the 1983 murder of Lynda Eastwood had been followed up without any result. It was only after Dawn Ashworth was killed in a similar manner that the authorities finally were able to locate a likely suspect. A rather peculiar seventeen-year-old boy, who worked in a local hospital as the kitchen porter, had been nearby when the second murder took place. Circumstantial evidence and his own previous history indicated that he could have been the murderer.

His arrest and questioning by the police were described in detail in Joseph Wambaugh's best-seller, *The Blooding*.[13] The testimony that he gave was confused, disjointed, and bizarre. He consistently denied any involvement with the murder of Lynda Eastwood. At one point he appeared to admit to the murder of Dawn Ashworth, but at other times he denied it. While in the midst of a confession, he finally made a sensible request: "I want a blood test."

It was not blood, but semen, however, that turned the tide. The boy's father had heard of the new technique: "I'd read somewhere, maybe in *Reader's Digest* or *Tomorrow's World*, about this DNA testing that the chappie in Leicester had discovered," he told Wambaugh. "I told my laddie's solicitor to look into it." By coincidence, the lab of Alec Jeffreys, the inventor of DNA profiling, was only six miles away.

Part of the results were as the police expected: The same person had murdered both Lynda Eastwood and Dawn Ashworth. That person was not the kitchen porter they had arrested, however. The DNA profile obtained from his blood did not match those obtained from the semen stains on the bodies of the victims. The porter became the very first accused murderer to be released as the result of a DNA profile.

The police, in despair because of the absence of any other strong leads, decided to turn the DNA technique to their own advantage. Blood samples were taken from over five thousand young men who lacked an alibi and lived in the area. No DNA patterns were found that matched the one from the semen stains, but the threat contained within the method provided a lead of equal value. A bakery employee, Colin Pitchfork, had persuaded a co-worker to provide a blood sample in his stead on the grounds that the police had treated him unfairly on past occasions. The co-worker discussed the switch in a conversation in a Leicester pub, however, and the news reached the police. When confronted, Pitchfork confessed to both murders. According to Wambaugh, the high court judge that presided in the case commented, "The rapes and murder were of a particularly sadistic kind. And if it wasn't for DNA you might still be at large today and other women would be in danger."

AMERICAN APPLICATIONS. The United States, more than Britain, suffers from an excess of rape, murder, and other crimes of violence. According to an article in *Chemical and Engineering News* by John I. Thornton, professor of forensic science at the University of California at Berkeley, homicide is the leading cause of death of Americans aged fifteen to twenty-three and the fourth leading cause for all Americans under sixty-five.[14] Two million Americans alive today will eventually be murdered. A woman born in the United States today has a twelve percent chance of being forcibly raped sometime in her life. Furthermore, a substantial minority of the crimes go unsolved.

Although it has been developed only to a small fraction of its ultimate potential, the technique of DNA profiling has already made some dent into the huge pileup of unsolved crimes. In 1988, for example, Victor Lopez, known in the press as "the Forest Hills rapist," was convicted of sexually abusing three women in the same area of Queens, New York, within a month. One juror commented, "The DNA was a kind of sealer on the thing. You can't really argue with science." The forewoman agreed, telling a *New York Times* writer, "That was the only thing that opened my eyes. That was the whole case in my opinion."[15]

DNA played an equally important role in an unusual case in Pierce County in the state of Washington. A fifty-seven-year-old woman who suffered from Alzheimer's disease was sexually assaulted on a bus, but could not remember the details. No witness was available. The bus driver was the most likely suspect, but he adamantly denied the charge. Conventional genetic tests were applied to a blood sample from the driver and a semen sample taken from the women. He was demonstrated to be a possible culprit by these tests, but twenty percent of the male population of the state also shared that distinction. When a DNA profile indicated that the odds were several million to one that he was the source of the semen, the driver pleaded guilty.

Events of this type have led writer Mark Thompson to comment, "Rapists might as well start leaving calling cards at the scene of their crime," and Professor John Thornton to conclude that "DNA typing is likely to revolutionize the analysis of biological evidence." In some cases, however, criminals have attempted countermeasures to neutralize the threat. Fortunately, they were clumsy. Daniel Garner of Cellmark Diagnostics, a commercial biotechnology company that offers DNA profile services, tells of a rapist who had the right idea and used a prophylactic, but left it at the scene of the crime! Another had the victim clean herself out with a washcloth; that, however, served to preserve the evidence, rather than destroy it.[16] Clumsiness can work in both directions, however. Ineptitude on the part of the technicians can damage the value in court of a particular DNA profile.

Geneticist-mathematician Eric Lander has pointed out that DNA profiles face difficulties in their crime-related applications that do not usually come up during medical diagnoses and paternity disputes. In the last two circumstances, unlimited samples are available. If a test should be mishandled, technically, another can be employed. If the results should prove ambiguous, another probe can be tried. (The full 6 billion letters of the human genetic text provide the ultimate backup.) By contrast, a sample of blood or semen left at the scene of a crime may be damaged by exposure or contaminated and extremely limited in amount. If errors were made in the analysis, a repeat might not always be possible. One particular murder case brought a situation of this type to widespread media attention.

THE BRONX, NEW YORK, FEBRUARY 2, 1987. Joseph Castro, a neighborhood handyman, broke into the apartment of Vilma Ponce, who was seven months pregnant and by his own confession (more than two years later) stabbed her and her two-year-old daughter, Natasha, to death. When brought to trial on a variety of evidence, Castro at first denied committing the crime. A small amount of Ms. Ponce's blood had spattered onto Mr. Castro's watch, however, and was sent to a DNA profile firm, Lifecodes, Inc., for analysis.

Unfortunately, the company did not use maximal care in conducting the analysis. The bands observed in the blood sample on Castro's watch matched those in Ms. Ponce's blood, but additional contaminants were accidentally introduced. The company dismissed them as artifacts and declared that a satisfactory match had been obtained. According to Lander, they also miscalculated the probability of a coincidental match, claiming the identification with more certainty than was justified by the mathematics.[17]

Although affirming the value of DNA profiling in general, the presiding judge concluded that the testing laboratory had "failed in several major respects to use the generally acceptable scientific techniques for obtaining reliable results."[18] He threw out the DNA evidence that identified the blood as belonging to Ms. Ponce. (I would disagree with this last step. The most likely result, by far, when two DNA samples are picked at random and compared is that they clearly will not match. If they come close enough so that contamination could explain the differences, the result has some meaning, although it is less decisive than a clear match.) Fortunately, the other evidence in the case was sufficient to convict Castro. Otherwise, the inadvertent errors of a technician might have led to the release of a murderer.

Eric Lander and the scientists who agreed with his data analysis did perform a valuable service. They made clear the need for competent supervision of DNA profile services under state-controlled licensing procedures. Protocols will be established that will help guard against false results due to contamination, another type of error called *band-shifting*, and other mistakes. The scientific basis behind the DNA profile methods is about as solid as one can get, but human error is possible in any procedure. In my own research, I have seen weeks of effort wasted when a student simply mislabeled a sample at the end of a complex separation that had been handled flawlessly.

Some of the media missed the point, however, even if the judge got it. *Nature* ran the headline "DNA Fingerprinting on Trial,"[19] and others reacted similarly. They might have concluded with equal validity that the physical laws governing optics had come into question when the mirrors of the Hubble telescope failed to come into focus properly. The media have shown much less tolerance toward the possibility of occasional error in the case of DNA profiles than they do for older, but equally fallible forms of evidence.

Consider for example the case of twenty-five-year-old Randall Lynn Ayers, as reported in the *New York Times* of July 23, 1990.[20] He had served eight years for the rape and robbery of a fifteen-year-old girl. In 1990, however, Robert Minton, a look-alike, confessed to the crime. Minton had already been charged with the killing of two other Cincinnati women. The rape victim could not differentiate between Ayers and Minton when confronted with both. Ayer's lawyer commented, "Basically the only evidence in the case was eyewitness testimony. You can see how eyewitness testimony can make a mistake." No spokesperson has come forward, however, to suggest that all such testimony be banned in court or that eyewitnesses pass a state examination.

Eric Lander has also pointed out that the technical disputes of today are temporary. Today's procedures will eventually be replaced by other ones based on direct DNA sequencing. With proper licensing and supervision in place, DNA misidentification will become a very rare event. As technical issues subside, however, others that involve privacy and civil liberties will come up.

THE DNA DATA BANK

We are rapidly approaching the day (we may be there already) when we will have the ability to reap an abundant harvest of DNA sequences from a hair; a flake of skin; or a tiny amount of blood, saliva, or semen. When substantial parts of the human genome have been read and interpreted, we will also learn much more from the data than whether or not two samples match. Special Agent Kenneth W. Nimmich of the Federal Bureau of Investigation (FBI) recently told the *New York Times*:

> Three [base] pairs [out of the billions in DNA] determine the color of hair, three pairs determine the color of eyes. It's just a matter of time until we find the right pairs to draw a physical profile from their DNA.[21]

Agent Nimmich was only scratching the surface. Much more will ultimately be learned from DNA than a few physical features. We mentioned genetic diseases and vulnerabilities in an earlier chapter, and in the next one I describe how relationships far more distant than paternity or grandpaternity may be determined from DNA sequences. For the information that lies within our DNA to be fully utilized, however, an extensive bank of DNA data drawn from the population at large will be needed. This prospect has already caused some observers to shudder at the thought and others to applaud in advance.

John Hicks of the FBI, nothing that 62.5 percent of those who commit serious crimes are rearrested within three years of their release from prison, has pointed out the value of keeping their DNA profiles in a permanent file. California and other states have already moved to set up such files. "There are lobbies both in the United States and the United Kingdom who wish to profile all individuals at birth," Peter Gill of the United Kingdom Home Office Forensic Service wrote in *Nature*.[22]

Some ethicists and lawyers, however, fear the uses to which a broad DNA data bank might be put by an intrusive and powerful government. Information on psychological and biological vulnerabilities might be used to harass and discredit political opponents and to deny employment to disliked individuals. Alan Westin of the Columbia University Department of Political Sciences has argued that "there is already a dangerous call from some government individuals to extend DNA testing to broad investigative data bank collections from . . . the general populace." He feels that such efforts should be rejected on privacy grounds: "Privacy is the claim of an individual to determine what information about himself or herself should be known to others."[23] Most ethicists who have considered the issue have endorsed the right of an individual, in principle, to keep his or her DNA profile to himself or herself in normal circumstances. (We will ignore for now the practical difficulties in such concealment that I mentioned earlier.)

Standing in counterpoint to the normal desire for privacy, however, are the many special circumstances when DNA information is essential to the well-being of others: unresolved paternity cases, unidentified remains, the Argentinian grandmothers, the presence of infectious viruses in sexual partners, and the identification of crime suspects, among others. In addition, more broadly based reasons exist to justify the collection of detailed genetic data from as many people as we can afford to handle.

THE ENVIRONMENTAL MUTAGEN THREAT. For the past century or more, the human race has been exposed to an ever-increasing number of synthetic chemicals, the great majority of which have been prepared for the first time in the history of this planet. Since the 1940s, it has also been clear that more than a few chemicals can cause mutations, changes in the DNA text, by reacting with DNA. As we have already seen, a single misspelling in an important gene area can damage a protein or lead to the production of too much or too little of it.

In normal circumstances, the malfunction of one cell out of the trillions we possess would not matter, but there are important exceptions. The uncontrolled reproduction of a cell may be triggered with cancer a result. Or, if the cell that suffers the DNA damage is a sperm or egg, an individual may be produced who has the spelling error in all the cells of his or her body. He or she may also pass that error on to the next generation.

When a chemical burns or poisons us, we usually recognize this quickly, but damage to DNA can be much more subtle. Cancer may take decades to manifest itself, whereas an increase in the number of carriers for a recessive disease such as cystic fibrosis might accumulate for generations until the actual death statistics revealed it.

One nightmare for geneticists, then, has been the possibility that a new chemical might react with our DNA and set up a genetic catastrophe that would not become evident for centuries. New drugs and food additives are tested, of course, using bacteria or mammalian cells in culture. In cases where the possible hazard justifies the expenditure, large-scale animal tests may be used. We cannot inspect all of the components of a smokestack or barbecue flame, however, or anticipate every alteration that the environment or our own bodies might make in a chemical. Many cases have been described where such changes have converted a harmless substance to a hazardous one.

As a safeguard against catastrophe, geneticists over the years have called for some early warning system to protect us from unrecognized chemical hazards. A national registry of birth defects was suggested, as was the analysis of the proteins in the blood of infants to search for abnormalities. A reporter from *Chemical and Engineering News* drew the following conclusion after a 1969 interview with Dr. James Neel of the University of Michigan:

> The detailed chemical analysis of the blood of hundreds of thousands, if not millions, of babies born in the U.S. would, of course, be a collossal undertaking. Yet, until elaborate studies of chemical abnormalities in vast numbers of humans are carried out, we are still "hopelessly mired in *terra incognita*."[24]

Such a program was not undertaken, undoubtedly because of the costs. We have reached a new era, however, where DNA spelling changes can be measured directly and immediately. The demand for data of this type is likely to expand: Specific sections of the DNA text of the unborn or newly born will be scanned routinely for disease-causing changes, just as their bodies are now routinely inspected for damage. It would not be too difficult to expand the analysis so that any above-average increase in the overall mutation rate for an individual would be signaled. A notable increase in mutations in a geographical area or among a group that shared some obvious life-style feature would provide the wanted warning flag to trigger a deeper investigation. In the same spirit, an individual who worked in a particular factory might have some of his or her white blood cell DNA monitored from year to year as a test for any exposure to mutation and cancer-causing agents.

THE BIOLOGICAL RECORD. In addition to these health concerns, there are many other reason why we would want to preserve a biological history of our present generation and future ones in the form of a DNA record. Historians may want to track movement of peoples, individuals will wish to construct genealogies, and evolutionists may want to compile overall changes in our DNA. All of these concerns argue for the accumulation of individual DNA sequences on the largest scale that we can afford.

Thus, the feeling of someone like Millennia that her DNA is her own business appears to run head-on into those of others who need the information. A doctor may need to know the mutation rate in the town and year of her birth or one hundred years later, her great-granddaughter may want to find out which of her genes came from Millennia. Fortunately, a way can be suggested to satisfy all of these feelings.

Imagine the following scheme: Some sequences will be read routinely for all newborns (unless there is parental objection) and all those who are fingerprinted today. In addition, the remainder of the population will be invited to provide voluntarily a saliva, hair, or blood sample for sequencing purposes. The information will be kept in computers for routine, but controlled access by academics, police, or others with a legitimate reason. No names would be listed with each sequence, however, but only a coded identification number. The information identifying particular individuals with their DNA code numbers would be stored elsewhere under strict guard. The computers holding such data should best be kept off-line with remote access impossible. The release of information from this file would require a court order or the equivalent.

Such files would be vulnerable, of course, to seizure by a totalitarian state. However, such a state could equally well compel individuals to provide DNA for sequencing or obtain samples surreptitiously. I feel that it is better to guard against the possibility of such governments by direct political action rather than follow a scorched earth policy in which we destroy in advance anything that might be useful to them and deprive ourselves of their benefits in the interim. With the hope that we can find some sensible way to both collect and protect DNA data, I have imagined some additional glimpses into tomorrow, to illustrate the power of the methods.

NEW YORK, 2020. Acting on a tip from a cruise ship captain, police recovered a badly decomposed human body from the East River. No facial features or fingerprints could be recorded, and the underwear band bore only the uninformative marking "J.C. Penney." The DNA profile of the individual was readily obtained, however. It was compared with the complete record of DNA profiles of missing persons stored in the New York Police Department computer, but no match was found. (In many cases, the missing individuals already had their profiles on record. In others, it was reconstructed with the aid of relatives. In a few instances, hairs or shed skin cells were recovered from the hairbrushes or clothing of the missing individuals.)

The DNA information of the corpse was then sent out over a network linking all police or equivalent departments in the United States, but again, no match could be made. The central DNA registry of the FBI was then consulted, and an exact match was found, originating in California. A court in that state then issued the necessary warrant to retrieve the name of the individual from the DNA identity registry.

The matching profile had been obtained from a Mr. Caliban of Los Angeles, California, when he had needed minor surgery two years earlier. He still maintained the same residence, but was, of course, not at home. The landlord reported that he had left some months ago on an extended vacation with instructions that the apartment not be disturbed. A police spokesperson indicated that some important leads concerning the victim's demise had been found in the apartment, but would not comment on their nature.

My inspiration for this tale came from the actual newspaper case I mentioned earlier, in which a corpse was identified by a coincidence involving an underwear band of very limited manufacture. The DNA method should work, however, as long as any flesh remains or perhaps even longer, as we will see in the next chapter.

Body parts, of course, work as well as bodies for DNA profile purposes. On the day before writing these words, I heard a radio account of the discovery by the police in Brooklyn of several bags that contained grisly contents of this type. Some months earlier, the Staten Island police uncovered several suitcases with similar items in them. In *Fingerprinting*, Eugene Block tells of Dr. Buck Ruxton, a surgeon who in 1935 murdered and dismembered his wife and another woman. Dr. Ruxton divided the remains among thirty parcels, which he abandoned in a gulley near Lancaster, England. He went to the trouble of removing his wife's dental work and fingertips to prevent her identification, but he was caught and hanged just the same.[25] In the Brooklyn case, the radio announcer added that the police did not know how many bodies there were or how to proceed in identifying them. That much, at least, will change as DNA profiles come of age.

MAMARONECK, NEW YORK, 2020. An abandoned newborn infant was discovered in a trash can near the Mamaroneck railroad station. (Again, this was inspired by a recent radio news item.) Local inquiries provided no clue, so the police turned to the national DNA data bank of the FBI. No complete match was found, nor was one expected. The computer was then instructed to scan its data base to find a sequence overlap of fifty percent, but no such match existed either. Neither parent had ever submitted a DNA profile. A search for any overlap close to twenty-five percent was more successful. Two such matches were located, both originating in the same town in Iowa.

Court permission was obtained to identify these individuals, who turned out to be a Des Moines couple. Their teenage daughter had run away from home several years ago; her present location was unknown. When informed that the abandoned infant was almost certainly their grandchild, they offered to take custody of it.

No moderately close relatives of the father of the infant were uncovered by the computer. Many text stretches were present on the paternal chromosomes, however, that were common in the eastern Mediterranean, with Greece the most likely site. Permission was requested of the Greek authorities and granted for a search of the data banks in that country. Routine screening of infants had been undertaken there for more than a decade, but fewer adults were on file. A boy in a provincial town was located, however, with a twenty-five percent DNA overlap to the infant. He was the youngest son of a large family with two older brothers employed as merchant seamen. One of them had disembarked briefly in New York nine months earlier. The American and Greek authorities agreed to work together to obtain a more secure paternity identification and possible child support from that sailor.

If the technique had arrived in Greece somewhat earlier, of course, it might have saved Oedipus from his fate. Rather than dwell longer on human affairs, I want to digress and consider the other inhabitants of this planet.

MAN'S BEST FRIEND AND SOME OTHERS

DNA profiling need hardly be limited to humans. As I mentioned earlier, the Human Genome Project itself will include some sequencing of bacteria, yeast, a worm, the fruit fly, and the mouse as models for the human effort. As sequencing technology improves, other groups may wish to add additional species to the list. First in line will be those that are important commercially and those we like to keep as pets.

According to one report, the American Kennel Club has launched a dog mapping project. Projects on cattle, pigs, sheep, and chicken have been considered by others. In the plant world, as we mentioned in Chapter 14, the first target will be a weed of academic interest, *Aribidopsis thaliana*, but important crop species will not be far behind.

Sooner or later, the sequencing of some text passages from almost every species is likely in the ultimate endeavor of taxonomy. With that information in hand, we may then pry into earthly biology as our whim dictates. The genetic secrets of every blade of grass on our lawn or every mosquito that intrudes into our bedroom will be open for inspection. Long before that time, however, DNA data will be used in animals as it has in humans: to determine identity. A few choice applications have already been made.

BIRDS AND THE NEST. To the religious, it may sometimes have seemed that birds were put on earth to model ideal family behavior for us. The loving parents cooperate to build their nest and later scurry about in search of food for the hungry mouths they have engendered. Loyalty and fidelity predominate. Alas, DNA profiles have alrady shattered that image. In many species, as many as thirty percent of the young in a nest are fathered by someone other than the resident male. Casual observation did not reveal this behavior: "Extra pair copulations are called sneakers and they really are," biologist Robert Montgomerie reported. "They're not easy to observe because the birds are very surreptitious about their behavior."[26] The birds do mimic certain human behavior after all, but not of the type we expected.

WHOOPING CRANES. The whooping cranes are among the most attractive members of endangered species. Although only about 150 of these elegant North American birds remain alive, this number still represents a considerable improvement over the eighteen that existed in 1938. Although matings among relatives are hard to avoid in this situation, would-be matchmakers still try to do the best they can in promoting as much diversity as possible. DNA profiles have been compared to try to find the least related pairs for breeding purposes. (I can't help thinking: But what if they don't *like* each other?)

THE FATE OF THE BLACK RHINOCEROS. DNA samples were taken from three widely separated populations of black rhinoceros in

Africa. They were very similar, almost as if the animals had been living and breeding together. The species is an endangered and dwindling one. The study showed that, if necessary, the survivors could be pooled together for better protection from poachers and to make it easier for males and females to meet at breeding time.

Many additional applications for animal DNA can be imagined. A DNA profile remains the ultimate identity mark, the brand mark that cannot be disguised. The steak in your freezer could be linked to a cow that at one time grazed on a particular acre of turf if the owner had troubled to keep such records. Birds and dogs that are sold as pets, fur coats, elephant tusks (provided that some scrap of flesh were still attached), and gorilla hands will all contain their own internal statement of origin. Hard times are clearly in store for poachers, dognappers, and old-fashioned cattle rustlers.

The DNA profile of an animal will reveal its parentage as well, where this is important. DNA will provide the ultimate guarantee for the owner of an animal with a pedigree, while those who prefer to rescue a dog or cat from the streets will be able to determine what mix of breeds has been combined to make their pet. Of course, some humans will be interested in this topic for themselves as well as their pets. It is time to return to our kind.

17

ROOTS, BRANCHES, AND TWIGS

We are the children of many sires, and every drop of blood in us betrays its ancestors.

—Ralph Waldo Emerson

Until our own time in history this "betrayal" had, with rare exceptions, only a poetic meaning. In some cases, a visible hereditary trait or defect marked a line of descent. A few of our many sires were flagged, but the others remained concealed by our bodies with only written or oral history acting as a guide to our forebears. Now that we have gained the ability to read the letters of our DNA, only the phrase "line of DNA text" need be substituted for "drop of blood" to make the above quote exactly correct.

Critics of the Human Genome Project have ridiculed the long stretches of junk, the Alus, CACAs, and other stretches of nonfunctional DNA, as unworthy of our attention. Perhaps a few portions of them need to be defined for the determination of paternity and the conviction of criminals, they say, but what need did we have for the billions of additional useless characters?

There is another commonplace number that runs in the billions: the total of living members of our species on this planet. Each of us is linked to all of the others by an intricate web of connections that extends back to the very first group of beings that we choose to call human. We each have had two parents, four grandparents, and eight great-grandparents. If this doubling were to continue for each generation back, the number of ancestors who were alive one thousand years ago, for each of us, would exceed in number the population of the earth today. Yet the count of humans alive now exceeds the total at any previous time in history. The roster of our forebears was filled out by having a much more limited number of ancestors appear in multiple positions in the scheme.

In the past, geographical and cultural barriers kept humans divided into separate breeding groups of races and nations much more than today. The dispersal of a presumed original pool of humans and the separate evolution of the separated branches have been a fascinating, but little known chapter in the history of our species.

Some humans today have well-documented histories of their ancestry at least for recent centuries, for example, Mormons and members of distinguished families. Others have little interest in such matters. For the remainder, our natural desire to understand our biological links to our forebears has been tempered by the effort needed to get such information or the hopelessness of the quest. Genetic science has now changed the odds.

The task of deciphering the connections that link billions of humans to the past is immense and formidable, of course. However, in the near future we shall gain a resource that measures up to the task: the six billion characters of information that reside in each of us. The data will be available for the taking, and computer power will be developed that can handle the information and extract meaning from it. To illustrate what can be done, I have selected a hereditary line that is more familiar than most: the royal family of Britain.

ROYAL GAMES

Charles Philip Arthur George Windsor, Prince of Wales, known more simply as Prince Charles, stands first in succession to the British throne. His position is secure, as a male and as the eldest child of the reigning monarch, Queen Elizabeth. Before Elizabeth ascended to the throne in 1952 upon the death of her father, King George VI, both Elizabeth and Charles could have been usurped in their direct accession to the throne. Had King George fathered a son before he died, that son would have taken precedence over Elizabeth and her sister Margaret, regardless of their ages. When that event did not occur, Elizabeth, the elder daughter, became queen, even though George VI had brothers and his brothers had sons.

The rules of the British succession attempt to keep a direct line from parent to child, then use sex and age to select a single heir to the throne when more than one child is present. Some rule is needed to prevent rivalry, for all children of the monarch are equally "royal," at least when heredity was thought related to blood. The advent of DNA profiles now allows more accurate distinctions to be made. We can, for our own amusement, invent some alternative tiebreakers based on actual heredity rather than age and sex.

Let us take a closer look at Prince Charles. Of his forty-six chromosomes, twenty-three came each from his mother and his father, Prince Philip, the Duke of Edinburgh. If he chose, he could undoubtedly arrange to have his own sequence read for any DNA passage that had already been recorded for any human. Let us use the beta-globin gene as an example. He would obtain two readings, one from each of his copies of chromosome 11. By comparing them with the sequences present in his parents, he could easily tell which passage came from his mother and which came from his father.

Of course, Elizabeth had two beta-globin genes of her own. Which of them was delivered to Charles? The selection was made at random in a process that I will call *chromosome countdown* (*meiosis*). In that process, Elizabeth's basic set of forty-six was reduced to twenty-three in the process of forming numerous eggs. If we name her two copies of No. 11 as A and B, then some eggs would get 11A and others 11B. Of those that got 11A, some might get 12A and others 12B, and so on. With twenty-three chromosomes to be distributed, several million egg possibilities would result.

In fact, an additional mechanism works during chromosome countdown to permit much more variety. It involves the process of crossing-over, or recombination, which we first met when we considered Thomas Hunt Morgan's fruit fly experiments. Although Charles could have received either 11A or 11B from Elizabeth, a more likely result was that he received a combination of both. For example, he might have taken both chromosome ends from 11A, while the middle area came from B or vice versa. This would be possible if two crossovers had taken place in the countdown of the egg that engendered him. This could actually be learned if a series of markers were determined along the length of his maternal chromosome 11, and they were compared to the same ones taken from both of Elizabeth's chromosomes.

Preliminary studies of this type have already been reported, although not for Elizabeth and Charles. A 1985 *Science* article by Dennis Drayna and Ray White[1] illustrated the relationship between the two X chromosomes of a woman and the single maternal X received by her four daughters and five sons. (The daughters also received a second X from their father, but this was not relevant, except that it made the analysis more laborious.) Using a larger sample of X chromosomes, the workers observed from zero to four crossovers with an average of 1.5 in a countdown process. In some cases the chromosomes received by two siblings were very different. One brother received most of the A chromosome with the tips from B, for example, whereas another received the reverse. Other pairs of siblings had received almost the identical X chromosome from their mother. Thus, with respect to the X chromosomes, two brothers could be virtually identical twins or totally unrelated.

In the 1985 study, only twenty markers on chromosome X were used, so crossover positions were approximate. The chromosome was sampled at positions separated by about five percent of its length. When the entire human genome has been sequenced, then it will be possible to take sequences from areas that are as close together as desired and to locate crossover points exactly when it is important to do so. For the routine study of human relatedness in the future, it will probably suffice to examine markers spaced about one hundred thousand base pairs apart. Ideally, each marker would be informative enough to tell any human apart from any other. Unless, of course, they shared an ancestor who had bequeathed that portion of their DNA to them. In practice, workers will settle for something less to keep costs down. I will call each marker of this type an *identity site*.

Some care will be needed in choosing the identity sites. Most likely, they will be picked from the junk areas, as variability among humans is greatest there. The repeating sequence on chromosome 1 developed by Alec Jeffreys that we described in Chapter 16 would be a candidate. On the other hand, costs would be diminished if each site were kept small and could easily be amplified by the Mullis technique. Various types of sequence each have their champion, even today: Alus, CACAs, repeating trimers, and others. Those who make the choice will have to avoid areas that change too rapidly every few generations or so, as that would ruin the point of an identity site. They could not even be secure of what a father had passed on to a son.

I will presume that these minor technical difficulties will be settled soon after the first human genome has been read, and some extra data have been gathered on a group of individuals to check the suitability of various sites. At the 1990 Genome II meeting in fact, I learned of a company, the Bios Corporation of New Haven, Connecticut, that had already started work in this direction. They were seeking informative markers on every chromosome, initially spaced about five hundred thousand base pairs apart, but ultimately closer together. As a data base, they were using the DNA of a reference set of several dozen large multigenerational families collected and made available by CEPH (Centre d'Etude du Polymorphisme Humain), an international gene mapping center located in Paris.

The Bios Corporation wished to use their markers as an aid to mapping, but their director of research, Richard Kouri, agreed in a phone interview[2] that they could also be used to study the relatedness of individuals, as I will describe below.

We will move ahead to a future date when suitable identity sites have been chosen one hundred thousand base pairs apart on every chromosome. Each might be one thousand bases or less so that, taken together, they might represent less than one percent of our DNA text. Further, procedures would have been automated so that they could be determined for an individual at no great cost. (When costs are still a problem, intermediate strategies will be used in which a lesser number of sites are examined and key areas then chosen for a closer study.) The procedure of Drayna and White could then be followed for all our chromosomes. We each could learn exactly which gene regions we had inherited from each of our parents.

Prince Charles, for example, could learn how each of his maternal chromosomes was derived from Queen Elizabeth's pair. She had obtained one of each pair from King George VI and her mother, Queen Mother Elizabeth, but may have shuffled them by crossing-over before passing them on to her son. Using identity site analysis, Charles could deduce where these crossing-over events had taken place. For each of his maternal chromosomes, he would learn which parts came from King George VI and which from the Queen Mother. (Some additional information would be needed. We will get to this in a moment.)

Please note that the contributions of his maternal grandparents, whether for one chromosome or all added together, need not be equal. Their sum total contribution to Charles' heredity would come to fifty percent, and a 25–25 split would be most likely, but other divisions, even very unequal ones, could occur. The same type of analysis could also be performed with the chromosomes he received from the parents of his father, Prince Philip, again provided that some extra information was obtained.

Direct sequencing would not tell us, right now, whether any particular chromosome of Queen Elizabeth came from her father or mother.

Some additional marks, or imprints, may actually keep maternal and paternal chromosomes distinct, but we have not learned how to read them yet. The most direct way to sort things out right now would be to run a comparison of Elizabeth's sequences with those of her parents. Her father has passed on, however. With one living parent, an analysis remains simple, as the unmatched sequences can be assigned to the other parent. Suppose that neither set was available. How would we then obtain genetic data on King George VI?

Fortunately, Elizabeth's kin on her father's side are both numerous and well documented. For example, I counted five first cousins in a genealogy of the mid-1980s. Each of them would, on the average, have a 12.5 percent text overlap with Elizabeth. We would look for runs of adjacent identity sites that were present in both Elizabeth and any of these first cousins. Any long text passages that are present in both Elizabeth and a cousin on her father's side were undoubtedly there in George VI as well. The data from five cousins taken together (if they cooperated) would serve to identify the origin of about half of Elizabeth's paternal DNA sequences. Additional assignments could be made by including more distant paternal relatives and by considering her mother's side. With effort, cooperation, and computers, most of the DNA text of King George VI could probably be reconstructed.

We can see that Charles could end up, after some work, with a complete assignment of his DNA to his four grandparents. He would undoubtedly also find a few spelling changes that were unique to him—the product of new mutations. The same analysis could also be made for his sister, Anne, and his brothers, Andrew and Edward. From this, each could learn how related he or she was to each of the other siblings.

The way that heredity works ensures that a child will get half of his or her DNA from each parent (the X versus Y difference will unbalance this slightly for males). Siblings all draw half of the gene pool available in their parents and so will *average* fifty percent relatedness to one another, but this need not be followed in any individual instance.

We saw above how two brothers could get X-chromosome combinations from their mother that had little overlap with each other. The same could happen for every chromosome, although this is very unlikely. Alternatively, siblings could come close to being identical twins. At the end of the DNA analysis, Charles would get a number indicating his percent relation to Anne, Andrew, and Edward. Values near fifty percent would be most likely, but some greater divergence could occur.

The same analysis could be done for any of us, of course. We could learn whether that brother or sister who seems so different (or close) really is so, genetically. If one grandparent on either side were alive or a preserved blood, hair, or saliva sample were available, the analysis would be simple and extended families would not be needed.

Let us return to the British succession. Queen Elizabeth's children each have a fifty percent relationship to her, but will vary in relation to King George VI. I will mischievously suggest a new rule based on maximal preservation of "royal" sequences in the ruling line. In this scheme, the grandchild of George and child of Elizabeth who had the greatest DNA relation to him would be designated the one to succeed Elizabeth. In this plan, all sequences, gene and junk, would weigh equally; total length of overlapping text would prevail.

Alternatives are possible, of course, in which some special text sections are given extra significance. In honor of the ancient tradition of "royal blood," I will nominate the alpha- and beta-globin gene areas for this honor. Together, the globins form hemoglobin, which gives blood its color, so the designation of these areas as royal would have great justification.

Adoption of any standard based on DNA would have the advantage of adding excitement to the succession. As each new candidate entered the scene, there would be a flurry of sequencing to determine whether the new entry had earned the right to follow the current monarch as the bearer of the royal DNA.

Considerations of this type may also enter into the lives of nonroyal individuals. For example, one of my aunts died several years ago leaving no will. She had neither husband nor children. The laws of the state in which she lived specified that her estate be divided equally among her brothers and sisters. If one of them had died, his or her children would divide that portion. Thus, genetics was given priority over any considerations of emotional closeness and involvement in my aunt's life. If this philosophy were carried to a logical conclusion, then when a closer genetic analysis becomes possible, actual degrees of relatedness might be taken into account when an estate is divided. Thus, a sibling with sixty percent overlap could be given priority over one with forty percent, and a thirty percent niece or nephew might outrank one with less genetic connection to the deceased.

In some cases, an individual may die without a will, and the whereabouts of close relatives may be unknown. A search for missing relatives will then be undertaken. Genetic screening might then provide a better guide to the relatedness of a claimant who turns up (unless the relationship is through marriage) than documentation.

THE QUEST FOR VICTORIA

Let us return to the genetics of the royal family. Because of the historical importance of this family, scholars have wanted to track the inheritance of hereditary conditions such as hemophilia and porphyria (a biochemical malfunction that can lead to insanity) through the generations. When a much more thorough genetic analysis becomes possible, the appetites of the historians will undoubtedly increase. One early goal of the not-yet-founded academic discipline of genetic history may be the reconstruction of the DNA text of Queen Victoria.

Many reasons make her an almost ideal target for such an inquiry. Her historical importance is profound. As the ruler of the British Empire for over sixty years at the time of its greatest influence, she lent her name to an entire era. Further, she gave rise to a prodigious number of children and grandchildren. In a recent survey of the royal succession, Alan Hamilton estimated that there were at least three hundred direct descendents alive today. He enumerated the first one hundred in line without exhausting the progeny of her first two sons (Victoria had nine children in all). Elizabeth, of course, is one, tracing her descent through Edward VII, George V, and George VI. Elizabeth's husband, Prince Philip, is another.[3]

The contribution of Victoria to the genetic text of Elizabeth and Philip should approximate six and a quarter percent. If heredity were transmitted in blood, this would represent a massive dilution. Text is involved, however, and a rough analysis can show that long stretches of Victorian DNA should lie uninterrupted in Elizabeth and Philip.

How will we locate those areas that belonged to Victoria? Before we answer this, we had better take another look at the way in which the DNA message gets scrambled over the generations. Crossing-over, or recombination, is the principal mechanism that shuffles the genetic deck within a chromosome. There is also a background of local mutations, text losses and duplications, and other events, such as gene conversion, that introduce changes that are not due to either parent. Fortunately, these take place at a rate of no more than one or two percent per million years. When sequences only a few generations apart are compared, they should produce only slight interference in our ability to follow how DNA messages are handed down. (Some special areas may mutate much more rapidly, but these can be excluded as identity sites.)

Little is known about the frequency and location of recombination in humans. It takes place somewhat more often in women than in men and may happen preferentially at favored "hot spots," whereas other areas, such as the chromosome hinge, are shunned. One estimate guessed that five thousand hot spots might exist with a crossing-over rate perhaps five or ten times the average. For now, I will ignore the complexities. On the basis of the data in hand, I have assumed that an average chromosome suffers about two recombinations in a generation. (This may be a bit high; if it were less, the analysis would be simpler.) If the crossover positions clustered, that might affect the detailed mathematics, but not the overall picture that I will present.

Let us track the fate of a Victorian chromosome, No. 11, for example, through the generations. The copy that Edward VII received from Victoria would have been divided by two crossovers into three separate areas en route to George V. Two additional crossover cuts would have been added by George V to the chromosome that he handed down to George VI. Yet two more splice points would be added as it went to Elizabeth. If we analyze Elizabeth's paternal chromosome 11 in terms of its authors in Victoria's generation, we would locate seven distinct text areas due to the six cuts. Each would be intact (apart from any changes by mutation), and each could be assigned to one of the eight great-great-grandparents that Elizabeth had on her father's side in Victoria's generation.

If eight authors have contributed only seven pieces, then one of the eight would not be represented in the copy of chromosome 11 that Elizabeth received. (If one author had "written" more than one section, then another would by necessity be excluded.) However, Elizabeth received twenty-three chromosomes from her father, so all eight ancestors are likely to be reflected somewhere. If each chromosome received six cuts, then 161 pieces in all were produced, about 20 of which would be derived from Victoria. The average length of each piece in chromosome 11 would be about 16 million letters, or one-seventh of its length. (Of course, there would be a range of sizes, with some that were larger or smaller.)

The job for genetic historians will be to identify the splice points and the authors of each text passage as they work backward generation by generation. I described earlier how the DNA of Elizabeth's father might be reconstructed. By including a greater range of more distant relatives, the analysis might be extended several generations further back. One assignment comes easily. Elizabeth's energy-cell (mitochondrial) DNA came to her directly in an all-maternal line. That trail leads not to Victoria, but a woman, Anne Caroline Salisbury, on the Queen Mother's side, who died in 1881. In a similar way, most of Prince Philip's Y chromosome can readily be followed back in an all-male line that leads back through the Greek and Danish royal families.

(Markers to do this have not yet been developed, but this should happen in the near future.) The remainder of the assignments will require more work.

Elizabeth's royal lineage leads back beyond Victoria, of course. If we continue for an additional eight generations into the past, we come to King James I of England (James VI of Scotland), who ruled in the early 1600s. Each of James's chromosomes would have suffered an additional sixteen cuts, on the average, in the passage through Victoria to Elizabeth. If Elizabeth's text were subdivided into portions that reflected the authors of James's generation, we would end up with twenty-three paternal chromosomes with each sliced, on the average, into twenty-three portions, or about 530 pieces in all. The individual segments would still average about 5 million letters each in length.

If we count the possible number of ancestors on Elizabeth's father's side in that generation, however, we get a surprise. The maximum number, 2048, is fourfold greater than the number of authors of her DNA text. If we presumed that this number of separate individuals was involved (which is untrue), there would be a three to one chance that King James, or any other ancestor of that time, would simply have dropped out of her ancestry: They would be unrelated.

James's chances have been improved above that minimal number by the same principle that operates when we purchase more than one lottery ticket. He appears multiple times in the queen's ancestry due to intermarriage. I do not have the data on the James I–Elizabeth II connection, but I have read that his mother, Mary Queen of Scots, appears twenty-two separate times in the ancestry of Prince Charles, Elizabeth's son. So the chances that James and Elizabeth share at least some DNA text look rather good.

As we move further and further back in time, however, the odds that a DNA bond connects an individual alive today with a specified ancestor grow more remote. Queen Elizabeth is a descendent of the first king of England, Egbert of Wessex, who ruled from 828 to 839 A.D., and William the Conquerer of the eleventh century, but she may have no DNA sequences at all that derive from them.

It would be more satisfying if these questions could be answered directly rather than in terms of probabilities. Is there a chance that analysis of the type we described for George VI and Victoria can be extended back to James I or even William the Conquerer? Mr. Gerald Paget, in his book *The Lineage and Ancestry of HRH Prince Charles, Prince of Wales*, is said to have counted and traced 262,142 ancestors of the prince.[4] It may be possible to locate an enormous number of distant relatives who are alive today and can be fit into the overall scheme. If they cooperated and computer algorithms were devised to analyze the data, then perhaps specific portions of Elizabeth and Charles's DNA could be attributed to James and Mary, Queen of Scots. Yet my feelings cringe at the magnitude of such a task, particularly when a tempting shortcut may be available. To learn more, we shall take a detour to Egypt.

MUMMY GENES

We can presume that the priest Nekht-ankh had a reasonably successful life, at least by the standards of the town Rifeh in Egypt, where he lived some four thousand years ago. He certainly had earned an elaborate embalment and burial after his death in the tradition reserved for notables of that era. His organs were removed from his body, and they and the body were prepared for mummification. According to the account of ancient Greek historians, one step was the immersion of the body and organs in *natron*, a mixture of salts that may be found on the surface of the desert in western Egypt.

Following mummification, Nekht-ankh's body was placed in an inscribed and painted coffin, and his organs were distributed among several jars. The coffin and jars, together with a chest, some statuettes, model boats, pottery vases, and bowls, were sealed into his tomb. At some point either earlier or later, the similarly treated remains of his brother, Khnum-nakht, were also placed in the tomb.

I cannot say what events Nekht-ankh may have anticipated following his burial, but the actual ones would have astonished him. For about four thousand years, nothing whatsoever happened except that his body tissues slowly disintegrated. His bones remained intact, though, and his skin, fingernails, and toenails were also well preserved. In all, the priest fared much better than corpses do that are put away without mummification. On this point we can accept the testimony of Shakespeare, as expressed in *Hamlet*, act 5, scene 1:

> *Hamlet:* How long will a man lie i' th' earth ere he rot?
> *Clown:* Faith, if 'a be not rotten before 'a die . . . , 'a will last you some eight or nine year. A tanner will last you nine year.
> *Hamlet:* Why he more than another?

> *Clown:* Why, sir, his hide is so tann'd with his trade that 'a will keep out water a great while; and your water is a sore decayer of your whoreson dead body.

In the case of Nekht-ankh, the exposure to salt utterly dried out at least the surface parts of his body and organs, preserving them for posterity. It was not until the early years of this century that posterity arrived, however, in the form of the expedition of Sir Flinders Petrie. The tomb was discovered and opened, and its contents were removed far away to a museum in Manchester, England. The remains were examined, cataloged, and stored there for eight additional decades, until they caught the attention of a young Swedish molecular biologist, Svante Pääbo.

I gathered these details while lunching with Svante in the sunny roof garden of a Berkeley restaurant in the spring of 1989. Tall, fair, lanky, bespectacled, reserved in speech, but relaxed in manner, he sounded and looked typically Swedish. He had first developed a passion for things Egyptian at age fifteen, when his mother and he had toured the King Tut exhibit in Cairo. After studying Egyptology for two years at a Swedish university, however, he was disappointed with the profession of archeologist: "It wasn't what I had dreamt it would be when I was a kid."[5] He altered his direction and eventually earned a Ph.D. degree in molecular biology at Uppsala University.

While he was still a graduate student, the thought occurred to him that he could combine his two interests. Many thousands of mummies in various states of preservation had been discovered in Egypt. What if he could rescue and sequence portions of their DNA using the newly developed techniques? If he could do that, then he might also be able to determine the degree that incest had been practiced among the royal populations of that time. Perhaps, as one geneticist wrote in *Nature*, "it would be possible to test the claim of the Coptic Christians of modern Egypt to be the lineal descendent of the ancients."[6]

The effort to rescue ancient DNA was difficult, and many mummies had to be sampled, but eventually he succeeded with the twenty-four-hundred-year-old mummy of a very young boy. Some sequences were cloned in bacteria, and he was able to identify a portion of an Alu family member. Little could be done with the information, but the possibility and potential of such studies had been demonstrated. To broaden his experience, Pääbo joined the research group of Allan Wilson at the University of California at Berkeley, where he used both cloning and DNA amplification techniques to extract DNA sequences from the liver of Nekht-ankh. Strings of A's, G's, T's, and C's that once functioned in the twelfth dynasty of Egypt were now reproducing happily in twentieth-century bacteria!

At this time, the work is still in a preliminary state, with technical improvements needed. Time does not treat the DNA of the deceased kindly. Much damage is inflicted in the initial days, as the still-functioning enzymes of the cell cut it into smaller pieces. When these are silenced by degradative processes of their own, slower but longer-lasting agents of destruction take over: water, air, and radiation. Further chain breakdown occurs, and the bases themselves suffer damage. The enormous redundancy of the message (two copies in almost every cell of our bodies), however, helps ensure its survival. Techniques such as DNA amplification can operate on relatively short stretches, selecting intact brief passages of DNA from the many mangled ones. Thus far, however, mostly energy-cell DNA sequences have been rescued from ancient samples. They have an additional advantage in survival in that thousands of copies of the message, rather than two, are present in every cell.

Despite these limitations, studies on ancient DNA are flourishing. Longevity can be ensured for our hereditary text by burial under dry or oxygen-free conditions, as well as by artificial mummification. DNA sequences have been obtained by Pääbo, Wilson, and John Gifford, for example, from a seven-thousand-year-old shrunken brain preserved in an oxygen-free peat bog in southwestern Florida. The energy-cell DNA text that was obtained differed from that seen in present-day native Americans. The nature of certain genes involved in the immune response has also been determined for one such individual.[7] Preliminary DNA work has also been performed on sources that vary from pre-Columbian Indian bodies buried in the sands of a dry desert in northern Chile to hospital samples of human tissues preserved in formaldehyde and set in paraffin wax that are a few decades old.

Recently, DNA amplification was used to retrieve text sequences from the paraffin-preserved tissues of a British sailor who died in 1959 of undiagnosed causes in a hospital near Manchester, England. The puzzling combination of symptoms that he displayed had motivated his doctors to publish the case in a prominent medical journal in 1960. What was more important is that they never forgot it. When new suspicions arose and new techniques became available, they were able to analyze his tissues and confirm their guess: The sailor became the earliest known victim of AIDS.[8]

Other preserved specimens beg for analysis. When I first visited the Soviet Union in 1960 and again in 1978, I was entranced by the sight of the embalmed and waxlike body of Lenin on display on his mausoleum in Red Square, Moscow. I later saw Mao Tse-tung in a similar condition in Beijing, China. Unless these figures really are made of wax and provided that the bodies survive their former owners' current fall from fashion, they should prove to be fascinating objects for analysis by future genetic historians.

Some of the more spectacular achievements in raiding the past have not involved humans at all, but animals and plants, particularly those that have become extinct. Naturalists have warned us of the imminent loss of endangered species, and some seem headed down that path despite our best efforts to save the survivors. Only thirty California condors, thirty-five hundred black rhinoceroses, and less than five thousand snow leopards remain, for example. These are not isolated instances. Harvard biologist Edward O. Wilson has estimated that we may be losing as many as 17,500 species each year. While they yet survive, we still have a chance to preserve samples of their DNA for future sequencing and study when technology permits it.

Efforts in that direction have already begun. The University of Queensland in Australia has set up a library, the Centre for Genetic Resources and Heritage, to preserve dried cells and DNA from endangered Australian species. The director, John Mattick, has complained that the Australian government "can't see the benefit of collecting genetic resources and sequencing key genomes." He himself feels that subsequent generations will ask why we did not do it when we had the chance.[9]

Such efforts alone cannot bring them back, but we can keep a record of exactly what it is that we have lost and preserve our options for the far-off future when we may be able to do something about it. In some cases, we may be able to do this after the species has vanished.

THE RETURN OF THE QUAGGA. Many curious types of beast have already marched to their extinction unaware of a coming future when their plan could be captured and recorded forever. Some noted examples that have departed within the last millennium include the moa, a ten-foot-tall bird of New Zealand; the dodo, a flightless bird that inhabited islands in the Indian Ocean; the passenger pigeon, which became extinct in 1914; and the quagga.

The last quagga was shot by a farmer somewhere in South Africa in the late nineteenth century. This beast resembled a zebra, stripes and all, in its frontparts, but a horse in its rear half. Evolutionists asked the obvious question: Was this a zebra, with some horse pretensions or the reverse? This might have remained a subject for idle argument, like the question of how Jack Dempsey would have fared if matched against Muhammad Ali, but Allan Wilson, Russell Higuchi, and their co-workers from Berkeley managed to rescue an answer.

Various quagga hides lay gathering dust and decay in museum cases in Europe. One in Mainz, Germany, had a scrap of flesh attached that was suitable for DNA analysis. Some sequences of energy-cell DNA could be rescued, and these declared the creature to be most closely related to the zebra, not the horse.[10] In reporting this achievement for *Nature*, Alec Jeffreys commented: "Friedrich Miescher, who discovered nucleic acids in 1868, could have saved Higuchi, et al., a lot of trouble if he had the foresight to make and store fresh quagga 'nuclein'."[11]

How much further back in time can the hand of science reach to retrieve vanished creatures? The end is not yet in sight. Russell Higuchi, now at Cetus, has rescued some DNA sequences from a forty-thousand-year-old woolly mammoth frozen in ice. He reported, not surprisingly, that the passages resemble elephant text.[12] Even this achievement pales before a recent report of the recovery of DNA text from the fossil remnants of a 17-million-year-old magnolia leaf that had been buried in oxygen-free, lake-bottom sediment in northern Idaho. Not much was learned—the same species thrives today. However, the achievement does demonstrate a new path that could, if we were lucky, give flashes of insight into the history of evolution.

The supply of mummified, desiccated, frozen, or otherwise preserved ancient samples cannot cover all of the cases we may be interested in, unfortunately. Whole new vistas would open up if the most common type of remains, bone, was shown to be a suitable source for the isolation of DNA.

BONES OF CONTENTION

It was only a one-page paper in *Nature* late in 1989.[13] Yet its content was striking enough to induce me to add a day in Oxford to my interview trip to England the next spring. Three individuals had reported the isolation and sequencing of DNA from long-buried human bones. They determined some of the DNA text of a thighbone found in a nearby seventeenth-century English Civil War cemetery and a 750-year-old upper arm bone from a medieval cemetery. The presence of DNA was demonstrated as well in a fifty-four-hundred-year-old thighbone obtained from a cave in the Judean desert, but sequencing was not reported.

Oxford has certainly had its share of history, and the collaboration reflected the mixture of medieval and modern themes that is present in that celebrated university town. Archeologist Robert Hedges led a unit that specialized in the technique of radiocarbon dating; his group had been one of three chosen internationally to determine the age of the Shroud of Turin. His unit was housed in comfortable brown buildings near the old colleges in the center of Oxford. Bryan Sykes ran a biochemical laboratory with a special interest in collagen, a protein that functions to give strength to tissues such as bone and cartilage. His group was located in the black-and-white building of the Institute of Molecular Medicine, part of the modern John Radcliffe Hospital complex at the edge of town. Hedges and Sykes combined forces to apply the latter's expertise to the problems that interested the former. They wanted to study the collagen in the bones of ancient peoples. A grant was obtained, and American-born, Cambridge-trained Erica Hagelberg was recruited to carry out the studies.

The strategic target of the work had changed, however, when Hedges visited the laboratory of Allan Wilson and learned about the newly discovered technique of DNA amplification from Russell Higuchi. Bone DNA would be the target, and amplification, the tool. According to them, the selection of that combination was the achievement. It was "not a complicated piece of research at all," said Hedges. "We tried it and it worked," commented Sykes, "anyone could have done it." Erica, who ran the experiments, emphasized that a good deal of trial-and-error experience was needed to make it work.[14]

The achievement was preliminary, and difficulties abounded. Impurities inhibited the DNA amplification reaction, and the danger of contamination by the DNA of the investigator or DNA from previous experiments was ever present. At first only one bone responded, then as their techniques improved, half of the samples gave results. Only short lengths of energy-cell DNA could be detected, however. (In unpublished work, Hagelberg has now amplified some chromosomal DNA sequences as well.)

Why should bones give results at all? Their cell and DNA content is much less than that in soft tissues, and DNA sequencing in very old specimens of such remains has succeeded only in cases where some unusual circumstance was operative, for example, mummification or burial in a bog. The Abingdon cemetery was rather typical for England, however, and the bones had been unearthed from the ground by routine excavation.

Only speculations could be provided by the authors. Perhaps the DNA came from *osteocytes*, specialized cells that are embedded deep within bone. The bone structure might have afforded them some additional protection from degradative enzymes within and the environment outside. In any event, it worked, at least for some bones. The best indicator of success in obtaining DNA appeared to be the general condition of the bone. A well-preserved specimen would give good results.

What conditions afforded the better bones? That was less clear. According to Hedges, one half of a site in England might yield well-preserved bones, and the other half, only decomposed ones. Temperature, acid, air, and sunlight were obvious adverse factors. The Mideast usually afforded poor conditions; they had worked with a specimen that was an exception, however; it was well preserved in a cave. Their experience to date, although limited, had left them optimistic. Hedges was the most reserved. Possibly the technique would work on only a few bones older than two thousand years. If so, he would be very disappointed, "but I think it would be very unlucky for that to be true," he felt. Sykes was less guarded: "Eventually you'll be able to get DNA out of all remains, all bones, all of them." Well, if not one hundred percent, ninety to ninety-five percent would do. The technique would become routine to workers in the new field of molecular archeology. Hagelberg also felt it possible that all bones could be made to work once enough experience had been obtained.

Of course, the results could be due to artifact or be limited to special cases despite the optimism of those involved. This outcome became less likely, though, when the isolation of DNA from bones was also reported by Satoshi Hirai and his colleagues in Japan.[15] They had worked with local specimens from sixty to six thousand years old and obtained from the latter some DNA text phrases not observed in modern Japanese. At the Human Genome II meeting in 1990, the sequencing of a 450-year-old bone of a native Hawaiian was also reported by a research group from the University of Hawaii.

What exciting applications could be made of the technique? Each of the Oxford trio had dreams or fantasies. Robert Hedges would like to determine DNA sequences from the bones of Neanderthals, brawny variants on the human theme that died out tens of thousands of years ago. The bones of fifty to one hundred individuals of that species had been found. Perhaps they could yield up some DNA samples to provide us with more accurate information on their makeup, and whether they interbred with humans before they became extinct.

Bryan Sykes wants to apply this technique to study the history of disease, as does Erica Hagelberg, who is now working independently. The black plague had been characterized by a description of its symptoms. How much better it would be if an accurate identification could be made by rescuing some DNA sequences of the supposed culprit organism from the bones of its victims! A four-thousand-year-old Egyptian skeleton in the British Museum also attracted him. From its appearance, its owner had suffered from a bone disease due to a collagen defect. Sykes had studied modern victims of this disease and identified in some cases the particular flaw in their DNA that was responsible. (It varied from case to case.) It would be a triumph to be able to perform the same diagnosis on an individual who had died millennia ago.

Determinations of identity will afford some of the most important applications of the DNA-in-bone technique. This procedure could open up a superb new highway into knowledge of particular mysteries of the past.

As we are all aware, the bones of unknown victims long outlast their flesh and serve for decades or centuries as memorials to the tragedies of the past. The Unknown Soldier rests with others of his kind in Arlington, Virginia. Many skeletons lie unidentified in mass cemeteries in Argentina, reminders of the murders of the military regime that ruled in the late 1970s. Other unidentified bones have rested for centuries under the chapel floor of the Tower of London, the victims of much earlier executions. Every cemetery may contain its quota of unidentified remains. DNA text reading could connect corpses to relatives living today.

Apart from the unknown, the other obvious targets for bone analysis would be the very well known. Queen Victoria's remains are buried at Windsor Castle, and those of her ancestors, Mary Queen of Scots and James I, lie in Westminster Abbey. How much simpler it would be to read their texts directly rather than reassemble them by extensive sampling of living relatives. The same applies to the remains of many buried monarchs, including those for whom the indirect method would prove impossible. Of course, the practice need not end with royalty. The DNA of those noted and famous for other reasons would become objects of scholarship and curiosity. When I discussed the possibilities with him, Fred Sanger quipped about getting "the sequences of the saints."

The first investigation on the DNA of a historical figure may soon be underway.[16] A committee appointed by the National Museum of Health and Medicine and headed by Dr. Victor McKusick of Johns Hopkins University is considering analysis of DNA from hair and bone fragment samples of Abraham Lincoln. The samples were taken at the autopsy of President Lincoln that followed his assassination and have been preserved in the museum's collection.

One particular question that could be settled if the analysis succeeds is whether Lincoln suffered from the inherited disease known as Marfan's syndrome. This condition causes weakness in the bones, joints, eyes, and circulatory system. Sufferers of this disease often are tall and gangly with long limbs and fingers, features that Lincoln possessed as well.

"The prospect of examining Lincoln's DNA is exciting," Dr. Mark E. Neely, director of the Lincoln Museum (Fort Wayne, Indiana) told the *New York Times*. "For more than 20 years scholars have debated the Marfan's issue and I would be relieved to have the issue put at rest. Lincoln spoke very little about his health but he was a remarkable specimen of frontier strength. Marfan's may have been quietly killing him, but it didn't affect him."[17]

Finally, many of those who led ordinary lives will still be of interest to their descendents. Those who wished to build DNA genealogies, but came from slender families, might find that the quickest route to the interpretation of their own DNA text was to be found by obtaining one last favor from their departed ancestors: the donation of a bone sample.

COUSIN, COUSIN

Two families living next door to each other on Long Island discovered that they were distantly related; their children who had been playmates for years, ran through the streets shouting, "We're cousins! We're cousins!"

—Dan Rottenberg in *Finding Our Fathers: A Guidebook to Jewish Genealogy*

Many of us, myself included, have family lines that terminate abruptly a generation or two into the past, due to migration (forced or otherwise), adoption, or other circumstances. What help will the new possibilities of DNA be to us without the support of historical records or well-preserved bones? The answer is a lot, provided that the people of today are interested enough to assemble a data bank of those who are here now. We could start by finding out who our living relatives are.

The most secure proof of relationship will become DNA sequence overlap. Because of the way heredity works, half of my DNA script comes from my mother and half from my father. My son has half of mine. As I mentioned earlier, siblings (my brother and I, for example) will have a lot of overlap, with fifty percent the most likely value, but it could be greater or smaller. In the same way, my overlap with a grandparent, uncle, aunt, niece, or nephew (and half-brother or -sister, had I any) will be in the vicinity of twenty-five percent. First cousins, great-grandparents, and a number of other connections will share about 12.5 percent. When we can actually measure these figures, some surprises will turn up to add spice to family Thanksgiving get-togethers. An uncle might turn out to be a closer connection than a brother, for example. Such events will be occasional, though, not frequent.

Some more practical use might also be made of such information. Governments vary in their marriage regulations with some forbidding unions between first cousins, for example, and others permitting them. Such rules exist to reduce the possibility that recessive disease-associated genes will come together in the same individual. When degrees of relatedness between individuals can be determined more exactly, some specified percent of relatedness rather than first or second cousinhood might be used as a criterion in approving such marriages. Later on, of course, when a fairly complete list of genetic diseases has been assembled, a detailed gene-by-gene analysis could be used in judging the advisability of marriages between close relatives.

Many adopted children will be very interested in learning what they can about their backgrounds through genetic analysis in cases where the identity of their parents has been kept secret. The actual DNA profiles of their parents, if stored in data banks, would undoubtedly be protected by legal safeguards just as other pertinent information is today. Many individuals in the population may want to locate their relatives and expand their geneology, however. Organizations may be founded whose members voluntarily exchange genetic data with one another for that purpose. Through such groups, adoptees might at least locate some more distant relatives and get some insight into their family and ethnic histories.

The gigantic size of our DNA text will allow much more distant connections to be turned up; the depth of the possible analysis is astounding. Let us suppose, for example, that two individuals matched in some three hundred thousand base pairs on a single chromosome. This would be found by noting that three adjacent identity sites in a row coincided in a one percent sampling of their genomes. The match could be confirmed by testing further markers in the area or even sequencing the entire stretch for both. If we assume that about two crossovers take place in each chromosome per generation, then two hundred generations might be needed to partition a chromosome into pieces of three hundred thousand base pairs. If we allow twenty-five years per generation, then the two individuals shared a single ancestor some five thousand years ago!

Not all descendents of this ancestor would be linked in this way. The gene in question would have to avoid being dropped for two hundred generations in each individual. The odds favoring this are very low, but it would happen for some genes in some individuals in the same way that some bridge hand will be dealt the next time you play, even though the odds against a particular hand are immense.

Of course, the five-thousand-year number I gave above is a soft one in scientific terms. If the average rate of crossing-over (or other mechanisms of gene switching) were lower, the distance that we can reach into the past would be more or vice versa. Further, spelling errors of various kinds will creep in over the millennia, causing two stretches that were once identical to differ from one another. Allan Wilson and others have actually used such differences in energy-cell DNA to estimate the time when races diverged in human evolution. Energy-cell DNA undergoes mutations much more rapidly than the rest of DNA, however. If we want to decide whether two individuals really did have a common ancestor some time back, then changes due to mutation would represent a "noise" background that would have to be subtracted so that the "signal" can come through. The further back we go, the greater the noise would be as mutations pile up, and the weaker the signal as the size of expected DNA text overlap shrinks. We have a chance, I think, of using DNA matching to explore links between individuals a few thousand years back, which provides an exciting prospect for anthropologists and historians.

Most of us will be more interested in exploring more recent links, ones that have deeper personal impact. I have created a few examples to illustrate some possible applications of the new science.

THE OTHER ROOTS. At the climax of his moving epic, *Roots*,[18] Alex Haley described a meeting with his kinsmen seven generations removed in the village of Juffure in the Gambia, West Africa. Haley's great-great-great-great-grandfather, Kunta Kinte, had been kidnapped from this village in 1767 and sold into slavery in America. Guided and inspired by an oral family tradition, Haley searched through old archives, census lists, maritime records, and newspaper accounts to locate the origin of his ancestor. His efforts were rewarded when he located an African *griot* (oral historian) who helped him culminate his search.

Ultimately, with the aid of the records of the insurance firm, Lloyd's of London, he was able to identify the ship *Lord Ligonier* that had brought Kunta Kinte to Annapolis, Maryland, in 1767. Haley had identified one of the sixty-four individuals of that generation who had contributed to his ancestry. The others, who were undoubtedly scattered over many other locations in Africa, America, and Europe, remained unidentified.

Haley's achievement was nonetheless notable for an American of African descent. The conditions of slavery in America were hardly encouraging to those who wished to keep family records, and the slave traders who took prisoners in West Africa were not interested in preserving the histories of their victims. Charles L. Blockson in his book *Black Genealogy* (written with Ron Fry) felt that "finding out where you came from might just make it easier to find out where you're going," yet had to conclude that "the vast majority of us, unfortunately, will *never* be able to add that final important piece—to trace our ancestors absolutely back to Africa."[19] The obstacles were formidable: You needed to wade through church, plantation, military, and census records to trace your family line back to a specific port of entry. After that, you had to identify the slave ship and somehow locate the African area from which it sailed and the tribe from which it had kidnapped slaves. Lacking that data, you could only fall back on soft information and guesses. For example, Blockson was told "you are from Dahomey" by a black serviceman who claimed that his facial bone structure was typical of that area.

Fortunately, African-Americans, like the rest of us, have preserved a more secure record of their ancestry to the eighteenth century and much farther back. It is inscribed in the DNA of the cells of their bodies. To make it fully interpretable, the human genome must be read, and suitable and extensive African and other ethnic data banks assembled. With that information in place, it may be possible to make not just one, but a host of connections back to Africa for every individual who wants to learn about the past.

The needed effort has already begun with population geneticists attempting to identify specific spelling changes with definite ethnic groups. For example, five different size variants were noted in a study of a repeat area on chromosome 6. The largest variant was present in thirty-five Central African Republic pygmies and two Zaire pygmies, but not in Caucasians, Chinese, and Melanesians. The next largest variant was observed in Melanesians, where it was the most predominant one, and Chinese, but not in either African group or Caucasians.[20]

Many other markers have been studied by various teams with less dramatic results. Ideally, with the entire human genome as a source of text, workers will be able to locate a list of characteristic identifiers scattered over the various chromosomes for every tribe or population group that has had a history of genetic isolation. Such regions would make superb identity sites. To do this, the groups should be sampled as soon as possible before further intermarriage and migration scramble the data.

Let us assume that this job has been done. The most useful collection of identity sites scattered across every chromosome has been recorded and stored in computers to form a global computer record of humanity. Alex Haley now has his own DNA read for those same sites. The various chromosome countdowns that passed since the generation of his ancestor Kunta Kinte would have divided Haley's chromosomes into five hundred pieces of DNA text, of which Kinte would be the author of about eight. Kinte's tribe still lives in the area, and many hundreds of descendents of his family are likely to be among them. Each of them would probably have some portion of Kinte's text within him or her, and for at least a handful, they should be the same pieces that Haley has.

This cluster of distant relatives would mark Juffure as the location of one of Mr. Haley's ancestral roots. He had this information anyway, of course. However, other clusters would also turn up in the analysis that flagged other African roots, as well as some in Europe that marked the home villages of Caucasian ancestors. The size of the areas of text overlap would indicate the length of the separation of each of Haley's ancestors from his homeland. An ancestor who was taken from Africa a generation later than Kunta Kinte would have left behind relatives whose DNA overlap with Haley would be about twice the size of those from Juffure. Needless to say, this analysis could be applied to any African-American, not just Alex Haley.

Not all questions concerning African-American heritage go back to Africa, of course. Many lost American relatives would turn up in such a search. Some fascinating questions originate in the United States. One of the most celebrated has been the question of the "unknown grandchildren" of Thomas Jefferson.

THE COURSE OF HUMAN EVENTS. Thomas Jefferson, the principal author of the Declaration of Independence and third president of the United States, may have had two families. One was legal and recognized: With his wife, Martha Wayles Jefferson, he fathered two daughters who survived to adulthood. The younger one lived only to age twenty-five, but the elder one had a large number of children of her own, preserving Jefferson's line for the future. A possible second family of Jefferson's has come to prominence largely through the efforts of the now-deceased historian Fawn Brodie. Her 1974 book, *Thomas Jefferson: An Intimate History*,[21] claimed that Jefferson had a thirty-eight-year illicit liaison with a slave, Sally Hemings, following the death of his wife. She bore him seven children, five of whom survived to adulthood. From these unacknowledged children sprang additional lines of descent from Jefferson.

Rumors of this connection were publicized by disappointed office-seeker James T. Callender and others during Jefferson's presidency. Anonymous ballads were dedicated to "Long Tom" and "Dusky Sally." Often set to the tune of "Yankee Doodle," they were published in newspapers controlled by Jefferson's political opponents. Some of Sally's children were said to have a remarkable resemblance to the president. In resurrecting these claims, Brodie relied on memoirs written by the children and the circumstances surrounding their upbringing. They were privately educated, given money, and allowed to run away or were freed in Jefferson's will. The time of their conception always coincided with Jefferson's presence in his estate, Monticello. Sally Hemings was said to have a privileged position there. She was freed by Jefferson's legal daughter after his death. Jefferson did not free her himself, according to Brodie, because "such a gesture for his celebrated concubine would have meant instant publicity, such as had been so humiliating during his presidency."[22]

Brodie's book was a Book-of-the-Month-Club selection. Her account was widely accepted by the media, including the *New York Times*. Further, it stimulated a number of claimed descendents of Sally Hemings and Thomas Jefferson to approach Brodie with scrapbooks, pictures, and genealogies. The results of her investigation of these claims were summarized in an article in *American Heritage Magazine* in October 1976.[23] Brodie felt that at least two lines of descent were authentic.

One of them represented the descendents of Sally's son Eston. Sally herself was only one-fourth African, and Eston and she were listed as "white" in an 1830 census. Later in life, he decided to hide his background and enter the Caucasian community. His descendents had kept a tradition of descent from Jefferson, but the nature of the connection remained obscure. They became prominent as company presidents, inventors, physicians, and hotel owners. A number of great-great-great-grandchildren of Eston were listed by Brodie. Madison, another son of Sally, chose to remain within the black community, as did some of his descendents. One of them became the first black legislator elected to the California State Assembly. Madison's line was represented in 1976 by several great-grandchildren.

Despite Brodie's certainty, the question of Jefferson's intimate relationship with Sally Hemings remains unsettled, with many historians unconvinced. In 1976, no obvious way to proceed with this historical problem may have been apparent. At exactly that time, however, Fred Sanger and Walter Gilbert were perfecting their two DNA readout methods in their laboratories in the two Cambridges. Can DNA evidence help settle the Jefferson–Hemings question? Perhaps, but this problem will be trickier than most.

The descendents of Madison Hemings cited above would, if the connection were true, have about six percent of their DNA from Jefferson; those of Eston Jefferson would have three percent. Their sequences could be compared to one another and to accepted descendents of Jefferson. A failure to match would be inconclusive, however. His accepted descendents come from one daughter; collectively, they could at best represent half of Jefferson's DNA text. Worse, even a match would be inconclusive. According to Brodie, Sally Hemings was herself the product of an illicit union. The complications exceed those found in most soap operas. Sally's mother, Betty, was the daughter of a sea captain, Hemings, and an African woman kidnapped into slavery. Sally Hemings was the illegitimate child of Betty and planter John Wayles. Wayles and his legal wife were the parents of Jefferson's wife, Martha. Sally Hemings was included in the inheritance that Martha Wayles Jefferson brought with her to Monticello. John Wayles, then, could be the common ancestor linking Sally Heming's descendents with those of Jefferson's legitimate line. DNA sequences shared by the latter and the former could be attributed to Wayles rather than Jefferson.

Another complication was introduced in the 1981 book *The Jefferson Scandals* by Virginius Dabney.[24] Dabney feels that Heming's children were probably fathered by Jefferson's nephew, Peter Carr. The same claim is made in another biography of Jefferson by Natalie S. Bober.[25] If this were true, again some overlap would be expected between Heming's and Jefferson's descendents. DNA data taken from those alive today could not resolve these possibilities.

One route out of this impasse remains open: direct analysis of the bones of Thomas Jefferson. If they retain his DNA, then his own text could be determined. If DNA sequences shared by Jefferson's legitimate descendents and those of Sally Hemings also turned up in Jefferson himself, then the possibility that they came from John Wayles would be eliminated.

The possibility would still remain open that Peter Carr had furnished those sequences to Hemings's descendents, as Carr would have about a twenty-five percent overlap with Jefferson. If this were the case, however, then Hemings's descendents of a given generation would have only one-fourth as much overlap with Jefferson as his own descendents of a given generation. This possibility could be explored by extensive comparison of sequences. If the remains of Peter Carr were available for DNA analysis as well, however, the determination would be greatly simplified.

Of course, some traditionalists may be perturbed at the prospect of teams of genetic historians lusting for the remains of Queen Victoria and other sovereigns, Thomas Jefferson, and virtually every other past celebrity who was not cremated or lost at sea. That hunt could well extend to the brain of surgeon Paul Broca, other human remains preserved in museums and hospitals, the clothing and private effects of the famous of past ages (they might retain some hairs or skin cells), and the bones of the saints. I cannot forsee whether activists will chain themselves to cemetery fences in defense of the privacy of the past, but lawsuits and legislation can be expected in many instances. In the case of Jefferson, both sides seem confident of their ground, so both should in theory welcome a resolution of the issue. (In practice, of course, things could turn out differently.)

Much more would be learned by disturbing for one last time the bones of this extraordinary individual than his relationship to Sally Hemings. He was a man of genius, accomplished in science, literature, and languages, as well as government. By studying his genetic text, we might ultimately learn what phrases in his own heredity helped produce this remarkable outcome. The *Encyclopedia Brittanica* describes Jefferson as a man who was "meticulous in preserving his letters and papers." He quite likely would have approved of this final addition to his works.

OF SHEM AND JAPHETH. While writing about the background of others, I could not help but get interested in my own. Prince Charles can trace his antecedents to the ninth century and Alex Haley made it back into the eighteenth, but my own documented connections do not reach outside the twentieth. Both of my parents spent their childhood in the land then ruled by Nicholas and Alexandra, now part of the Soviet Union, near the towns of Kiev, in my father's case, and Pinsk, in my mother's. Political events forced their migration. Foremost among them was the extended family quarrel that broke out among the grandchildren and grandchildren-in-law of Queen Victoria (George V of England, Kaiser Wilhelm of Germany, Emperor Nicholas of Russia, and some others) and that ultimately was called World War I. This cataclysm, of course, disturbed or ended many lives everywhere. Eventually, students from Thomas Morgan's laboratory had to put down their fruit flies and help settle it, along with hundreds of thousands of their fellow nationals.

After the war ended, the former Russian Empire became hazardous for those of my own ethnic background, the Ashkenazi Jews, and many of them, including my parents and one grandparent, migrated to the United States. Those who remained were scattered or annihilated by the Nazis who, in addition to their other crimes and shortcomings, had totally perverted the teachings of modern genetics to construct a philosophy that justified their actions.

As a consequence, my knowledge of my forebears extends back only two generations (with the second one back rather incomplete), supplemented by some shadowy recollections about one generation further back. Some rumors also remain. My mother's maternal grandmother had a distinct oriental cast to her eyes, which persists slightly in some cousins. My father was said to include a famous Rabbi Levi Isaac (1740–1810), "the Berdichev rabbi," among his forebears. (My wife shares that distinction as well, with her mother claiming direct female descent.) My father's father had the family name Rabinowitz, which indicates descent from Rabbis. For personal reasons, he adopted his mother's maiden name, Shapiro. According to Dan Rottenberg's book on genealogy,[26] that name ultimately derives from the town of Speyer on the Rhine, south of Mainz, Germany. A Jewish community lived there prior to their expulsion in 1435.

Conventional research is unlikely to connect me to earlier ancestors, as few records were kept of Eastern European Jewry before World War I, and most of those that existed were destroyed in subsequent wars. For further understanding, I will also have to fall back on the resources in my own DNA. When an extensive DNA data bank is compiled, text matching will undoubtedly furnish me with a list of unsuspected cousins in this country, as well as in other lands, such as the Soviet Union, Israel, and Argentina. As I have not had the time to keep in contact with most of the cousins that I do know about, I doubt that I could do much with an extended list. Unlike most African-Americans, I know the immediate locations on the other side of the ocean from which my forebears came. Such locations have less significance to me than to, say, Alex Haley as my ancestors left voluntarily and were glad to do so. The question abut my background that interests me most has a broader context: What were the ethnic streams that came together to provide me with my particular constitution?

The history of the Jewish people has been one of expulsion and migration. The kingdom founded by King David about 1000 B.C. was conquered by the Babylonians, and the population sent into exile in 586 B.C. They returned after about sixty years, but were dispersed more permanently by the Romans after a revolution was put down in 70 AD. After this, a division into two great branches took place, the Sephardim and the Ashkenazim. Both branches have existed to modern times, with the Ashkenazim in the large majority.

According to tradition, the Sephardic Jews migrated through the Arab lands of the Mediterranean, settling in Spain in the Middle Ages. They developed their own dialect, *ladino*, a mixture of Spanish and Hebrew elements. After their expulsion from that country at the end of the fifteenth century, they resettled in a variety of Mediterranean lands and in the Ottoman Empire. Some were among the earliest Jewish settlers in the United States.

The origin of the Ashkenazim has been the subject of more dispute. The more conventional and majority view among scholars has featured a migration from Mediterranean areas to France, England, and Germany, where their settlements were documented by the ninth century A.D. From 1290 into the fifteenth century, they were expelled from one place after another and forced to migrate to the east to Poland and Russia. They retained their language, Yiddish, which is a combination of German and Hebrew elements, and some German-derived names, like my own. With the eruption of anti-Semitism on an enlarged scale in the nineteenth and twentieth centuries, many migrated to the United States and Israel.

One alternative theory, well publicized by writer Arthur Koestler in his book *The Thirteenth Tribe*, would alter some parts of the history of the Ashkenazim. According to Koestler and some others, Jews also migrated north and east from Palestine after the disaster of 70 A.D., and contacts were made with the peoples in those areas. One unexpected result was the conversion of the upper classes of a medieval kingdom of Khazar to Judaism in 740 A.D.

The Khazars were a blue-eyed, reddish-haired people that migrated from the steppes of Asia to the region of the Caucasus Mountains (now in the Soviet Union, near Turkey) from the fifth century on. They were classified among the Turkic peoples, a distinction that reflects their language group more than their ethnic composition. Their ethnic relatives were supposedly the Huns of Attila and the Magyar ancestors of the Hungarians. In their traditions, they traced their ultimate origins to Japheth, third son of Noah.

At its peak, the Khazar kingdom extended from the Ukraine and Black Sea to the Volga River, Caucasus Mountains, and Caspian Sea. At one point, according to legend, the ruling king, called the *Kagan*, tired of their religion of primitive shamanism. He called representatives of Christianity, Islam, and Judaism and asked them to make presentations. At the end, the story continues, he took note that the first two were the religions of powerful neighbors that bordered his land. To avoid taking sides between them and offending one, he selected Judaism.

At any event, Judaism became the state religion, and after a time, the Hebrew alphabet was adopted. After some centuries, the kingdom started to decline. Kiev was lost to their enemies in 862, and the capital, Itil, was destroyed in 965. After the year 1245, history makes no mention of them. Their land was overrun by the Golden Horde of Genghis Khan, and the population assimilated or dispersed.

The narration to this point would cause little furor among Jewish historians. Koestler and his sources believe, however, that Khazar refugees moved west into Russia and Poland, providing the bulk of the population that came to be called the Ashkenazim. According to Koestler:

> The evidence adds up to a strong case in favor of those modern historians—whether Austrian, Israeli or Polish—who, independently from each other, have argued that the bulk of modern Jewry is not of Palestinian, but of Caucasian origin. The mainstream of Jewish migrations did not flow from the Mediterranean across to France and Germany, to the east and then back again. The stream moved in a consistently westerly direction, from the Caucasus through the Ukraine into Poland and thence into Central Europe.[27]

They mingled to some extent with Jews from the West and adopted some of their Germanic traditions, so that "the numerical ratio of the Khazar to the Semitic and other contributions is impossible to establish." The main bulk of the population came from the Khazars, though.

If so then, in biblical tradition, the Ashkenazim would trace their line back not to Noah's first son, Shem, ancestor of Semitic peoples, but to Japheth.

The controversy remains open. The *Encyclopedia of Judaism*, simply states flatly: "The notion that Ashkenazi Jewry is descended from the Khazars has absolutely no basis in fact,"[28] without debating the issue. For myself, I have cause to wonder. I am six feet tall, with greenish-grey eyes and hair that is now brown with some grey, but was blond earlier in life. My brother has a similar appearance. I find relatively little resemblance between myself and many Jews from the Mideast or even Sephardic background. My parents on the other hand were both under five feet six inches with quite dark hair. I suspect the contribution of recessive genes.

Anthropologist and historian Raphael Patai has written that "contrary to popular opinion, there is no Jewish race." In his book *The Myth of the Jewish Race*, written with Jennifer Wing, he cites theories that the ten to fifteen percent blond component found among Ashkenazic Jews can be attributed "to admixture with blond Slavs in Eastern Europe during the Middle Ages."[29] He suggests that blue eyes and reddish hair might also have been brought into the population of eastern Jews by Khazar infusion.

Jews in other regions often resemble the native populations. The American Museum of Natural History in New York has a display on Jews of Asia. The Chinese Jews depicted there look remarkably like native Chinese, and the Indians like natural Indians. Their appearance suggests a heavy genetic admixture with the native population, even though the display claimed that their forebears traveled from the Mideast by land and sea, respectively.

Population geneticists L. Luca Cavalli-Sforza and Dorit Carmelli have discussed several possible bases for this effect, with interbreeding only one of them. Random genetic drift was another, and natural selection, the last. There might be a genetic advantage in fairer skin in a northern climate, and Mediterranean people who migrated might have evolved in that direction. Cavalli-Sforza came to no firm conclusion on the issue, but estimated the admixture of other populations into the Ashkenazim at perhaps forty percent (with a thirty-five percent margin of error).[30] Other investigators have attempted to interpret fragmentary data from blood and other conventional markers in terms of a largely Mediterranean origin for Ashkenazic Jews.

The good news, I feel, is that DNA will provide the answer. Using many identity points over a large sample, we should be able to figure out that the Ashkenazim, collectively, are more closely related to

1. The Sephardic Jews and other branches of ancient Palestinian Judaism who were dispersed far and wide.

2. Other descendents of the Turkic peoples of the steppes, such as the Hungarians.

3. The current non-Jewish population of the land that they inhabited most recently.

4. Some mixture of all of the above.

An intensive analysis using large populations will be needed at a time after the needed collection of human DNA identity sites has been assembled. Some work could be done immediately, however, using energy-cell DNA. Again, this small circle of sixteen thousand base pairs is inherited from your mother, and her mother before her, and so on in a strictly maternal line. This parallels exactly a Jewish tradition that one is a Jew if one's mother is a Jew and so on.

The set of females who inhabited the Jewish kingdom in Palestine before the dispersal of 70 A.D. were likely to have a set of energy-cell DNA sequences that differed in composition from those of other ancient peoples. These sequences should have been carried over into both the later Sephardic and Ashkenazic Jewish communities if the majority view of Jewish population history is correct. If any relationship persists today, it should still be reflected there.

I have seen no reports on this subject, but would expect some before too long. For population geneticists have been using energy-cell DNA to construct trees and establish relationships for other ethnic groups that extend back to a single common ancestor of us all two hundred thousand years ago. That target topic is well worth our attention.

ENERGY-CELL EVE AND ADAM-Y

The Garden of Eden had all of the conventional trimmings: serpent and apple tree in a parklike setting with one nude human couple. As this was the cover of the January 11, 1988, *Newsweek* and not *Playboy*, skin exposure was minimal. Only the upper parts of the man and woman were displayed, and her breasts were covered by strands of hair and a bar-code emblem. It was the color rather than the amount of the exposed skin that was newsworthy. The couple's skin was light brown, their features were African, and their overall appearance was well groomed, clean shaven (no facial or underarm hair), and thoroughly modern. The headline that accompanied this picture, "The Search for Adam and Eve," implied that they might be found in a fashionable nudist resort outside of New York City or some other modern urban center.

The story within was somewhat different. Eve alone was discussed: "more likely a dark-haired, black-skinned woman, roaming a hot savanna." Quite muscular, "she might have torn animals apart in search of food."[31] This woman differed in several other ways as well from either the Eve of the Bible or the *Newsweek* cover. She probably lived in a community and had sisters, brothers, and parents. Further, she may not even have been a modern human, but a type that preceded us in evolution, perhaps with a brawny physique and a large protruding face, with the forehead receding behind prominent brow ridges. Her most important characteristic was the time and place of her existence in sub-Saharan Africa some two hundred thousand years ago.

Why should she be called Eve then, and why was she worthy of *Newsweek*'s attention? Her existence had been inferred by a detailed analysis of energy-cell DNA samples taken from 147 individuals and performed by Allan Wilson and his colleagues, Rebecca Cann and Mark Stoneking. They had persuaded women from a variety of locations around the world, Europe, Asia, Africa, Australia, and New Guinea, to donate placentas to science and used these as a source of DNA.

Spelling differences among the energy-cell DNAs of the above women were determined using text cutters, and the differences were compared. They were used to construct a "tree" in which the modern spellings formed the tips of the twigs and the underlying branches were deduced by studying the variations among the twigs. The process may become clearer if we again look at an analogy in English. I have kept the spaces in this time to make the differences more readily visible.

1. THE RAIN IN SPAIN STAYS MAINLY ON THE DRAIN
2. THE RAIN ON SPAIN SLAYS MAINLY ON THE PLAIN
3. THE PAIN IN SPINE STOPS MAINLY IN THE BRAIN

The three phrases above are obviously related. They could have been derived from a common ancestor (THE RAIN IN SPAIN STAYS MAINLY IN THE PLAIN) by mutations involving the loss, gain, or replacement of letters. With only three sentences in our sample, several possible ancestors could be written. Wilson and his co-workers had a larger group that constrained them more than that. In our case, to save work, we will rely on faith and our memories of the play *My Fair Lady*, which contained the original line.

In the analysis, the fewer the changes that are needed to make two phrases identical, the more closely related they are assumed to be. Lines 1 and 2 above, for example, differ from one another in four letters. Both can be derived from the presumed ancestor by first changing the second IN to ON and then introducing two extra substitutions in each. This may have been the course. They may have had a joint ancestor (THE RAIN IN SPAIN STAYS MAINLY ON THE PLAIN) that was not shared by the third variant.

Line 3 differs from the first two more extensively. It would be placed on a different branch of the tree leading back to the common ancestor. It differs from that ancestor more greatly than the first two, so presumably it branched off at an earlier time.

The tree that Allan Wilson and his colleagues constructed by a more extensive analysis of this type was put forward as a model of the actual path of evolution of energy-cell DNA in human history. This DNA, once again, is passed exclusively from mother to daughter without any shuffling due to crossing-over or any male input. Only mutations cause spelling changes. The tree then follows the separation of maternal lines through time. If my twig is close to yours in the tree then you and I shared a common female ancestor not too long ago. If we are far apart, then our peoples separated further back. The fact that the tree comes together to a single trunk indicates that everyone on this planet had a common super-great-grandmother at some point in history. This person was given the somewhat misleading name of Eve, or the more forbidding "mitochondrial Eve." In this book, she will be called Energy-Cell Eve.

Population geneticists tell us that in cases where no genetic scrambling occurs, one line of descent will triumph eventually and dominate the population. Consider surnames, for example, which pass in an all-male line. If one hundred males with different last names and an equal number of females lived on an island, and nobody came or left, only ninety surnames would remain after twenty generations. The other ten would not necessarily have ended with childless couples; they could also disappear when families had only daughters. Such families would still have descendents, but the descendents would have different last names. After enough generations had passed, all of the settlers of the island would bear the same surname; one of them would have won. Energy-Cell Eve was the winner of such a sweepstakes. Many other women and men of her generation may also have been our ancestors, but all of them had families without daughters somewhere along the line. Their DNA survives in our chromosomes. Only Eve, however, has presented us with our energy-cell DNA.

Every area in our genome that is free of recombination would conduct its own sweepstakes of the above type with a single winner likely. Most of the Y chromosome, for example, is handed down in an all-male line. There must therefore exist an Adam-Y counterpart to Energy-Cell Eve.[32] Technical problems in locating the most suitable identity points have slowed down the construction of a tree in this area, but by 1991 there were promising signs that one might be forthcoming. The cover of *Newsweek* notwithstanding, our most recent common paternal ancestor, Adam-Y, will not necessarily have been the husband of Maternal Eve, nor need he even have lived in the same generation with her. The romantically inclined among us, however, can still cling to the hope that some couple, somewhere, gave rise to us all.

The mother and grandmother of Energy-Cell Eve, as well as the females before them, were also the ancestors of all of us. Eve was just the last one before a great split occurred. The same analysis applies to Adam-Y, his father, and the earlier relatives in his all-male line. At some earlier point, particularly if the number of humans or prehumans was very small, we may actually have had an authentic Adam and Eve couple.

So far, so good. However, the picture gets messier when we recall that other DNA areas, the chromosome hinges, for example, may also be free of crossing-over. As sequencing work progresses, identity sites could be selected for any hinge area, and that information could also be used to build a human family tree. We might find again that the data point to a single last common ancestor in the distant past, but the line of descent would pass through both male and female lines. Naming this individual would present a vexing problem to the media. Somehow, the name Hinge-22 Harry or Hilda (or using the technical term, Centromere Charlie) lacks both recognition and zing.

These games remain for future geneticists. Thus far, substantial work has only been done with energy-cell DNA, so let us see what harvest has come from those efforts.

Energy-Cell Eve had one distinction that was not shared by her mother and grandmother and earned her the name. Her daughters (she must have had at least two) were not the mothers of us all. Each engendered only a part of the human race. We do not know what actually happened, but we may imagine that one of them, together with some friends, simply left town and was not heard from again.

In the original energy-cell DNA tree of Wilson and his colleagues, the first branch from the original trunk separated some Africans from other Africans and from the remainder of the humans sampled. Energy-Cell Eve, then, presumably lived in Africa. (An implied consequence would then be that some Africans, Nigerians, for example, would be more closely related to Europeans, Asians, and Aboriginal Australians than they would be to other Africans, for example, the click-speaking !Kung people of southern Africa. So much for stereotypes.) Some of her descendents migrated extensively and gave rise to most populations of the globe, while others stayed near home. One of Eve's daughters, who we will call Afra, stayed near her mom, while the other one, Globa, hit the road.

The analysis so far tells us the order in which separations took place, but nothing of the time scales involved. Dates can be attached to events if we assume that the spelling changes that were studied happened at a fairly constant rate. The more differences between two DNA sequences, the longer the time that has passed since they were united in a common ancestor. Spelling changes, or mutations, thus provide a "molecular clock."

The molecular clock concept was devised by Linus Pauling and Emile Zuckerkandl in the early 1960s and has been used extensively since then, but many quarrels have come up over its applications. Different areas of DNA can evolve at different rates. For example, energy-cell DNA changes five to ten times faster than the bulk of DNA in the chromosomes. Each clock must be calibrated by linking its branch points of the tree associated with it to other events that have been dated in a different way. Geological events such as the separation of land masses and fossils whose age has been established by radioactive dating have been used for this purpose.

Some triumphs have been recorded through the use of this mutation-based clock. The separation of humans, chimps, and gorillas in evolution was set at 5 to 7 million years ago with this technique. The time is considerably less than the one previously accepted by anthropologists, but has now come to be generally accepted, even by those who first opposed it. The bickering that remains concerns whether gorillas split off first, leaving the chimp and human together, or whether humans were the first to diverge from the trio. The latter idea was put forth earlier, but the former has now gained an advantage. Sequencing data from the globin gene cluster and now from energy-cell DNA have suggested that chimps are our closest relatives.

Matters have not gone so smoothly in the case of Energy-Cell Eve. Allan Wilson and his collaborators originally estimated her time to be about two hundred thousand years ago. More recently they remeasured it by comparing the divergence of humans in energy-cell DNA to the human–chimp differences that exist there. The human–chimp difference was a bit more than twenty times the separation in sequence between individual humans. If we parted from the chimps 5 million years ago, then Maternal Eve lived 238,000 years before the present.

That conclusion need not vex you nor me, but then we have not spent our professional careers developing an alternative theory of human evolution, one that requires a very different date. According to the media, those who had done so were in a state of outrage. Milford Wolpoff of the University of Michigan, for example, commented to *Discover* that "our public enemy number one is Allan Wilson."[33]

Wolpoff has been described in an article in *Mosaic* as the person who "is said to have looked at more hominid fossils than any other individual."[34] He and similar-minded colleagues have examined humanlike fossil remains from Europe, Africa, Asia, and Australia and developed a very different theory of human origins. In their account, *Homo erectus*, a predecessor of our own species, *Homo sapiens*, migrated from Africa at least 850,000 years ago. *Homo erectus* colonized the various continents, and the separate communities developed in parallel into the modern human race. Some genetic exchanges between the emerging races kept us together as one species despite the fact that the different groups evolved in widely separated locations.

In the Wilson scenario, the separation and diversification of the human species is a much more recent affair. Energy-Cell Eve's daughter Globa did not set sail immediately. Her descendents may not have left the African continent until one hundred thousand years ago or later. It may have taken them sixty thousand more years to reach the more remote corners of the Eurasian landmass, such as Australia, and the Americas were not colonized until twelve thousand years ago.

The differences in the DNA of certain ethnic groups are too profound to reflect mutations of the last forty thousand years alone. In the case of the Australian aborigines, for example, fifteen basic patterns exist in their energy-cell DNA. These types probably diverged before Australia was settled. It seems likely, then, that these varieties were present in at least fifteen individuals who first colonized that continent. Similar circumstances have been described in other migrations. Most of the evolution of energy-cell DNA sequences, then, took place before the divergence of the races.

If the Wilson view of human evolution should be correct, then more ancient fossils such as Java and Peking represent examples of evolutionary sidelines that died out. Questions have arisen concerning the circumstances of their demise. Were they killed by the new migrants in a prehistoric holocaust, or did they perish on their own through changes in climate or loss of food supplies? Such items may have to wait until the main event is settled: Which overall theory is correct?

Quarrels still abound today about the best method of tree construction and the homeland of Energy-Cell Eve as well as the chronology.

How will such issues be settled? If some fossils representing modern humans were found (especially in a non-African site) that dated at more than a few hundred thousand years of age, it would bode ill for the Wilson theory. However, no obvious strategy exists to move us dramatically ahead in the collection of new types of fossil. Rebecca Cann has written: "All in all, it is too much to hope that the trickle of bones from the fossil beds of east Africa will, in itself, provide a clear picture of human evolution any time soon."[35]

Floods of data will be coming from another direction, though. The techniques used in the Eve investigation have already improved dramatically in three years. Newer studies now use single plucked hairs thanks to DNA amplification rather than placentas or blood samples. (Some individuals in certain tribes feel queasy at the thought of giving blood. I have shared that feeling myself.) Whole sequences are being read instead of individual letters at the sites recognized by text cutters. The energy-cell DNA of much larger groups of people will be sampled.

Even more important will be the extension of such studies to the chromosomes themselves. As the Human Genome Project proceeds, other areas free from crossing-over will be included in the analysis. Each study can proceed separately of the others with its own individually calibrated clock. If all should agree on a particular time scale for human evolution, then we could feel secure about the results.

Of course, such studies need not be limited to the course of human evolution. The ones I have discussed have focused on the one percent or so of DNA sequences in which humans differ from one another. If we open the width of our lens to encompass the other ninety-nine percent and include other species in our sequencing efforts, then the whole vista of evolution of life on this planet comes into view.

Such efforts are already underway. I mentioned them briefly when we spoke of globin evolution and again with the human–chimp–gorilla divergence. This book concerns humans; the remainder of the tale will be told elsewhere. For now we can note that the discoveries that began with Mendel and Meischer and culminated with Sanger and Gilbert have provided us with a telescope that can peer aeons into the past, back when life on this planet existed only in the form of simple one-celled organisms. Tomorrow we shall know much more about yesterday and the day before that.

Another prospect begs for attention. I have made some guesses about events to come. The consequences of our capture of our own blueprint extend into the indefinite future. I would like to peer further forward.

PART IV
AFTER TOMORROW

We are the generation of the text. It falls to us to be the first to read the DNA script that has accumulated for billions of years and evolved to provide the blueprint for our species. We will acquire the full contents of a single human script and pass it on to our descendents as a legacy. As a reward, we shall harvest the first benefits that come with knowledge of the text. Maladies that have plagued humans for millenia will be fully comprehended and controlled or eliminated.

The following generation will have the opportunity to comprehend the message more fully and appreciate how its variation from person to person leads to differences between us. More subtle genetic malfunctions will be diagnosed and dealt with, and human genetic history and relationships will be explored. When that time has also passed, and humans have become more familiar and comfortable with the blueprint and legacy stored in their DNA, additional questions may come up:

1. Shall each of us have a chance to preserve for the future a record of our total biological identity and life history with the expectation that we, or at least the plan that gave rise to us, will someday be recalled?

2. Should the human race or some parts of it introduce

deliberate changes in the text in the hope of improving the human condition?

These feats will not be possible very soon. We may gain some vision of our directions and a sense of purpose concerning the future, however, if we consider them now.

18

OUTDOING OZYMANDIAS

And on the pedestal, these words appear:
My name is Ozymandias, King of Kings,
Look on my works, ye Mighty, and despair!
Nothing beside remains. Round the decay
Of that colossal Wreck, boundless and bare
The lone and level sands stretch far away.

—Percy Bysshe Shelley in "Ozymandias"

The gentleman at least left us with his name and inscription, which betters the accomplishment of most humans who have lived on this planet. We come, struggle, and achieve and then vanish. At many funerals I have attended, the rabbi or minister has pointed out that the departed remains among us in the works that he or she has done and the memories of those who have loved or appreciated him or her. This solution holds in the short range, but one day all those who have known the person will also pass on. The works, unless they are pyramidlike in their proportions, also leave ripples that are absorbed by the greater flow of human events and vanish. What hint remains of the charities that were supported or school curricula reformed in ancient Ur or Babylon?

Despite this erosion, philanthropists attach their names to university buildings and charitable foundations in an attempt to have their mark endure for a while at least. Dictators still name cities for themselves and have their own statues erected. The name of Richard M. Nixon is inscribed on metal plates attached to the base of the landers that were left on the moon. Free from the ravages of erosion, they may endure for millions of years.

Yet a name does not reflect the essence of a person, nor does a statue capture very much of him or her. Without fanfare, modern science has been preparing the basis for more comprehensive and enduring memorials for all of us. The ultimate description of our body plan lies in the letters of our DNA. I have guessed that by 2020 society may find it advantageous to store as much as one percent of the DNA text of individuals for reasons of public health, history, and identification. As the uses of sequences are extended and technology improves, costs will eventually drop to the point where everyone may readily preserve his or her body plan on a small disc or the even more compact forms of data storage that will undoubtedly be developed in the twenty-first century.

A DNA script, of course, does not define a living individual. Yet, by taking a bit more space, each of us could add our recorded reminiscences, written letters, documents of any form, and digital equivalents of photographs. Our personal discs would be placed on the most durable material available and reproduced in a number of copies for distribution to descendents, as well as for storage in public archives. It is not hard to imagine the data for all those alive today stored efficiently in a single library of reasonable size. Our great-great-grandchildren or an interested browser ten thousand years from now would have the chance to learn how we were constructed, how we felt, and what we did. Ozymandias would be jealous.

Of course, we humans are curious and sentimental creatures. Given the plan for something that once was and had meaning to the people of that time, we are tempted to bring it back again. Ancient Cretan palaces are rebuilt, nineteenth-century villages are reconstructed, and musical instruments of the Renaissance are assembled to their original specifications. The same impulses may impel some of us to bring back the beings of the past when technology permits. The recapture of extinct species would be an obvious early target for such efforts.

RECALL OF THE WILD

When the last individual of a race of living things breathes no more, another Heaven and another Earth must pass before such a one can breathe again.

—William Beebe, New York Zoological Society

I found that quote recently in the rebuilt elephant house of the society, located in the Bronx Zoo in New York. I grew up within walking distance of that zoo and wandered over on many an idle afternoon of my childhood. I particularly remember the large groups of elephants, together with hippos, rhinos, and other massive beasts, that inhabited the indoor and outdoor enclosures of the earlier version of the elephant house. In its rebuilt form, such beasts are barely represented and hardly visible, while much of the space is given over to exhibits and conservation.

It would be sad if such magnificant animals or even the less familiar ones on the long list of endangered species were to be no more. Should some of them depart before the human race can get its act together, perhaps we will not have to wait as long as Mr. Beebe suggests. We may even take the offensive and try to recall some of those who departed long ago. Some dreamers, at least, are already laying their plans.

In his 1983 text *Genetic Principles: Human and Social Consequences*, Gordon Edlin wrote of one of his "favorite fantasy experiments that is actually a reasonable extension of experiments that have already been performed."[1] His fantasy was the resurrection of the woolly mammoth. He would extract cell nuclei from a well-preserved, thousands-of-years old, frozen specimen. They would be inserted in elephant eggs that had had their own nuclei removed, and the product (after it had performed a few cell divisions) implanted in a female elephant that had been treated with hormones to induce pregnancy. When she delivered, a baby mammoth might be born.

Some parts of the scheme are already in place. In a news story in *Science* in 1988,[2] Jean Marx reported that nuclei could be removed from sheep eggs. The nuclei could then be fused with cells taken from sheep embryos at the eight- or sixteen-cell stage, and the product reimplanted in a sheep mother-to-be. This procedure could afford identical copies, or clones, of individual sheep. The *New York Times* of January 2, 1990, reported cases where a common species could carry the offspring of a closely related, but rare one.[3] In the Cincinnati Zoo, the fetus of an endangered Indian cat was carried to term by a domestic cat, and a rare bongo (an African antelope) was born of a more common eland.

These developments do not yet bring us close to rescuing the mammoth. Cells that have already differentiated, or committed themselves to becoming a particular tissue, do not work in the above scheme. In many species, differentiation starts at the stage of sixteen or fewer cells, and adult cells would be unlikely to work. More serious yet would be the condition of the frozen mammoth cell nucleus. We have seen that DNA can survive in such specimens, but it is hardly intact or functional. Other cell components are undoubtedly degraded as well. A more ambitious plan will be needed, one that cannot be executed now, but is conceivable in the future.

Mammoth cell DNA exists in many pieces, but the fragments can be rescued and sequenced. So many copies of DNA exist in an organism that it may be possible, when sequencing technology has speeded up considerably, to determine the entire DNA text of the beast. If we could do this then we would be ready for the next task: to reassemble mammoth chromosomes. This might prove less formidable than we think.

Hopefully, we would have saved some samples of all the mammoth text fragments used for sequencing. They would have to be spliced together in the proper order, the type of task that can be done, on a smaller scale, even today. Necessary proteins would then be added to make up functional chromosomes. Artificial yeast chromosomes are being used in DNA work today. If some special ingredients were needed for the mammoth, we could find the recipe within the DNA text. Finally, the chromosomes would have to be packaged in a nucleus of some other form that an elephant egg would accept. It sounds formidable to me, but in 1986, Cambridge University zoologist R. A. Laskey published an article, "Prospects for Reassembling the Cell Nucleus," in the *Journal of Cell Science*. He noted: "It is becoming possible to reconstitute whole nuclei from purified DNA."[4] What was becoming true by 1986 should be old hat by the mid-twenty-first century.

We would undoubtedly have to go to some trouble to make sure that the cell was in an undifferentiated state, which might require a number of bothersome adjustments to the final product. The question of how cells differentiate interested Thomas Hunt Morgan, but he was premature. It is a hot topic in many laboratories today. Sooner or later, we will understand it and learn how to control it.

Possibly only shreds of mammoth DNA have survived in the Siberian freezer—not enough to put together the original text. If so, we are out of luck. If the information survives, though, and we care enough to make the effort, I would not be surprised if someday the first reissued mammoth were not welcomed by a future zoo. For species recently perished or on the way out, of course, prospects are even better. We need not wait until another heaven and earth come by again.

The re-creation of dinosaurs would be much tougher, despite a recent novel, *Jurassic Park*, by Michael Crichton.[5] The survival of a complete DNA text over 60 million years, whether in old bones or blood swallowed by a preserved insect, is very chancy. The problems involved in putting flesh around such a text, even if recovered, would dwarf those involving species where a close relative survives. This item can be marked down for the quite distant future or never.

THE FREEZER AND THE FILING CABINET

What we can do for animals could of course be done for individual human beings. Much more experience and wisdom will be needed before we set out in this direction, of course: We are willing to make blunders with mice that we would not tolerate with humans. The possibilities are centuries away, but the issues involved are very relevant today.

"Some people want to achieve immortality through their works or their descendants; I want to achieve immortality through not dying." Many of us have shared this feeling as described by Woody Allen. We are taught that death is an inevitable part of life, yet, given the choice, we would just as soon not go. Many religions offer us a better existence in the hereafter. The Buddhists would return us to this realm by reincarnation, but in some other form. No guarantee obviously accompanies these offers, and faith is required. Even if one of the above were true, I and many of those I have spoken to would still want the option of exploring this existence, in this body, until we were bored with one or both.

About twenty-five years ago, a group of my academic friends organized a discussion group to consider cryonics, a concept in which people would be frozen upon death and stored in liquid nitrogen. When the skills needed to defrost and reanimate a corpse had been learned by science, they would be thawed and restored to life. Whatever the damage was that had killed them, surely at some point in the distant future it could be repaired, along with the injuries inflicted in freezing and the decay during storage. Sooner or later, a benign future society would take pity on those who were frozen and bring them back to resume their existence. This idea had been publicized in a 1964 book, *The Prospect of Immortality* by physicist Robert W. Ettinger.[6]

As academics, my friends and I were more given to thought and speech than action, but other individuals took steps, and several societies for the above purpose were founded. We had assumed in the 1960s that the idea, once publicized, would sweep the country. If it worked, you had gained another life, perhaps an indefinite one, whereas if it failed, you had lost nothing. The unreligious who followed this scheme could still cling to some hope on their deathbeds, whatever the outcome. My friends and I were naive, however. By 1989, a total of exactly eight deceased individuals were preserved in liquid nitrogen awaiting future resurrection, according to an article by gerontologist Steven B. Harris in *Free Inquiry*.[7] (Dr. Gerald Feinberg has estimated the number as thirty, citing conversations with officials at the Alcor Foundation in California.) Many obstacles have come up to discourage would-be "cryonauts."

The practical problems were formidable. One had to arrange to have trained technicians available at an instant's call and to make sure to die in a place that they could get to quickly. To pass away from a lingering disease in a hospital would be ideal, to perish in a violent plane crash would be totally impractical. Further, funds to arrange for the perpetual maintenance of the frozen remains had to be set aside in advance and secured in an ironclad insurance policy. The person who was to be frozen would have to presume not only that science would advance, but that the society to which he or she had entrusted his or her remains would remain stable and honor contracts over many centuries.

An interesting alternative that avoids these hassles and hazards is afforded by our DNA text. For many reasons, society is likely to want to keep a portion of that text on record. At some future point it will not cost too much to have the remainder of it recorded as well. The genetic plan for my body would then be stored indefinitely in some filing cabinet or disc within a computer system at negligible cost and in several copies for safety. Those who die before the data bank DNA sequencing becomes a routine affair or the data banks are set up can fall back on an interim measure. A sample of blood, hair, or saliva could be preserved in the cold until sequencing can be done. Some of the Argentinian grandmothers have already done this for the benefit of their lost grandchildren, and American facilities are being set up as well for storage purposes. In a few years, you or I should be able to arrange for the preservation of our DNA blueprints without disrupting our lives or bankrupting our savings to any great extent. The DNA text could be supplemented with written memoirs and documents to provide a fuller record of our lives.

What would we have accomplished? At the least, we would have left a lasting memorial. If we borrow one assumption from the cryonics group, we would also have gained the hope that some group in the indefinite future would choose to recreate us out of curiosity or benevolence.

Viewed from this perspective, a human being at death would resemble a play that had ended its run. No more performances would be given for a while, but the script and stage directions would always be with us. When circumstances warranted, a revival might be staged.

19

A MATTER OF CHOICE

What would this Man? Now upward will he soar, And little less than Angel, would be more.

—Alexander Pope, *An Essay on Man*, 1733

As we gradually get familiar with our genetic script, both the general plan for the species and our own personal variations on the theme, new opportunities and temptations will arise. We humans, certainly in this culture, are seldom inclined to accept stoically whatever is given to us. Think about the way we behave when we buy a house. Few of us can afford to build exactly to specifications. We select from what is available, according to what we can afford. Once we move in and get somewhat familiar with what we have, we begin to repair the obvious defects. After that, we may rearrange the furnishings to suit us more. Finally, if we have the resources, we may redesign the place, altering the layout of rooms and building additions to suit our needs and tastes.

In the exploration of the blueprint for our bodies, we are definitely still taking stock and exploring what we have. Even so, we have started, tentatively, to repair some defects. Even as I write this, the first experiment to introduce new genes into the body cells of a human has received the necessary government approval and is under way. The DNA text of certain cells called *T cells* in the blood of a four-year-old girl have been altered artificially so that they will produce an important enzyme that she lacks. Whether this initial effort succeeds or not, others will follow. Physicians will attempt to treat a variety of human genetic defects, including sickle-cell anemia, hemophilia, and cystic fibrosis, by inserting or replacing the genes in the most appropriate body tissues. Formidable technical difficulties have slowed this work, and responsible guidelines have been put into place to protect the health of those treated from damaging side effects. These measures have delayed the date when gene therapy will become common medical practice, but its role in the future seems assured.

The offspring of those who have received gene therapy will still be at risk for the disease. Of course, they will be able to get such treatment on their own. Alternatively, a would-be parent who carried the disease could still obtain genetic counseling. Marriages between carriers could be avoided, as certain Ashkenazic Jews do in the case of Tay–Sachs disease. Abortion or even embryo selection are other measures that would protect future generations from the disease-carrying gene.

It would be a nuisance, of course, if one generation after another had to undertake complex medical procedures or selection measures. Most of us would like the option of marrying whom we choose without fear of adverse genetic consequences, and some religious and ethical groups oppose abortion and embryo selection as a matter of principle. An obvious alternative would be to apply germ-line therapy to a willing individual, in which the defect is repaired in his sperm or her eggs at the same time that it is corrected in his or her body. George Annas, who directs the Boston University law, medicine, and ethics program, put it simply: "If you can get rid of a genetic disease, why not do it rather than cure all the kids as they are born?"[1] *Nature* suggested that "royal families with haemophilia would no doubt have jumped at the technique."[2]

Will germ-line therapy be a practical alternative in the near future? Some scientists have argued that we need not debate the issue right now because its practical applications, particularly to the more complex human traits, lie in the distant future. Robert Williamson, for example, asserted in 1982:

> When we come to such vague concepts as "aggression" or "intelligence" or "good looks," we are off in the land of the fairies. If there are any inherited determinants for these attributes, we are so far from understanding, isolating or measuring them that it's barely worthwhile even consid-

> ering that an attempt at gene therapy will be made with any prospect of success.[3]

Our past experience indicates that the person who takes such positions runs a lot of risk of being wrong. Jonathan Glover, in *What Sort of People Should There Be*, cites one such instance:

> Sir MacFarlane Burnet, writing in 1971 about the use of genetic engineering to cure disorders in people already born, dismissed the idea that a virus could carry a new gene into the human body to replace a faulty one: "I should be willing to state in any company that the chance of doing this will remain infinitely small to the last syllable of recorded time."[4]

Burnet expressed himself beautifully, but may have set a record for the extent to which one can be incorrect in a prediction.

Another instance, reported by biochemist Christian Anfinsen,[5] emerged in a 1977 lecture by Nobel laureate Ernst Chain, the discoverer of penicillin:

> There exists no method, at present, nor is there the likelihood that one will be discovered in the forseeable future, by which it would be possible to alter the nucleotide sequence, and thereby the genetic properties, in any gene of any mammalian cell in a controlled manner which could be called "genetic engineering." Any speculations that such a process may be near at hand and could influence the heredity of man must be dismissed as science fiction.

A dozen years later, genes are being delivered to specific positions in the DNA of animals.

Sooner or later, germ-line therapy of humans applied to complex traits as well as simple ones will become possible. Scientists first learned how to manipulate genes within bacteria in the 1970s, and events have rushed forward since then.

Genetically altered bacteria are now used to manufacture such useful biochemicals as human growth hormone on an industrial scale. The genetic manipulation of mammals and plants has become an active frontier of science. The text that I use in my New York University biochemistry class, for example,[6] carries an account of the experiment of the early 1980s in which a rat growth hormone was microinjected into fertilized mouse eggs. The eggs were then implanted into the uteri of surrogate mouse mothers. Some of the progeny grew much more rapidly than normal mice and attained double the normal size.

Such transgenic, or genetically altered, mice and other animals can now be produced with much more precision, although much scientific ingenuity has been needed to move ahead. As Mario Capecchi, a geneticist at the University of Utah, put it: "Essentially you're trying to go into what amounts to 1,000 volumes worth of genetic information and make one little editing change."[7] Yet the promise of a new technique called *gene targeting* has been so high that the same individual also declared: "We now have the potential to generate mice of virtually any desired genotype."[8] This new advance will prove helpful on many different fronts.

Biochemists and physicians will use them to learn what newly discovered genes do in the body. Farmers will want to have animals that afford better yields of milk, beef, and pork and more disease-resistant crops. Ultimately, some humans will want to improve their own lot or that of their children.

One likely scenario will involve the treatment of an embryo of a few cells in a test tube, followed by reimplantation in the womb. As in the case of gene therapy, exquisite care will be needed to safeguard the subjects, which is not a major concern in animal research. Again, such precautions will slow down possible applications, but not prevent them.

Considerations of safety are important, but they have come up and been handled in many other human activities, from brain surgery to skydiving. A deeper concern for many has been the ethical choices. If only the correction of Huntington's disease and a handful of other dreadful diseases in the afflicted family lines were involved, then the divisions among ethical thinkers would be much less. However, once the elimination of awful diseases has been accepted, why not treat less severe ones as well? Activist Jeremy Rifkin pictured a gradual slide in this direction in his book *Algeny*:[9] He argued that there was no logical stopping place, once society had decided to cure disease by altering human genes. Once sickle-cell anemia and cancer had been cured, why not go on to fix color-blindness, left-handedness, and even an undesirable skin color? Rifkin and some others have opposed the apparently benevolent uses of germ-line therapy and even gene therapy for this reason. Ethicist Arthur Caplan put it succinctly: "They fear the camel's nose. They think that if gene therapy comes in under the tent, germ therapy is not far behind."[10]

Some scientists and ethicists have suggested that any debate on such issues be postponed until we are closer to the technical possibility. I feel that the questions have profound significance for the human future and that we will need all the time that we can get to think them over. I will present several very different views and then hazard a guess about the eventual outcome.

Certain thinkers have had strongly negative feelings about the prospect of designed human genetic change. Walter Charles Zimmerli, a German philosopher, commented at a CIBA Foundation Symposium: "Human beings *must* under *no* conditions whatever tamper with human germlines!"[11] In a similar vein, the Nobel laureate N. Tinbergen wrote to the British newspaper, the *Guardian*, "I find it morally reprehensible and presumptuous for anybody to put himself forward as a judge of the qualities for which we should breed."[12]

In 1988, an essay contest conducted by *The Humanist* was won by Deborah Dalton, a Boston University freshman. She wrote:

> Evolution has been a slow process, and biotechnology threatens to artifically accelerate and direct its course, thus totally destroying the structure of nature. . . . Can we trust an individual doctor or scientist to choose the course of human evolution instead of allowing natural selection to continue? . . . Many believe that it is immoral to bring a deformed baby into the world. But do we as humans have the authority to decide who should be born and who should not? Are we not trying to create a superior race that parallels Hitler's Aryan race?[13]

The eugenics movement, of course, was founded by Francis Galton for the purpose of improving the heredity of the human species, and Ms. Dalton's comment reflects the abominations ultimately inflicted in the name of that cause.

Memories of past horrors and fears of future ones have motivated the opponents of germ-line therapy to pool their efforts. One of Jeremy Rifkin's most remarkable achievements was the bringing together of an unusual coalition in June 1983 to convene a press conference and issue a resolution. Included in his group were the cleric and politician Pat Robertson; Jerry Falwell, the founder of the Moral Majority organization; and Nobel laureate George Wald of Harvard, as well as twenty-one Roman Catholic bishops and a number of Protestant and Jewish religious leaders. The group members may have disagreed previously on almost every issue from Vietnam to abortion, but they came together to advocate that no attempts should be made to "engineer specific traits into the germline of the human species."[14]

They did not suggest that the issue should be discussed, but included the following arguments for their position:

> Since part of the strength of our gene pool consists of its very diversity, tampering with it might ultimately lead to the extinction of the human race. . . . No individual, group of individuals or institutions can legitimately claim the right or authority to make decisions on behalf of the rest of the species alive today or for future generations. Society should oppose human genetic engineering with the same courage and conviction as we now oppose the threat of nuclear extinction.[15]

Of course, those who would forbid human germ-line therapy are doing just what the above group deplores: presuming to make decisions on behalf of the rest of the species alive today. Many of us would question their conclusion that human genetic engineering would necessarily lead to extinction of the species or even to a loss of gene diversity.

The *New York Times*, hardly a radical or impulsive institution, devoted an editorial, "Whether to Make Perfect Humans," to this topic on July 22, 1982. They said, in part:

> Repairing a defect is one thing, but once that is routine, it will become much harder to argue against adding genes that confer desired qualities, like better health, looks and brains. There is no discernable line to be drawn between making inheritable repairs and improving the species. . . . The question of whether the human germline should be declared inviolable deserves close attention.
>
> Such a restriction will probably prove unjustifiable. But deliberate manipulation of the human germline will constitute a watershed of history, perhaps even in evolution. It should not be crossed surreptitiously, or before a full

> debate has allowed the public to reach an informed understanding of where scientists are leading. The remaking of man is worth a little discussion.[16]

In any such discussion, the question Who decides? will rate a high priority. Opponents of germ-line therapy have assumed that any use of the technique must involve decisions by authority that would be forced upon individuals. A limited number of approved models would be selected, and human genetic variety would decrease. In *Brave New World*, George Orwell's *1984*, and the actual Nazi and Stalinist dictatorships, technology was used to suppress human choice. The decisions and enforcement, however, were made by politicians or people trained in science but functioning in the political realm. Science and technology, by extending human control over the environment, increase human choices.

Jet aircraft, penicillin, and many other items exist today that were unknown when Thomas Hunt Morgan was growing up. They have been placed at our disposal because many visionary individuals like Morgan devoted their lives to hard work and study. No one is compelled to use these items in a free society. We may still cross the Atlantic in a sailing vessel or perish from pneumonia if we wish, but most of us prefer the alternatives that have been provided.

In his book *What Sort of People Should There Be?*, Oxford philosopher Jonathan Glover has described a future in which humans also retain the right of individual choice about germ-line therapy. He considers the arguments of those who would forbid genetic intervention, but does not find them compelling: "Preserving the human race as it is will seem an acceptable option to all those who watch the news on television and feel satisfied with the world." He is optimistic about the possible gain if we go ahead: "If we decide on a positive programme to change our nature, this will be a central moment in our history, and the transformation might be beneficial to a degree we can hardly imagine." Yet he understands the concern about central control: "It is a protest against a particular group of people, necessarily fallible and limited, taking decisions so important to our future." He endorses, as an alternative, the idea of a "genetic supermarket."[17]

In this scheme, which Glover attributes to philosopher Robert Nozick, prospective parents could select which traits their child could gain and which defects would be corrected from a list of genetic options. A couple in the future might want their child to have two copies of a gene associated with musical talent, but to lose one that conferred greater susceptibility to environmental cancer. Other couples would have their own individual agendas. No central planning by any controlling board on behalf of the entire population would be involved, and variety would be as likely to increase as decrease. The question of which changes would provide "better" human beings would be left for each to decide, and the answers would vary, as individual life-styles do today.

In a recent talk, Nancy Wexler mentioned the worries over "Barbara Walters couples" who would go "shopping for the perfect baby."[18] (She was referring to a recent television special on that possibility.) Such parents might be concerned with appearances, height, weight, eye color, and, of course, intelligence. However, why not, Wexler pointed out, "shop for a sense of humor, empathy, wisdom, compassion and passion?" Obviously, many different choices will be made.

The role of government would be more defensive to protect children from being harmed by adverse or peculiar selections by parents, just as children are protected by law from cruelty today. Occasionally, some intervention on behalf of society might be necessary to prevent the ratio of the sexes from becoming too unbalanced, but this could be done by subsidy or lottery rather than by compulsion. The particular options that would be available and choices that would be vetoed could vary from one nation or locality to another, as they do now.

I have brought up this scheme to show that societies opting for germ-line therapy need not resemble *Brave New World* and may remain as open as they are today. I doubt that people who have been staunchly opposed to germ-line therapy will be converted by this suggestion, although it may help those who are undecided if the possibilities for the future are expanded and made more explicit. At the end, however, I expect that there will still be disagreement. The dispute concerns values, not technical possibilities, and humans have not reconciled such differences very often through civilized discussions.

After centuries of debate and combat, no consensus has been reached as to which is the best religion or economic system for humankind, nor can experts agree on more limited ethical questions such as the morality of abortion. The question of the biological future of humankind is basic to many ideologies, and the differences are unlikely to be bridged by a dialogue. Fortunately, there is no compelling reason why we need to do so.

When I was a child, the idea of one world was very popular. Humanity would some day coalesce in one global confederation where we recognized our common interests and worked together under the same rules of justice. I no longer think that this is either likely or desirable. In the world today, the number of existing confederations that seem to be pulling apart seems at least as large as those that are coming together. No obvious move to full unity exists, nor is this necessarily a bad thing. With experience, I have found the alternative of local control much more attractive, as it allows for the expression of the authentic differences that exist in humanity.

Certain issues, for example, the prevention of nuclear war and the preservation of the ozone layer, require cooperation between the various societies on this planet to avoid the destruction of us all. Others do not, and the editing of our genetic text belongs to this category. For moral reasons, or because they value themselves as they are, some cultures may wish to keep their germ lines inviolate. Others may decide to make modifications, but only to eliminate genetic diseases. Yet others may allow individuals to introduce such "improvements" as they choose. The options permitted will vary from place to place.

The cardinal virtue in this situation will be tolerance rather than cooperation. As stated by geneticist Albert Jacquard in his book *In Praise of Difference*, "Genetic richness comes from diversity . . . we need to understand that others are precious to us insofar as they are different from us."[19] Efforts of one group to impose their values on another by force run the risk of violent conflict. Even if successful at first, perpetual enforcement would be exceedingly difficult. Very little technology is needed for genetic experimentation in comparison to that needed for constructing nuclear arms. An effective global police state, the very evil that many who oppose germ-line therapy wish to avoid, might be needed to enforce genetic prohibitions.

Further, when a particular change has been made and found to work, it will be as "natural" as any other. The opinions of Jeremy Rifkin notwithstanding, the history of life on this planet appears to involve the continual extension, shuffling, and editing of the DNA texts of species. Humans have participated in this for millennia by domesticating and breeding plants and animals, for example, and arranging marriages. The uninformed experiments that Jacob performed on his flock, as described in Genesis, were the forerunners of the more skilled germ-line therapy efforts of tomorrow. Anyone who declares that certain DNA sequences are forbidden because they have come into being through intelligent human intervention would by definition be advocating genocide.

If we accept the idea that various groups and nations will follow their separate ideals with respect to germ-line therapy, a unique vision of the future emerges. The races and ethnic groups that we constitute today were produced largely by geographic and language separations.

These barriers are being reduced today by the ease of travel and communication, resulting in greater mobility and intermarriage. The options made available by germ-line therapy may produce a new division of the human race, however, based upon freely selected values.

Some portions of humanity may feel it unsafe or immoral for humans to modify their own genetic blueprint and that evolution should continue in the future as it has in the past. They should certainly be allowed to act upon their own beliefs within their own nations, although not to enforce their opinions upon others. It seems very appropriate, both as a safeguard and for sentimental reasons, that the current versions of the human race continue into the future.

For others, the possibility of planned human genetic change offers the brightest prospect for our future. They look beyond the cure of disease to the possibiility of improving the human condition.

Biologist Robert Sinsheimer has been quite eloquent in advocating human genetic improvement. He wrote earlier:

> Indeed, this concept marks a turning point in the whole evolution of life. For the first time in all time, a living creature will have the power to direct its own evolution. . . . Today we can envision that chance—and its dark companion of awesome choice and responsibility.[20]

He added, on another occasion,

> Man has always been doubly tethered to his past: through his genes and through his culture. . . . The genetic tether has been more tight and inflexible. . . . In the end we may come to see the improvement of our biological inheritance to be as natural and as important as the improvement of our cultural heritage. . . . Would a man be less human if he knew less hate or rage or envy or terror? Perhaps in a strange sense we might be less human but more humane.[21]

From this viewpoint, with which I agree, the history of life on this planet has concerned its struggle to survive and evolve to more advanced levels within an indifferent and sometimes inhospitable environment. The slow and clumsy method of natural selection was the only tool at its disposal in this long-term quest. Finally, after 4 billion years of evolution by this method, human consciousness was produced, the finest achievement to date. (I admit to being a human chauvinist.) By the application of our minds, using the methods of science, we have moved from a primitive existence and a relatively short human life span to a condition where most of us live longer, and a good fraction have a comfortable existence.

By applying our consciousness to evolution, we should be able to make our position and that of life, in general, not only more secure, but ultimately much better. Many miseries of our bodies that we have accepted as inevitable, even aging itself, may now be challenged. Some feel it unnatural that we do this, but others disagree, holding that such developments were implied in the very evolution of consciousness. As psychoanalyst Willard Gaylin, president of the Hastings Center said, "I not only think that we will tamper with Mother Nature, I think Mother wants us to."[22]

As we experiment with the passages of our genetic text, we may find that certain selections (or none) work better than others and move together in a common direction. As I have suggested, it seems more likely that we will disagree on the relative values of certain changes. Our descendents, the products of those changes, may vary even more in their preferences. If we continue in this direction, then the human race may separate eventually into a number of subspecies (perhaps separated by national borders) each following its own vision as to the best biological future for our species. For this future of multiple possibilities to work, it will be crucial that the differing versions of humankind recognize their common heritage and respect each other's parallel quests.

A future of this type would certainly harmonize with the history of life on this planet. It has not been a record only of extinctions. Species, when successful, have often diverged and followed different evolutionary paths to occupy the various ecological niches that became available. In the coming centuries and millennia, humankind will have the opportunity to expand beyond the confines of our planet and colonize portions of the larger universe. A multiplicity of worlds may become available in the long run to house the many human possibilities for the future.

It may be only coincidental that the first instance, after billions of years, when living beings on earth were able to read their own genetic text came within a decade of the time when they first set foot on another world. These events, even if put together accidentally in their first instance, may contain the essence of the human future.

EPILOGUE

The knight put down the last scroll and looked wearily around the ancient library. His labor had been long, but less tiresome and arduous than he feared when he had first started the task. With experience, he had learned many shortcuts in deciphering the scrolls, and after a time the messages had so held his attention that he did not take notice of the passing months and years. At last, he had completed the work and could move on.

Much of what the voice had promised him had been delivered. He understood his identity and his past. He had learned how the very muscles and organs within him functioned. To test his knowledge, he had taken his sword from the corner where it rested and swung it with much more power and skill than he ever had before. He trusted that his new abilities would be needed in the adventures that lay before him.

The answers to some key questions still eluded him: Where was he going? What was his quest? The voice only suggested that the answers still lay ahead. Now that he had mastered the scrolls, he needed to move onward.

A massive door stood at the far end of the library. The very sight of it had intimidated him until now. He now found that he could, with his new strength, open it easily. A dimly lit corridor lay beyond. After a few paces, however, it split into several passages. Which one should he choose? He hesitated and waited for guidance.

After some time, the answer arose within him, although he did not understand it fully. It was his destiny to explore all of them, and each would change him in some way. Puzzled, he selected the one that seemed most attractive and took a few tentative steps down it. A loud noise that resounded ahead of him suddenly changed his mood. With a smile, he adjusted his armor and unsheathed his sword. Here was a challenge that he knew how to face. There would be dragons.

NOTES

CHAPTER 1

1. Hans Stubbe, *History of Genetics—From Prehistoric Times to the Rediscovery of Mendel's Laws*, translated by T.R.W. Waters (Cambridge, Mass.: MIT Press, 1972).
2. Conway Zirkle, "Mendel and His Era," in Roland M. Nardone, ed., *Mendel Centenary: Genes in Development and Evolution* (Washington, D.C.: Catholic University of America Press, 1968).
3. A. Jacquard, *In Praise of Difference: Genetics and Human Affairs*, translated by Margaret M. Moriarty (New York: Columbia University Press, 1984).
4. Zirkle, "Mendel and His Era."
5. Francois Jacob, *The Logic of Life: A History of Heredity*, translated by Betty A. Spilman (New York: Pantheon, 1973).
6. D. H. Kenyon and G. Steinman, *Biochemical Predestination* (New York: McGraw-Hill, 1969).
7. Stubbe, *History of Genetics.*
8. Jacob, *Logic of Life.*
9. Ibid.
10. Quoted in Stubbe, *History of Genetics.*

CHAPTER 2

1. Hugo Iltis, *Life of Mendel*, translated by Eden and Cedar Paul (New York: Norton, 1932).
2. Ibid.
3. Madan K. Battarchayya, Alison M. Smith, T. H. Noel Ellis et al., "The Wrinkled-Seed Character of Pea Described by Mendel Is Caused by a Transposon-like Insertion in a Gene Encoding Starch-Branching Enzyme," *Cell* 60 (1990): 115.
4. Iltis, *Life of Mendel.*
5. Ibid.
6. V. Orel, *Mendel*, translated by Stephen Fines (Oxford, England: Oxford University Press, 1984).

7. Iltis, *Life of Mendel.*
8. Orel, *Mendel.*
9. Iltis, *Life of Mendel.*
10. John Farley, *The Spontaneous Generation Controversy From Descartes to Oparin* (Baltimore: Johns Hopkins University Press, 1974).
11. John B. Jenkins and P. Michael Conneally, "The Paradigm of Huntington's Disease," *American Journal of Human Genetics* 45 (1989): 169.
12. George Huntington, "On Chorea," *The Medical and Surgical Reporter* 26 (1872): 317.
13. Franklin H. Portugal and Jack S. Cohen, *A Century of DNA* (Cambridge, Mass.: MIT Press, 1977).
14. Frances Galton, *Hereditary Genius: An Inquiry into Its Laws and Consequences* (London: Macmillan, 1869).

CHAPTER 3

1. Hugo Iltis, *Life of Mendel*, translated by Eden and Cedar Paul (New York: Norton, 1932).
2. Barbara Tuchman, *The Guns of August* (New York: Macmillan, 1962).
3. Ibid.
4. Robert K. Massie, *Nicholas and Alexandra* (New York: Atheneum, 1967).
5. James Herrick, "Peculiar Elongated and Sickle-Shaped Red Blood Corpuscles in a Case of Severe Anemia," *Archives of Internal Medicine* 6 (1910): 517.
6. Eugene B. Block, *Fingerprinting* (New York: David McKay, 1969).
7. Ian Shine and Sylvia Wrobel, *Thomas Hunt Morgan, Pioneer of Genetics* (Lexington: University of Kentucky Press, 1976).
8. Ibid.
9. Ibid.
10. Thomas Hunt Morgan, *Experimental Zoology* (New York: Macmillan, 1907).
11. Garland E. Allen, *Thomas Hunt Morgan, the Man and His Science* (Princeton, N.J.: Princeton University Press, 1978).
12. T. H. Morgan, A. H. Sturtevant, H. J. Muller et al., *The Mechanism of Mendelian Heredity* (New York: Holt, 1915).
13. Allen, *Morgan, the Man and His Science.*

14. Shine and Wrobel, *Morgan, Pioneer of Genetics.*
15. Alfred H. Sturtevant, *A History of Genetics* (New York: Harper & Row, 1965).
16. Shine and Wrobel, *Morgan, Pioneer of Genetics.*
17. Allen, *Morgan, the Man and His Science.*
18. Ibid.
19. Shine and Wrobel, *Morgan, Pioneer of Genetics.*
20. Daniel J. Kevles, *In the Name of Eugenics: Genetics and the Uses of Human Heredity* (New York: Knopf, 1985).
21. Charles B. Davenport, *Heredity in Relation to Eugenics* (New York: Holt, 1911), cited in Kevles, *In the Name of Eugenics.*
22. Madison Grant, *The Passing of the Great Race* (New York: Charles Scribner's Sons, 1916), cited in Mark H. Haller, *Eugenics: Hereditarian Attitudes in American Thought* (New Brunswick, N.J.: Rutgers University Press, 1963).
23. Gordon Edlin, *Genetic Principles: Human and Social Consequences* (Boston: Jones & Bartlett, 1983).
24. Benno Muller-Hill, *Murderous Science,* translated by George R. Fraser (Oxford, England: Oxford University Press, 1988).
25. Ibid.

CHAPTER 4

1. Elizabeth Longford, *The Queen: The Life of Elizabeth II* (New York: Knopf, 1983).
2. Zhores Medvedev, *The Rise and Fall of T.D. Lysenko,* translated by I. Michael Lerner (New York: Columbia University Press, 1969).
3. Valery N. Soyfer, "New Light on the Lysenko Era," *Nature* 339 (1989): 415.
4. James D. Watson, *The Double Helix* (New York: Atheneum, 1968).
5. Francis Crick, *What Mad Pursuit: A Personal View of Scientific Discovery* (New York: Basic Books, 1988).
6. Watson, *Double Helix.*
7. Erwin Chargaff, "A Quick Climb Up Mount Olympus" (review of *The Double Helix*). *Science* 159 (1968): 1448.
8. Oswald T. Avery, Colin M. MacLeod, and Maclyn McCarty, "Induction of Transformation by a Deoxyribonucleic Acid Fraction Isolated from Pneumococcus Type III," *Journal of Experimental Medicine* 79 (1944): 137.

9. Erwin Chargaff, "Preface to a Grammar of Biology. A Hundred Years of Nucleic Acid Research," *Science* 172 (1971): 639.
10. Erwin Schrödinger, *What is Life?* (Cambridge, England: The University Press, 1944).
11. Horace Freeland Judson, *The Eighth Day of Creation* (New York: Simon & Schuster, 1979).
12. Ibid.
13. Francis Crick, "The Double Helix: A Personal View," *Nature* 248 (1974): 766.
14. Crick, *What Mad Pursuit.*
15. Ibid.
16. Watson, *Double Helix.*
17. Judson, *Eighth Day.*
18. Crick, *What Mad Pursuit.*
19. Ibid.
20. Watson, *Double Helix.*
21. Judson, *Eighth Day.*
22. Watson, *Double Helix.*
23. Erwin Chargaff, *Heraclitean Fire, Sketches from a Life Before Nature* (New York: Rockefeller University Press, 1978).
24. Erwin Chargaff, "Building the Tower of Babble," *Nature* 248 (1974): 776.
25. Watson, *Double Helix.*
26. Judson, *Eighth Day.*
27. Ibid.
28. Watson, *Double Helix.*
29. Judson, *Eighth Day.*
30. Ibid.
31. Watson, *Double Helix.*
32. Crick, *What Mad Pursuit.*
33. Anthony Liversidge, "Interview, James D. Watson," *Omni* 6 (1984): 75.
34. Judson, *Eighth Day.*
35. J. D. Watson and F. H. C. Crick, "Molecular Structure of Nucleic Acids. A Structure for Deoxyribose Nucleic Acid," *Nature* 171 (1953): 737.
36. J. D. Watson and F. H. C. Crick, "Genetical Implications of the Structure of Deoxyribonucleic Acid," *Nature* 171 (1953): 964.
37. Judson, *Eighth Day.*
38. Crick, *What Mad Pursuit.*
39. Judson, *Eighth Day.*

40. D. Harel, R. Unger, and J. L. Sussman, "Beauty Is in the Genes of the Beholder," *Trends in Biochemical Sciences* 11 (1986): 155.
41. Judson, *Eighth Day.*
42. Gunther S. Stent, "Molecular Biology and Metaphysics," *Nature* 248 (1974): 779.
43. "Twenty-one Years of the Double Helix," *Nature* 248 (1974): 721.
44. Liversidge, "Interview, James D. Watson."
45. Chargaff, "Building the Tower of Babble."
46. Frederick Sanger, "Sequences, Sequences and Sequences," *Annual Review of Biochemistry* 57 (1988): 1.
47. Interview with Frederick Sanger, Cambridge, England, March 14, 1990.
48. H. Franklin Bunn and Bernard G. Forget, *Hemoglobin: Molecular, Genetic and Clinical Aspects* (Philadelphia: Saunders, 1986).
49. Crick, *What Mad Pursuit.*
50. Judson, *Eighth Day.*

CHAPTER 5

1. Michael Rogers, "The Pandora's Box Congress," *Rolling Stone* 189 (1975): 36.
2. Horace Freeland Judson, *The Eighth Day of Creation* (New York: Simon & Schuster, 1979).
3. Rogers, "Pandora's Box Congress."
4. Erwin Chargaff, "On the Dangers of Genetic Meddling," *Science* 192 (1976): 938.
5. Nicholas Wade, "Recombinant DNA: New York State Ponders Action to Control Research," *Science* 194 (1976): 705.
6. Rogers, "Pandora's Box Congress."
7. Erwin Chargaff, "A Slap at the Bishops of Asilomar," *Science* 190 (1975): 135.
8. Frederick Sanger, "Sequences, Sequences and Sequences," *Annual Review of Biochemistry* 57 (1988): 1.
9. Ibid.
10. Ibid.
11. Ibid.
12. Ibid.
13. Interview with Alan Coulson, Cambridge, England, March 14, 1990.
14. Sanger, "Sequences."

15. Ibid.
16. Interview with Frederick Sanger, Cambridge, England, March 14, 1990.
17. Coulson interview, March 14, 1990.
18. Sanger, "Sequences."
19. F. Sanger, G. M. Air, B. G. Barrell et al., "Nucleotide Sequence of Bacteriophage Phi-X-174," *Nature* 265 (1977): 687.
20. F. Sanger, S. Nicklen, and A. R. Coulson, "DNA Sequencing With Chain-Terminating Inhibitors," *Proceedings of the National Academy of Sciences USA* 74 (1977): 5463.
21. Sanger interview, March 14, 1990.
22. Sanger, "Sequences."
23. Ibid.
24. Thomas H. Jukes, "Members of the Club" (review of *The Eighth Day of Creation*), *Nature* 281 (1979): 505.
25. Interview with Walter Gilbert, Cambridge, Mass., July 6, 1990.
26. Stephen S. Hall, *Invisible Frontiers, The Race to Sequence a Human Gene* (New York: Atlantic Monthly Press, 1987).
27. Ibid.
28. Gilbert interview, July 6, 1990.
29. Frederick Sanger, "Determination of Nucleotide Sequences in DNA," *Science* 214 (1981): 1205.
30. Walter Gilbert, "DNA Sequencing and Gene Structure," *Science* 214 (1981): 1305.
31. Ibid.
32. Gilbert interview, July 6, 1990.
33. Gilbert, "DNA Sequencing."
34. Judson, *Eighth Day*.
35. Ibid.
36. Anthony Liversidge, "Interview, James D. Watson," *Omni* 6 (1984): 75.
37. David Pendelbury, "Cold Spring Harbor Tops Among Independent Labs," *The Scientist*, March 19, 1990, 20.
38. Leslie Roberts, "Cold Spring Harbor Turns 100," *Science* 250 (1990): 496.
39. Judson, *Eighth Day*.
40. Theodore Melnechuk, "The Dream Machine," *Psychology Today* 17 (November 1983): 22.

CHAPTER 6

1. Renato Delbucco, "A Turning Point in Human Cancer Research: Sequencing the Human Genome," *Science* 231 (1986): 1055.
2. James D. Watson, "Directors Report," in *Annual Report, 1988, Cold Spring Harbor Laboratory* (Cold Spring Harbor, N.Y., 1989).
3. Erwin Chargaff, "What Really Is DNA? Remarks on the Changing Aspects of a Scientific Concept," *Progress in Nucleic Acid Research and Molecular Biology* 8 (1968): 297.
4. Carl W. Schmid and Warren R. Jelanek, "The Alu Family of Dispersed Repeated Sequences," *Science* 216 (1982): 1065.
5. "Banking DNA Sequences," *Nature* 285 (1980): 59.
6. James D. Watson, "Organization—Different Views of Current and Future Science and Procedures," presented at Human Genome I Meeting, San Diego, October 4, 1989.

CHAPTER 7

1. "Banking DNA Sequences," *Nature* 285 (1980): 59.
2. "Boring but Better Sequences," *Nature* 312 (1984): 667.
3. Horace Freeland Judson, *The Eighth Day of Creation* (New York: Simon & Schuster, 1979).
4. Susan Greenberg, "City of Hope Scientist Strikes Accordant Note Between DNA and Music," *Genetic Engineering News*, July–August 1986, 49.
5. Hiroshi Nakamura, "SV40 DNA—A Message from Epsilon Eri?" *Acta Astronomica* 13 (1986): 573.
6. Sydney Brenner, "The Decipherment of Linear G," *Nature* 329 (1987): 490.
7. Warren B. Leary, "Scientists Cite Gains on Sickle Cell Disease," *New York Times*, 28 November 1989, C1.
8. Natalie Angier, "Girl, 4, Becomes First Human to Receive Engineered Genes," *New York Times*, 15 September 1990. 1.
9. Alan N. Schechter, Constance Tom Noguchi, and Griffin P. Rodgers, "Sickle Cell Disease," in George Stamatoyannopoulos, Arthur W. Nienhuis, Philip Leder et al., eds., *The Molecular Basis of Blood Diseases*, (Philadelphia: Saunders, 1987), 179.
10. Matt Clark, Mariana Gosnell, Daniel Shapiro et al., "Medicine: A Brave New World," *Newsweek*, March 5, 1984, 64.
11. Sandra L. Martin, Karen A. Vincent, and Allan C. Wilson,

"Rise and Fall of the Delta Globin Gene," *Journal of Molecular Biology* 164 (1983): 513.

12. Ibid.

13. Nick Proudfoot, "Globin Gene Monkey Business," *Nature* 231 (1986): 730.

14. Josée Pagnier, J. Gregory Mears, Olga Dunda-Belkhodja et al., "Evidence for the Multicentric Origin of the Sickle Cell Hemoglobin Gene in Africa," *Proceedings of the National Academy of Sciences USA* 81 (1984): 1771.

15. Stylianos E. Antonarakis, Corinne D. Boehm, Graham R. Serjeant et al., "Origin of the Beta^{s} Gene in Blacks: The Contribution of Recurrent Mutation or Gene Conversion or Both," *Proceedings of the National Academy of Sciences USA* 81 (1984): 853.

16. C. Monteiro, J. Rueff, A. B. Falcao et al., "The Frequency and Origin of the Sickle Cell Mutation in the District of Coruche/Portugal," *Human Genetics* 82 (1989): 255.

17. Ibid.

18. Ibid.

19. Andrew Murray, "All's Well That Ends Well," *Nature* 346 (1990): 747.

CHAPTER 8

1. Paul M. Quinton, "Cystic Fibrosis: A Disease in Electrolyte Transport," *FASEB Journal* 4 (1990): 2709.

2. Leslie Roberts, "The Race for the Cystic Fibrosis Gene," *Science* 240 (1988): 141.

3. Leslie Roberts, "Race for Cystic Fibrosis Gene Nears End," *Science* 240 (1988): 282.

4. Roberts, "Race for the Cystic Fibrosis Gene."

5. Gina Kolata, "Rush Is on to Capitalize on Testing for Gene Causing Cystic Fibrosis," *New York Times*, 6 February 1990, C3.

6. Roberts, "Race for the Cystic Fibrosis Gene."

7. Roberts, "Race for Gene Nears End."

8. Ibid.

9. Benita Sirkin, "Utah Institute Isolates and Studies DNA Markers For Familial Disorders," *Genetic Engineering News* (October 1986): 13.

10. Roberts, "Race for the Cystic Fibrosis Gene."

11. Roberts, "Race for Gene Nears End."

12. Roberts, "Race for the Cystic Fibrosis Gene."

13. Johanna M. Rommens, Michael C. Iannuzzi, Bat-Sheva Kerem et al., "Identification of the Cystic Fibrosis Gene: Chromosome Walking and Jumping," *Science* 245 (1989): 1089; John R. Riordan, Johanna M. Rommens, Bat-Sheva Kerem et al., "Identification of the Cystic Fibrosis Gene: Cloning and Characterization of Complementary DNA," *Science* 245 (1989): 1066; Bat-Sheva Kerem, Johanna M. Rommens, Janet A. Buchanan et al., "Identification of the Cystic Fibrosis Gene: Genetic Analysis," *Science* 245 (1989): 1073.

14. Natalie Angier, "Team Cures Cells in Cystic Fibrosis by Gene Insertion," *New York Times*, 21 September 1990, A1.

15. Paul M. Quinton, "Righting the Wrong Protein," *Nature* 347 (1990): 226.

16. Kolata, "Rush Is on to Capitalize on Testing."

17. Amy Virshup, "Perfect People? The Promise and Peril of Genetic Testing," *New York Magazine*, July 27, 1987, 26.

18. Jeremy Rifkin, *Algeny* (New York: Viking Penguin, 1983).

19. Ibid.

20. Virshup, "Perfect People?"

21. Dorothy Nelkin and Laurence Tancredi, *Dangerous Diagnostics* (New York: Basic Books, 1989).

22. Virshup, "Perfect People?"

23. A. H. Handyside, E. H. Kontogianni, K. Hardy et al., "Pregnancies from Biopsied Human Preimplantation Embryos Sexed by Y-specific DNA Amplification," *Nature* 344 (1990): 769.

24. Marc Lappe, *Genetic Politics* (New York: Simon & Schuster, 1977).

25. Ibid.

CHAPTER 9

1. Interview with Nancy Wexler, New York City, June 21, 1990.

2. Gina Kolata, "Huntington's Disease Gene Located," *Science* 222 (1983): 913.

3. Marion Steinman, "In the Shadow of Huntington's Disease—Nancy Wexler's Quest for a Cure," *Columbia* (November 1987): 14.

4. *Experiences of a Huntington's Disease Patient* (New York: Huntington's Disease Society of America, 1977).

5. Kolata, "Huntington's Disease Gene Located."

6. Mary B. Hans and Arnulf H. Hoeppen, "Huntington's Chorea: Its Impact on the Spouse," *Journal of Nervous and Mental Disease* 168 (1980): 209.

7. Marjorie Guthrie, *A Family Member Speaks About Huntington's Disease* (New York: Huntington's Disease Society of America), reprinted from *Counseling in Genetics*, Y. E. Hsia, K. Hirschhorn, R. L. Silverberg et al., eds. (New York: Liss, 1979).

8. Julia Bell and J. B. S. Haldane, "The Linkage Between the Genes for Colour-blindness and Haemophilia in Man," *Proceedings of the Royal Society of London, B* 123 (1937): 119.

9. Susan Greenberg, "Conference Highlights the Benefits of Human Genome Mapping Initiative," *Genetic Engineering News* (July–August 1989): 30.

10. Robin McKie, *The Genetic Jigsaw* (Oxford, England: Oxford University Press, 1988).

11. Alan Newman, "The Legacy on Chromosome 4," *Johns Hopkins Magazine* (April 1988): 31.

12. Nancy Wexler, "Ethical, Social, and Legal Issues Concerning the Human Genome Project," paper presented at Human Genome II Conference, San Diego, October 22, 1990.

13. PBS's "Nova," *Confronting the Killer Gene*, broadcast March 28, 1989.

14. Leon Jaroff, "The Gene Hunt," *Time*, March 20, 1989, 62.

15. "Nova," March 28, 1989.

16. Wexler interview, June 21, 1990.

17. Steinman, "In the Shadow of Huntington's."

18. Leslie Roberts, "Huntington's Gene, So Near, Yet So Far," *Science* 247 (1990): 624.

19. Matt Clark, Mariana Gosnell, Daniel Shapiro et al., "Medicine: A Brave New World," *Newsweek*, March 5, 1984, 64.

20. Interview with Robert Moyzis, San Diego, October 3, 1989.

21. M. Bucan, M. Zimmer, W. L. Whaley et al., "Physical Maps of 4p16.3, the Area Expected to Contain the Huntington Disease Mutation," *Genomics* 6 (1990): 1.

22. Roberts, "Huntington's Gene."

CHAPTER 10

1. Robert K. Massie and Suzanne Massie, *Journey* (New York: Knopf, 1973).

2. Ibid.

3. Robert K. Massie, *Nicholas and Alexandra* (New York: Atheneum, 1967).

4. Ibid.

5. Robert F. Weaver and Philip Hedrick, *Genetics* (Dubuque, Ia.: Wm. C. Brown, 1989).

6. Andrew H. Sinclair, Phillippe Berta, Mark S. Palmer et al., "A Gene from the Human Sex-determining Region Encodes a Protein with Homology to a Conserved DNA-binding Motif," *Nature* 346 (1990): 240; John Gubbay, Jérôme Collignon, Peter Koopman et al., "A Gene Mapping to the Sex-determining Region of the Mouse Y Chromosome Is a Member of a Novel Family of Embryonically Expressed Genes," *Nature* 346 (1990): 245; David C. Page, Elizabeth M. C. Fisher, Barbra McGillivray et al., "Additional Deletion in Sex-determining Region of Human Y Chromosomes Resolves Paradox of X,t(Y;22) Female," *Nature* 346 (1990): 279.

7. Natalie Angier, "Scientists Say Gene on Y Chromosome Makes a Man a Man," *New York Times*, 19 July 1990, A1.

8. Anne McLaren, "What Makes a Man a Man?" *Nature* 346 (1990): 216.

9. Kevin Davies, "The Essence of Inactivity," *Nature* 349 (1991): 15.

10. P. J. Willems, I. Dijkstra, B. J. Van der Auwera et al., "Assignment of X-linked Hydrocephalus to Xq28," *American Journal of Human Genetics* 45 (Suppl.), (1989): A167.

11. Weaver and Hedrick, *Genetics*.

12. Julia Bell and J. B. S. Haldane, "The Linkage Between the Genes for Colour-blindness and Haemophilia in Man," *Proceedings of the Royal Society of London, B* 123 (1937): 119.

13. Jane Gitschier, William I. Wood, Theresa M. Goralka et al., "Characterization of the Human Factor VIII Gene," *Nature* 312 (1984): 326; William I. Wood, Daniel J. Capon, Christian C. Simonsen et al., "Expression of Active Human Factor VIII from Recombinant DNA Clones," *Nature* 312 (1984): 330; Gordon A. Vehar, Bruce Keyt, Dan Eaton et al., "Structure of Human Factor VIII," *Nature* 312 (1984): 337.

14. Marsha F. Goldsmith, "Target: Sexually Transmitted Diseases," *Journal of the American Medical Association* 264 (1990): 2179.

15. George C. Brownlee and Charles Rizza, "Clotting Factor VIII Cloned," *Nature* 312 (1990): 307.

16. John Maddox, "Who Will Clone a Chromosome?" *Nature* 312 (1984): 306.

17. Brownlee and Rizza, "Clotting Factor VIII Cloned."

CHAPTER 11

1. David V. Goeddel, Herbert L. Heyneker, Toyohara Hozumi et al., "Direct Expression in Escherichia Coli of a DNA Sequence Coding for Human Growth Hormone," *Nature* 281 (1979): 544.

2. Gina Kolata, "New Growth Industry in Human Growth Hormone?" *Science* 234 (1986): 22.

3. Thomas H. Murray, "Human Growth Hormone and Abuse Potential," *Genetic Engineering News* (May–June 1984): 6.

4. Natalie Angier, "Human Growth Hormone Reverses Effects of Aging," *New York Times*, 5 July 1990, A1; "Growth Hormone and the Drive for a More Youthful State," *New York Times*, 6 July 1990, A9.

5. Kolata, "New Growth Industry"; Angier, "Growth Hormone and the Drive for a More Youthful State."

6. Harold M. Schmeck, Jr., "Cell Growth Factors Emerge as Potent Therapies," *New York Times*, 28 March 1989, C1.

7. Norman H. Carey and P. E. Crawley, "Commercial Exploitation of the Human Genome: What Are the Problems?" in *Human Genetic Information: Science, Law and Ethics*, Ciba Foundation Symposium 149 (Chichister, England: Wiley, 1990), 133.

8. Erwin Chargaff, "Engineering a Molecular Nightmare?" *Nature* 327 (1987): 199.

CHAPTER 12

1. Roger Lewin, "Proposal to Sequence the Human Genome Stirs Debate," *Science* 232 (1986): 1598.

2. James D. Watson, "The Human Genome Project: Past, Present and Future," *Science* 248 (1990): 44.

3. Edmund H. Kelly, "The Human Genome Initiative: A Different Kind of Research," *FASEB Journal* 4 (1990): 1423.

4. Pamela S. Zurer, "Molecular Biologists Backing Effort to Map Entire Human Genome," *Chemical and Engineering News*, March 14, 1988, 22.

5. Fred Hoyle, "High Hopes for the Space Telescope," *Nature* 344 (1990): 808.

6. Jean Marx, "Dissecting the Complex Diseases," *Science* 247 (1990): 1540.

7. Robert A. Weinberg, "Positive and Negative Controls on Cell Growth," *Biochemistry* 21 (1989): 8263.

8. Daniel E. Koshland, Jr., "Sequences and Consequences of the Human Genome," *Science* 246 (1989): 189.

9. Nancy Wexler, "Ethical, Social, and Legal Issues Concerning the Human Genome Project," paper presented at Human Genome II Conference, San Diego, October 22, 1990.

10. Lewin, "Proposal to Sequence the Human Genome Stirs Debate."

11. James Trefil, "Beyond the Quark, the Case for the Supercollider," *New York Times Magazine*, 30 April 1989, 24.

12. Robert A. Weinberg, "To Sequence or Not to Sequence?" *Scientific American* (November 1988): 150.

13. Robert A. Weinberg, "The Case Against Genetic Sequencing," *The Scientist*, November 16, 1987, 11.

14. Sydney Brenner, "The Human Genome: the Nature of the Enterprise," in *Human Genetic Information: Science, Law and Ethics*, Ciba Foundation Symposium 149 (Chichister, England: Wiley, 1990), 6.

15. James Bruce Walsh and Jon Marks, "Sequencing the Human Genome," *Nature* 322 (1986): 590.

16. Laura Manuelidis, "A View of Interphase Chromosomes," *Science* 250 (1990): 1533.

17. Robert Sinsheimer, "The Santa Cruz Workshop—May 1985," *Genomics* 5 (1989): 954.

18. Nathalie Angier, "Great 15-Year Project to Decipher Human Genes Stirs Opposition," *New York Times*, 5 June 1990, C1.

19. Joshua Lederberg, "Does Scientific Progress Come from Projects or People?" Address to the National Association of State Universities and Land Grant Colleges, Washington, D.C., November 9, 1987, reprinted in *Current Contents, Life Sciences* November 27, 1989, 4.

20. Angier, "Great 15-Year Project."

21. Dan M. Cooper, "Human Genome Project" *Science* 246 (1989): 874.

22. Charles DeLisi, "The Human Genome Project" *American Scientist* 76 (1988): 488.

23. J. Corbett McDonald and Alison D. McDonald, "Asbestos and Carcinogenicity," *Science* 249 (1990): 844.

24. Joseph Palca, "Grants Squeeze Stirs up Lobbyists," *Science* 248 (1990): 803.

25. David Baltimore, "Genomic Sequencing: A Small Scale Approach," *Issues in Science and Technology* (Spring 1987): 48.

26. Joshua Lederberg, "The Genome Movement Holds Promise,

but We Must Look Before We Leap," *The Scientist*, March 20, 1989, 10.

27. DeLisi, "The Human Genome Project."

28. Leslie Fink, "Whose Genome Is It, Anyway?" *Human Genome News* 2 (National Center for Human Genome Research, National Insititutes of Health) (July 1990): 5.

29. Roger Johnson, "The Public Image of Science and Its Funding," *FASEB Journal* 4 (1990): 2431.

30. Norton D. Zinder, "The Genome Initiative: How to Spell 'Human'," *Scientific American* (July 1990): 128.

31. Timothy Ferris "The Space Telescope: A Sign of Intelligent Life," *New York Times*, 29 April 1990, sec. 4, 1.

32. Richard Seltzer, "Impact of Apollo 11 Moon Landing Probed," *Chemical and Engineering News*, July 31, 1989, 12.

33. Francois Jacob, *The Logic of Life: A History of Heredity*, translated by Betty A. Spilman (New York: Pantheon, 1973).

CHAPTER 13

1. Tabitha M. Powledge, "Shall We Peddle Human Genes?" *The Scientist*, March 23, 1987, 12.

2. Leslie Roberts, "Who Owns the Human Genome?" *Science* 237 (1987): 358.

3. Ibid.

4. Robert Kanigel, "The Genome Project," *New York Times Magazine*, 13 December 1987, 44.

5. Powledge, "Shall We Peddle Human Genes?"

6. Kanigel, "Genome Project."

7. Ibid.

8. Stephen S. Hall, *Invisible Frontiers: The Race to Sequence a Human Gene* (New York: Atlantic Monthly Press, 1987).

9. Ibid.

10. Kanigel, "Genome Project."

11. Hall, *Invisible Frontiers*.

12. Interview with Walter Gilbert, Cambridge, Mass., July 6, 1990.

13. Office of Energy Research, Office of Health and Environmental Research, *The Human Genome Initiative of the Department of Energy*, Document no. DOE/ER-0382 (National Technical Information Service, Springfield, Va.).

14. Robert Mullan Cook-Deegan, "The Alta Summit, December 1984," *Genomics* 5 (1989): 661.

15. Interview with Charles Cantor, New York City, June 1, 1989.
16. Leslie Roberts, "Agencies Vie Over Human Genome Project," *Science* 237 (1987): 486.
17. Ibid.
18. G. Christopher Anderson, "Genome Project Planners Vie for Leadership," *The Scientist*, June 12, 1989, 12.
19. Cantor interview, June 1, 1989.
20. "There's No Place Like Home," *The Scientist*, October 10, 1989.
21. Cantor interview, June 1, 1989.
22. Ibid.
23. Ibid.
24. Interview with Cassandra Smith, New York City, June 1, 1989.
25. Paul Selvin, "Charlie Cantor Gets Kicked Upstairs," *Science* 249 (1990): 1238.
26. Ibid.
27. Interview with Cassandra Smith, San Diego, October 23, 1990.
28. James D. Watson, "Directors Report," in *Annual Report, 1988, Cold Spring Harbor Laboratory* (Cold Spring Harbor, N.Y., 1989).
29. James D. Watson, "The Human Genome Project: Past, Present and Future," *Science* 248 (1990): 44.
30. Leslie Roberts, "Cold Spring Harbor Turns 100," *Science* 250 (1990): 496.
31. Colin Norman, "Science Budget: Growth Among Red Ink," *Science* 251 (1991): 616.
32. Interview with James Watson, Cold Spring Harbor, N.Y., November 22, 1989.
33. Anthony Liversidge, "Interview, James D. Watson," *Omni* 6 (1984): 75.
34. James Watson, "The Human Genome," paper presented at the 40th Annual Meeting of the American Society of Human Genetics, Baltimore, Md., November 15, 1989.
35. Robert Wright, "Achilles' Helix," *The New Republic*, July 9 & 16, 1991, 21.
36. Watson, "The Human genome."
37. Ibid.
38. Watson interview, November 22, 1989.
39. Ibid.
40. Ibid.
41. James D. Watson, "Organization—Different Views of Cur-

rent and Future Science and Procedures," paper presented at Human Genome I Meeting, San Diego, October 4, 1989.

42. Christopher Anderson, "Genome Project to Tackle Mass Screening," *Nature* 348 (1990): 569.

43. Watson, "The Human Genome Project."

44. Leslie Roberts, "Plan for Genome Centers Sparks a Controversy," *Science* 246 (1989): 204.

45. Watson interview, November 22, 1989.

46. Ibid.

47. Roberts, "Plan for Genome Centers Sparks Controversy."

48. Ibid.

49. Watson interview, November 22, 1989.

50. Ibid.

51. Roberts, "Plan for Genome Centers Sparks Controversy."

52. Ibid.

53. Watson, "The Human Genome."

54. Barbara J. Culliton, "Jim Watson on the Budget," *Science* 242 (1988): 1376.

55. Ibid.

56. Watson interview, November 22, 1989.

57. James D. Watson, "The NIH Perspective," paper presented at Human Genome II Conference, San Diego, October 22, 1990.

58. Victor A. McKusick, "The Human Genome Organization: History, Purposes and Membership," *Genomics* 5 (1989): 385.

59. David Dickson, "Watson Floats a Plan to Carve up Genome," *Science* 244 (1989): 521.

60. Anderson, "Genome Project Planners Vie for Leadership."

61. Watson interview, November 22, 1989.

62. Peter Coles, "A Different Approach," *Nature* 347 (1990): 701.

63. Bertrand R. Jordan, "The French Human Genome Program," *Genomics* 9 (1991): 562.

64. "Greens Against Genes," *Nature* 332 (1990): 667.

65. Leslie Roberts, "Watson versus Japan," *Science* 246 (1989): 576.

66. Watson interview, November 22, 1989.

67. Dickson, "Watson Floats a Plan."

68. Watson, "The Human Genome."

69. Roberts, "Watson versus Japan."

70. David Swinbanks, "Japan's Project Stalls," *Nature* 349 (1991): 360.

CHAPTER 14

1. Interview with Alan Coulson, Cambridge, England, March 14, 1990.
2. Leslie Roberts, "A Sequencing Reality Check," *Science* 242 (1988): 1245.
3. T. Friedmann, "Rapid Nucleotide Sequencing of DNA," *American Journal of Human Genetics* 31 (1979): 19.
4. Daniel E. Koshland, Jr., "The Molecule of the Year," *Science* 246 (1989): 1541.
5. Ibid.
6. Ruth Levy Guyer and Daniel R. Koshland, Jr., "The Moleclule of the Year," *Science* 246 (1989): 1543.
7. Abigail Grissom "PCR Expands, Creates Revolution," *The Scientist*, July 10, 1989, 14.
8. Daniel Sinnett, Jean-Marc Deragon, Louise A. Sinard et al., "Alumorphs–Human DNA Polymorphisms Directed by Polymerase Chain Reaction Using Alu-Specific Primers," *Genomics* 7 (1990): 331.
9. Koshland, "Molecule of the Year."
10. John Bell, "Out of Chains," *Nature* 341 (1989): 196.
11. Interview with Henry Erlich and Randall Saiki, Emeryville, Calif., May 26, 1989.
12. Interview with Kary B. Mullis, San Diego, October 4, 1989.
13. Ibid.
14. Kary B. Mullis, "The Unusual Origin of the Polymerase Chain Reaction," *Scientific American* (April 1990), 56.
15. Randall K. Saiki, Stephen Scharf, Fred Faloona et al., "Enzymatic Amplification of Beta-Globin Genomic Sequences and Restriction Site Analysis for Diagnosis of Sickle-Cell Anemia," *Science* 230 (1985): 1350.
16. Mullis interview, October 4, 1989.
17. Marilyn Chase, "Biotech Battle, Du Pont and Cetus Fight Over Patents on New Genetic Tool," *Wall Street Journal*, 18 December 1989, A1.
18. Jeffrey S. Price, "PCR Origins," *Nature* 342 (1989): 623.
19. Mullis interview, October 4, 1989.
20. Walter Gratzer, "Adventures of Ben Trovato," *Nature* 342 (1989): 307.
21. Interview with Walter Gilbert, Cambridge, Mass., July 6, 1990.
22. Helen Donis-Keller, Philip Green, Cynthia Helms et al., "A Genetic Linkage Map of the Human Genome," *Cell* 51 (1987): 319.

23. Leslie Roberts, "Flap Arises Over Genetic Map," *Science* 238 (1987): 750.
24. Leslie Roberts, "Whatever Happened to the Genetic Map?" *Science* 247 (1990): 281.
25. Interview with David Schwartz, New York City, March 5, 1990.
26. Interview with Charles Cantor, New York City, June 1, 1989.
27. Harold M. Schmeck, Jr., "Burst of Discoveries Reveals Genetic Basis for Many Diseases," *New York Times*, 31 March 1987, C1.
28. Lois Wingerson, *Mapping Our Genes* (New York: Dutton, 1990).
29. James D. Watson, "The Human Genome Project: Past, Present and Future," *Science* 248 (1990): 44.
30. Leslie Roberts. "The Worm Project," *Science* 248 (1990): 1310.
31. Ibid.

CHAPTER 15

1. Interview with Walter Gilbert, Cambridge, Mass., July 6, 1990.
2. Francisco J. Ayala, "Roundtable Forum, The Human Genome Initiative: Issues and Impacts," in Avril D. Woodhead and Benjamin J. Barnhart, eds., *Biotechnology and the Human Genome* (New York: Plenum, 1988), 93.
3. L. Luca Cavalli-Sforza, "Opinion: How Can One Study Individual Variation for 3 Billion Nucleotides of the Human Genome?" *American Journal of Human Genetics* 46 (1990): 649.
4. Allan C. Wilson, "Will Sequencing the Human Genome Revolutionize Biology?" *The New Biologist* 2 (1990): 1.
5. Deborah Franklin, "What a Child Is Given," *New York Times Magazine*, 3 September 1989, 36.
6. Woodhead and Barnhart, eds., *Biotechnology and the Human Genome*, 103.
7. Joshua Lederberg, "Medical Science, Infectious Disease, and the Unity of Humankind," *Journal of the American Medical Association* 260 (1989): 684.
8. Aubrey Milunsky, *Choices, Not Chances* (Boston: Little, Brown, 1989)
9. P. A. Kitchin, Z. Szotyori, C. Fromholc et al., "Avoidance of False Positives," *Nature* 344 (1990): 201.

10. Franklin, "What a Child is Given."
11. Robert Plomin, "The Role of Inheritance in Behavior," *Science* 248 (1990): 183.
12. Thomas J. Bouchard, Jr., David T. Lykken, Matthew McGue et al., "Sources of Human Psychological Differences: The Minnesota Study of Twins Reared Apart," *Science* 250 (1990): 223.
13. Thomas H. Murray, "Roundtable Forum, The Human Genome Initiative: Issues and Impacts," in Avril D. Woodland and Benjamin J. Barnhart, eds., *Biotechnology and the Human Genome* (New York: Plenum, 1988), 97.

CHAPTER 16

1. Robert K. Massie, *Nicholas and Alexandra* (New York: Atheneum, 1967).
2. Eugene B. Block, *Fingerprinting* (New York: David McKay, 1969).
3. Robert D. McFadden, "Year-Old Mystery Solved as Police Identify Body Found in East River," *New York Times*, 14 May 1988.
4. Alec J. Jeffreys, Rita Neumann, and Victoria Wilson, "Repeat Unit Sequence Variation in Minisatellites: A Novel Source of DNA Polymorphism for Studying Variation and Mutation by Single Molecule Analysis," *Cell* 60 (1990): 473.
5. Jeffrey Schmalz, "Couple Suffer a Double Loss in Daughter's Death," *New York Times*, 27 September 1988, A18.
6. "Tests Indicate That Girl, 10, Is Child of Claimants," *New York Times*, 20 November 1989, A21.
7. Alec J. Jeffreys, John Y. K. Brookfield, and Robert Semeonoff, "Positive Identification of an Immigration Test-case Using Human DNA Fingerprints," *Nature* 317 (1985): 818.
8. Interview with Mary-Claire King, Berkeley, Calif., May 26, 1989.
9. Clyde C. Snow, "The Investigation of the Human Remains of the 'Disappeared' in Argentina," paper presented at symposium, *Defending Human Rights with Genetics and Forensic Evidence*, Annual Meeting of the American Association for the Advancement of Science, San Francisco, January 16, 1989.
10. Jared M. Diamond, "Abducted Orphans Identified by Grandpaternity Testing," *Nature* 327 (1987): 552.
11. Ibid.
12. King interview, May 26, 1989.

13. Joseph Wambaugh, *The Blooding* (New York: Morrow, 1989).

14. John I. Thornton, "DNA Profiling," *Chemical and Engineering News*, November 20, 1989, 18.

15. George James, "Man Convicted of Rape on DNA Evidence," *New York Times*, 20 October 1988, B1.

16. Daniel Garner, "DNA Forensics: Science, Law and Public Policy Issues," presented at American Society of Human Genetics, 20th Annual Meeting, Baltimore, Md., November 12, 1989.

17. Eric S. Lander, "DNA Fingerprinting on Trial," *Nature* 339 (1989): 501.

18. Alun Anderson, "Judge Backs Technique," *Nature* 340 (1989): 582.

19. Lander, "DNA Fingerprinting on Trial."

20. "Look-Alike's Confession Sets Man Free After 8 Years in Jail," New York Times, 23 July 1990, A11.

21. Andrew H. Malcolm, "F.B.I. Opening Door to Wide Use of Genetic Tests in Solving Crimes," *New York Times*, 12 June, 1989, A1.

22. Peter Gill, "Case Study," *Nature* 344 (1990): 394.

23. Alan F. Westin, "A Privacy Analysis of the Use of DNA Techniques as Evidence in Courtroom Proceedings," in Jack Ballantyne, George Sensabaugh, and Jan Witkowski, eds., *DNA Technology and Forensic Science* (Banbury Report 32) (Cold Spring Harbor, N.Y.: Cold Spring Harbor Laboratory Press, 1989).

24. Howard J. Sanders, "Chemical Mutagens. The Road to Genetic Disaster?" *Chemical and Engineering News*, May 19, 1969, 51; June 2, 1969, 64.

25. Block, *Fingerprinting*.

26. Natalie Angier, "Mating for Life? It's Not for the Birds or the Bees," *New York Times*, 21 August 1990, C1.

CHAPTER 17

1. Dennis Drayna and Ray White, "The Genetic Linkage Map of the Human X Chromosome," *Science* 230 (1985): 753.

2. Telephone interview with Richard Kouri, October 30, 1990.

3. Alan Hamilton, *The Royal 100* (London: Pavilion Books, 1986).

4. Ibid.

5. Interview with Svante Pääbo, Berkeley, Calif., May 21, 1989.

6. J. S. Jones, "Mummified Human DNA Cloned," *Nature* 314 (1985): 576.

7. David A. Lawlor, Cynthia D. Dickel, William W. Hausworth et al., "Ancient HLA Genes From 7,500-year-old Archeological Remains," *Nature* 349 (1991): 785.

8. Lawrence K. Altman, "Puzzle of Sailor's Death Solved after 31 Years: The Answer is AIDS," *New York Times*, 24 July 1990, C3.

9. Tania Ewing, "Preserving the Present," *Nature* 345 (1990): 463.

10. Russell Higuchi, Barbara Bowman, Mary Freiberger et al., "DNA Sequences from the Quagga, an Extinct Member of the Horse Family," *Nature* 312 (1984): 284.

11. Alec J. Jeffreys, "Raising the Dead and Buried," *Nature* 312 (1984): 198.

12. Svante Pääbo, Russell G. Higuchi, and Allan C. Wilson, "Ancient DNA and the Polymerase Chain Reaction: The Emerging Field of Molecular Archeology," *Journal of Biological Chemistry* 264 (1989): 9707.

13. Erika Hagelberg, Bryan Sykes, and Robert Hedges, "Ancient Bone DNA Amplified," *Nature* 342 (1989): 485.

14. Interviews with Robert Hedges, Bryan Sykes, and Erika Hagelberg, Oxford, England, March 15, 1990.

15. Satoshi Horai, Kenji Hayasaka, Kumiko Murayama et al., "DNA Amplification from Ancient Human Skeletal Remains and Their Sequence Analysis," *Proceedings of the Japan Academy* 65, series B, (1989): 229.

16. Warren E. Leary, "A Search for Lincoln's DNA," *New York Times*, 10 February 1991, 1.

17. Ibid.

18. Alex Haley, *Roots* (New York: Doubleday, 1976).

19. Charles L. Blockson with Ron Fry, *Black Genealogy* (Englewood Cliffs, N.J.: Prentice-Hall, 1977).

20. Rowena K. Track, Florence K. Ricciuti, and Kenneth K. Kidd, "Information on DNA Polymorphisms in the Human Gene Mapping Library," in Jack Ballantyne, George Sensabaugh, and Jan Witkowski, eds., *DNA Technology and Forensic Science* (Banbury Report 32) (Cold Spring Harbor, N.Y.: Cold Spring Harbor Laboratory Press, 1989).

21. Fawn M. Brodie, *Thomas Jefferson: An Intimate History* (New York: Norton, 1974).

22. Ibid.

23. Fawn M. Brodie, "Thomas Jefferson's Unknown Grandchildren—A Study in Historical Silences," *American Heritage* (October 1976): 28.

24. Virginius Dabney, *The Jefferson Scandals. A Rebuttal* (New York: Dodd Mead, 1981).

25. Natalie S. Bober, *Thomas Jefferson: Man on a Mountain* (New York: Atheneum, 1978).

26. Dan Rottenberg, *Finding Our Fathers: A Guidebook to Jewish Genealogy* (New York: Random House, 1977).

27. Arthur Koestler, *The Thirteenth Tribe* (New York: Random House, 1976).

28. *The Encyclopedia of Judaism*, Geoffrey Wiggoner, ed., s.v. "Khazars" (New York: Macmillan, 1989), 414.

29. Raphael Patai, *The Myth of the Jewish Race* (New York: Charles Scribner's Sons, 1975).

30. L. Luca Cavalli-Sforza and Dorit Carmelli, "The Ashkenazi Gene Pool: Interpretations," in R.M. Goodman and A.G. Motulsky, eds., *Genetic Diseases Among Ashkenazi Jews* (New York: Raven Press, 1979), 93.

31. John Tierney, Linda Wright, and Karen Springen, "The Search for Adam and Eve," *Newsweek*, January 11, 1988, 40.

32. Ann Gibbons, "Looking for the Father of Us All," *Science* 251 (1991): 378.

33. James Shreeve, "Argument Over a Woman," *Discover* (August 1990): 52.

34. Arthur Fisher, "The More Things Change," *Mosaic* 19, (Spring 1988): 22.

35. Rebecca L. Cann, "In Search of Eve," *The Sciences*, (September–October 1987): 30.

CHAPTER 18

1. Gordon Edlin, *Genetic Principles: Human and Social Consequences* (Boston: Jones & Bartlett, 1983).

2. Jean L. Marx, "Cloning Sheep and Cattle Embryos," *Science* 239 (1988): 463.

3. Jon R. Luoma, "Hope for Threatened Species Seen in Effort to Breed Captive Animals," *New York Times*, 2 January 1990, C2.

4. R. A. Laskey, "Prospects for Reassembling the Cell Nucleus," *Journal of Cell Science*, Supplement 4 (1986): 1.

5. Michael Crichton, *Jurassic Park* (New York: Knopf, 1990).
6. Robert W. Ettinger, *The Prospect of Immortality* (New York: Doubleday, 1964).
7. Steven B. Harris, "Many Are Cold but Few Are Frozen," *Free Inquiry* (Spring 1989): 19.

CHAPTER 19

1. Gina Kolata, "Why Gene Therapy Is Considered Scary But Cell Therapy Isn't," *New York Times*, 16 September 1990, sec. 4, p. 5.
2. "Are Germ-lines Special?" *Nature* 331 (1988): 100.
3. Robert Williamson, "Gene Therapy," *Nature* 298 (1982): 416.
4. Jonathan Glover, *What Sort of People Should There Be?* (Harmondsworth, England: Penguin, 1984)
5. Christian Anfinsen, "Bio-Engineering: Short-Term Optimism and Long-Term Risk," in Robert Esbjornson, ed., *The Manipulation of Life*, Nobel Conference XIX (San Francisco: Harper & Row, 1984).
6. Lubert Stryer, *Biochemistry*, 3rd ed. (New York: Freeman, 1988).
7. Natalie Angier, "Gene Swap in Mice Holds Hope for Medical Gains," *New York Times*, 19 April 1990, B8.
8. Mario Capecchi, "Tapping the Cellular Telephone," *Nature* 344 (1990): 105.
9. Jeremy Rifkin, *Algeny* (New York: Viking Penguin, 1983).
10. Kolata, "Why Gene Therapy is Considered Scary."
11. Walter Charles Zimmerli, "Discussion," in *Human Genetic Information: Science, Law and Ethics*, Ciba Foundation Symposium 149 (Chichister, England: Wiley, 1990), 91.
12. Glover, *What Sort of People Should There Be?*
13. Deborah Dalton, "Genetic Gerrymandering," *The Humanist* (January–February 1988): 16.
14. Daniel J. Kevles, "Unholy Alliance," *The Sciences* (September–October 1986): 25.
15. Kenneth A. Briggs, "Clerics Urge U.S. Curb on Gene Engineering," *New York Times*, 9 June 1983, A1.
16. "Whether to Make Perfect Humans," *New York Times*, 22 July 1982, A22.
17. Glover, *What Sort of People Should There Be?*
18. Nancy Wexler, "Ethical, Social, and Legal Issues Concerning the Human Genome Project," paper presented at Human Genome II Conference, San Diego, October 22, 1990.

19. A. Jacquard, *In Praise of Difference—Genetics and Human Affairs*, translated by Margaret M. Moriarty (New York: Columbia University Press, 1984).

20. Robert Sinsheimer, "The Prospect for Designed Genetic Change," in Marc Lappe and Robert S. Morrison, eds., *Ethical and Scientific Issues Posed by Human Uses of Molecular Genetics* (New York: New York Academy of Sciences, 1975).

21. Robert Sinsheimer, "Genetic Engineering: The Modification of Man," *Impact of Science on Society* 20 (1970): 279.

22. Willard Gaylin, "What's So Special about Being Human?" in Robert Esbjornson, ed., *The Manipulation of Life*, Nobel Conference XIX (San Francisco: Harper & Row, 1984), 51.

BIBLIOGRAPHY

More than one thousand technical articles, reviews, monographs, and newspaper and magazine accounts were consulted in preparing this book. A number of them are cited in the notes. I have compiled a list here of additional selections that I have chosen for their value as general references.

Beard, Timothy F., and Denise Denong, *How to Find Your Family Roots* (New York: McGraw-Hill, 1977).

Brownlee, G.G., "The Molecular Genetics of Haemophelia A and B." *Journal of Cell Science*, Suppl. 4 (1986): 445.

Bunn, H. Franklin, and Bernard G. Forget, *Hemoglobin: Molecular, Genetic and Clinical Aspects* (New York: Saunders, 1986).

Cann, Rebecca L., Mark Stoneking, and Allan C. Wilson, "Mitochondrial DNA and Human Evolution." *Nature* 325 (1987): 31.

Congress of the United States, Office of Technology Assessment, *Mapping Our Genes; Genome Projects: How Big, How Fast?* (Baltimore: Johns Hopkins University Press, 1989).

Edwards, J.H., "The Importance of Genetic Disease and the Need for Prevention." *Philosophic Transactions of the Royal Society of London, B* 319 (1988): 211–227.

Ford, Simon, and William C. Thompson, "A Question of Identity: Some Reasonable Doubts About DNA Fingerprinting." *The Sciences* (January–February 1990): 37.

Galton, David J., *Molecular Genetics of Common Metabolic Disease* (New York: Wiley, 1985).

Gornick, Larry, and Mark Wheelis, *The Cartoon Guide to Genetics* (New York: Barnes & Noble Books, 1984). This work contains historical material and simplified principles in an amusing illustrated form.

Honig, George R., and Junius G. Adams III, *Human Hemoglobin Genetics* (Vienna: Springer-Verlag, 1986).

Judson, Horace Freeland, *The Eighth Day of Creation* (New York: Simon & Schuster, 1979). This history offers a rich and detailed account of the development of molecular biology through the 1970s.

Kaufman, R.J., "Genetic Engineering of Factor VIII." *Nature* 342 (1989): 207.

Kidd, Charles, and Patrick Montague-Smith, *Debrett's Book of Royal Children* (New York: William Morrow, 1982).

Kirby, Lorne T., *DNA Fingerprinting, An Introduction* (New York: Stockton Press, 1990).

Levenson, Corey H., and Norman Arnheim, "Polymerase Chain Reaction." *Chemical and Engineering News*, 1 October 1990, 36. A review of the applications of DNA amplification is presented.

Lewin, Benjamin, *Genes IV* (Oxford, England: Oxford University Press, 1990). This text provides greater depth into the details of DNA biochemistry and is updated periodically. (The next one presumably will be *Genes V*.)

Maddox, John, "Understanding Gel Electrophoresis." *Nature* 345 (1990): 381. A brief overview of pulsed-field gel electrophoresis is presented.

McCarty, Maclyn, *The Transforming Principle: Discovering That Genes Are Made of DNA* (New York: Norton, 1985). This book presents a first-hand account by the surviving member of Avery's team.

McKusick, Victor, *Mendelian Inheritance in Man*, 9th ed. (Baltimore: Johns Hopkins University Press, 1990). This reference provides endless information, both technical and anecdotal, on human genetic conditions.

———, "Mapping and Sequencing the Human Genome." *New England Journal of Medicine* 320 (1989): 910. A brief history of the project is given.

Murray, Thomas H., "Ethical Issues in Genome Research." *FASEB Journal* 5 (1991): 55.

National Research Council, *Mapping and Sequencing the Human Genome* (Washington, D.C.: National Academy Press, 1988).

Roberts, Leslie, "New Game Plan for Genome Mapping." *Science* 245 (1989): 1438. The STS (sequence-tagged site) mapping strategy is described in a nontechnical manner.

Sargeant, Graham R., *Sickle Cell Disease* (Oxford, England: Oxford University Press, 1985).

Stephens, J. Clairborn, Mark L. Cavanaugh, Margaret I. Gradie et al., "Mapping the Human Genome: Current Status." *Science* 250 (1990): 237. This summary is accompanied by a large fold-out map of human chromosomes.

U.S. Department of Health and Human Services and the U.S. Department of Energy, *Understanding Our Genetic Inheritance,*

The U.S. Human Genome Project: The First Five Years, FY 1991–1995 (Springfield, Va.: National Technical Information Service).

Verma, Inder M., "Gene Therapy." *Scientific American* (November 1990): 68.

Vogel, F., and A. G. Motulsky, *Human Genetics, Problems and Approaches*, 2nd ed. (Berlin: Springer-Verlag, 1986).

Watson, James D., and John Tooze, *The DNA Story* (San Francisco: Freeman, 1981). Many articles and original documents related to the recombinant DNA controversy are collected here.

Weatherall, D.J., *The New Genetics and Clinical Practice*, 2nd ed. (Oxford, England: Oxford University Press, 1985).

Weaver, Robert F., and Philip Hedrick, *Genetics* (Dubuque, Iowa: Wm. C. Brown, 1989). This text provides a capable introduction to the subject at the college level.

White, Ray, and Jean-Marc Lalouel, "Chromosome Mapping with DNA Markers." *Scientific American* (February 1988): 40.

Wintrobe, Maxwell W., *Blood Pure and Eloquent* (New York: McGraw-Hill, 1980).

INDEX